Foundations of Economic Psychology

Kazuhisa Takemura

Foundations of Economic Psychology

A Behavioral and Mathematical Approach

Kazuhisa Takemura
Department of Psychology
Waseda University
Tokyo, Japan

ISBN 978-981-13-9048-7 ISBN 978-981-13-9049-4 (eBook)
https://doi.org/10.1007/978-981-13-9049-4

This Springer imprint is published by the registered company Springer Nature Singapore Pte Ltd.
The registered company address is: 152 Beach Road, #21-01/04 Gateway East, Singapore 189721, Singapore

Preface

This book provides an overview of the idea of economic psychology from behavioral and mathematical perspectives and related theoretical and empirical findings. Economic psychology is described briefly as the general term for descriptive theories to explain the psychological processes of microeconomic behaviors and macroeconomic phenomenon. As the studies of G. Katona who started economic psychology, G. H. A. Simon who won the Nobel Prize for economics in 1978, D. Kahneman, and R. Thaler who won the prize in 2002 and 2017, respectively, suggest, however, the psychological methodology and knowledge of economic psychology have been applied widely in such fields as economics, business administration, and engineering, and are expected to become useful in the future. This book will explain various behavioral and mathematical models of economic psychology related to microeconomic and macroeconomic phenomenon. Numerous models have been proposed to explain the psychological processes related to such economic phenomenon. This book will also introduce some new models that are useful to explain human economic behaviors. It ends with some speculation about the future of modern economic psychology while referring to their relation with fields related to neuroscience, such as neuroeconomics, that have been developed in recent years.

Economic psychology is recently widely recognized in many countries since Daniel Kahneman received the Nobel Prize in 2002. In fact, there are many researchers and students who are studying economic psychology in Europe and North America. Interest in economic psychology is increasing in Asian countries such as China, Taiwan, Singapore, Malaysia, Korea, and Japan. I think that many people are becoming concerned with economic psychology because they have recognized that standard economic policies and standard marketing policies sometimes fail to predict consumer's behaviors and decisions.

This book covers a range from classical to relatively recent major studies related to economic psychology. It comprises nine chapters—What is Economic Psychology (Chap. 1), Rational Choice and Revealed Preference (Chap. 2), Expected Utility Theory and Economic Behavior (Chap. 3), Nonlinear Utility Theory and Prospect Theory (Chap. 4), Mental Accounting and Framing (Chap. 5),

Multi-attribute Decision making Process in Economic Behavior (Chap. 6), Deployment on the Consumer's Interaction Research (Chap. 7), and Consumers' Preference Construction, Affects, and Neuroscientific Research (Chap. 8).

Chapter 1 introduces the concepts of economic psychology and historical perspectives of the research frameworks, such as the hermeneutic paradigm, utility theory paradigm, psychoanalytic paradigm, macroeconomic psychological paradigm, social psychological paradigm, behavioral decision theory paradigm, and behavioral economics paradigm, and explains that all of these are closely related to one another. Chapter 2 describes that a rational choice is based on the premise that at least the best option can be selected and the options can be ordered in descending order of preference from axiomatic properties such as completeness and transitivity, and provides an explanation of revealed preference concepts relation to the rational choice while presenting examples that do not satisfy transitivity, which is the premise of the ordinal utility concept. Chapter 3 highlights the relation to expected utility theory, which is often used in economics and psychology, and presents empirical research findings related to the axioms of expected utility theory and the Allais and Ellsberg Paradoxes. Chapter 4 introduces basis of the general expected utility of Choquet integral, and the prospect theory that Kahneman and Tversky proposed, and illustrates the economic phenomenon that is explainable using this theory.

Chapter 5 focuses on the economic psychology of mental accounting especially in price judgment where people organize and understand price information, and interprets the problems of mental accounting and framing by using the mental ruler model proposed by the author. Chapter 6 outlines the theoretical framework that describes the process of consumers' multi-attribute decision making and the method for its analysis, showing empirical and theoretical research findings. Chapter 7 explains game theory as a theoretical viewpoint to examine consumer interaction and discusses the research on happiness in relation to people's social welfare. Finally, Chap. 8 explains the consumer's preference construction and the influences of affects such as mood on the decision making process from the viewpoint of prospect theory and related models especially in the fields of economic psychology and neuroeconomics as neuroscientific research of economic behavior.

Reading this book requires no advanced expertise. Nonetheless, introductory knowledge of psychology, business administration, and economics and approximately high school graduate-level mathematics should improve a reader's comprehension of the content. In addition, each chapter includes corresponding references, which can be referred to when studying more details related to economic psychology.

This book is based on the Japanese book entitled "Economic Psychology: Psychological Foundations of Behavioral Economics" published by Baifukan Co., Ltd. in 2015. The information provided in this book has been also used for lectures at Waseda University, Gakushuin University, Rikkyo University, The University of Tokyo, Tokyo Institute of Technology, Nagoya University, Kansai University, Osaka University of Human Sciences, Osaka University, Kobe University, University of Tsukuba, Saint Petersburg State University, Russia, National Cheng

Kung University, Taiwan and Venice International University, Italy. Questions and answers exchanged with students at all of those places have contributed greatly to the compilation of this book. Particularly, I have received highly valuable opinions from graduate students taking the Takemura Seminar at Waseda University and from researchers in decision making studies through usual discussions. Above all, Mr. Hajime Murakami, Mr. Keita Kawasugi, Mr. Rammaru Watanabe, and Ms. Momo Tamura of Waseda University helped with some of the proofreading, corrections, and editing files of figures and tables. Ms. Tayeko Kondo of Baifukan gave me helpful comments to the Japanese book. The Springer team, especially the editor, Yutaka Hirachi, has been helpful from the inception of this book project.

The research discussions and workshops for the Experimental Social Science Project (headed by Prof. Tatsuyoshi Saijo at Kochi University of Technology) and Economic Behavior Project conducted under a Grant-in-Aid for Scientific Research on Priority Areas of The Ministry of Education, Culture, Sports, Science and Technology (No. 19046007), Grant-in-Aid for Scientific Research A (No. 24243061, No. 16H02050, No. 18H03641, and No. 19H00601), Grant-in-Aid for Scientific Research B (No. 16H03725, and No. 16H03676), Grant-in-Aid for Scientific Research C (No. 17K03637), Grant-in-Aid for Challenging Exploratory Research (No. 18K18570, and No. 18K18701), and Waseda University Grand-in-Aid for Research have allowed me to exchange opinions with researchers from various fields including experimental psychology, behavioral economics, social psychology, consumer psychology, and experimental economics. Professor Satoshi Fujii at Kyoto University, who has been conducting joint research on economic decision making for nearly two decades, also provided me with extremely informative advice and suggestions on a regular basis. A part of our joint research is introduced in this book. Professor Hidehiko Takahashi at Tokyo Medical and Dental University, Prof. Yutaka Nakamura at University of Tsukuba, Prof. Yoichiro Fujii at Meiji University, Prof. Takayuki Sakagami, Prof. Toshiko Kikkawa, the late Mr. Shigetaka Ohkubo at Keio University, Prof. Kaori Karasawa at the University of Tokyo, Prof. Naoko Nishimura at Shinshu University, Prof. Yutaka Matsushita at Kanazawa Institute of Technology, Prof. Mieko Fujisawa at Kanazawa University, Prof. Tsuyoshi Hatori at Ehime University, Prof. Yumi Iwamitsu at Kitasato University, Prof. Mikiya Hayashi at Meisei University, Prof. Takashi Ideno at Tokuyama University, Prof. Yuki Tamari at University of Shizuoka, Prof. Henry Montgomery at Stockholm University, Prof. Marcus Selart at the Norwegian School of Economics and Business Administration, Prof. Michael Smithson at the Australian National University, Prof. Yuri Gatanov at Saint Petersburg State University, Prof. Baruch Fischhoff at Carnegie Mellon University, Prof. Colin Camerer at California Institute of Technology, and Prof. Cheng-Ta Yang at National Cheng Kung University have given me useful comments for our joint research on economic decision making through daily practice, which also benefitted this book.

I have been participating in the 30-year-old Cognitive and Statistical Decision Making Research SIG (headed by Prof. Kazuo Shigemasu at Keio University) from its inception. Moreover, I continue to learn much from researchers in economic psychology and decision studies such as Prof. Gerrit Antonides of Wageningen University, Prof. Tommy Gärling of Gothenburg University, Prof. David Leiser at Ben-Gurion University of the Negev, Prof. Tomasz Zaleskiewicz of SWPS University of Social Sciences and Humanities, Prof. Ola Svenson of Stockholm University, Prof. Shuzo Abe, Prof. Hiroo Sasaki, Prof. Mamoru Kaneko, Prof. Yukihiko Funaki, Prof. Tsuyoshi Moriguchi, Prof. Naoto Onzo, Prof. Kazumi Shimizu, Prof. Kenpei Shiina of Waseda University, Tetsuo Sugimoto of Sophia University, Prof. Mitsuro Nagano of Kyoto Tachibana University, Prof. Taizoh Motonishi, Prof. Yasuhiro Ukai of Kansai University, Prof. Manabu Akiyama of Kobe Gakuin University, and Prof. Makoto Abe of University of Tokyo.

I am most appreciative of the guidance and encouragement offered by predecessors such as the late Prof. Sotohiro Kojima (Doshisha University), Prof. Osamu Takagi (Kansai University), Prof. Kazuo Shigemasu (Keio University), Prof. Nozomu Mastubara (Seigakuin University), and Prof. Tomio Kinoshita (International Institute for Advanced Studies).

Finally, this book is the fruit of valuable advice from numerous people with whom I have become acquainted but whose names have not been put into print here. I am truly grateful for all of their support.

Tokyo, Japan Kazuhisa Takemura

Contents

About the Author

Kazuhisa Takemura is a Japanese psychologist and educator. Besides serving as a professor at the Department of Psychology, Waseda University, he is also director of the university's Institute for Decision Research, a professor at the Waseda MBA School, and a research fellow at the Waseda Research Institute for Science and Engineering.

He received his B.A. and M.A. from the Department of Psychology, Doshisha University, in 1983 and 1985, and received his Ph.D. (System Science) from Tokyo Institute of Technology in 1994, and an additional Ph.D. (Medical Science) from Kitasato University in 2013.

He has also worked abroad as a visiting researcher at: James Cook University, La Trobe University, and Australian National University (Australia); the Tinbergen Institute (the Netherlands); Gothenburg University and Stockholm University (Sweden); the University of Konstanz (Germany); and National Cheng Kung University (Taiwan). He was also a Fulbright Senior Researcher at the Department of Social and Decision Science, Carnegie Mellon University (USA) from 1999 to 2000, and a Visiting Professor at the Department of Psychology, St. Petersburg State University (Russia) in 2008.

His main research area is human judgment and decision making, especially the mathematical modeling of preferential judgment and choice. He received a Hayashi Award (Distinguished Scholar) from the Behaviormetric Society in 2002, an Excellent Paper Award from Japan Society of Kansei Engineering in 2003, Book Awards from the Japanese Society of Social Psychology (in 2010) and the Behaviormetric Society (in 2016), and a Fellow Award from the International Association of Applied Psychology in 2018.

In the course of his career, he has taught extensively on behavioral decision theory, marketing, economic psychology, consumer behavior, social psychology, and psychometrics at many universities (Tokyo University, Osaka University, University of Tsukuba, Kobe University, Nagoya University, Tokyo Institute of Technology, Gakushuin University, Rikkyo University, Tokyo International University, and St. Petersburg State University).

List of Figures

List of Tables

Chapter 1
What Is Economic Psychology? The Perspective of Economic Psychology and the Research Framework

Keywords Economic psychology · Economic behavior · Consumer behavior · Marketing · Economics

Economic psychology studies the psychological aspects behind various economic phenomena and the mechanisms of people's judgment, decision making, and behavior in an economic situation. Traditional economics is the study of how limited resources are used for production and distribution, with the aim of exploring economic phenomena through theoretical and empirical analyses. However, traditional economics has mainly studied economic phenomena by assuming the rationality of individuals and assuming the concept of "homo economicus," who makes a rational choice. With this background, an area called "experimental economics," in which theoretical hypotheses on economic phenomena are examined through experiments similar to those conducted in psychology, has emerged in recent years. Moreover, the term "behavioral economics" has also spread since Daniel Kahneman, a psychologist, received the Nobel Prize in economics in 2002. Recently, R. Thaler who proposed mental accounting concept of consumer psychology also received the Nobel Prize in Economics in 2017. Meanwhile, economic phenomena related to consumer behavior have been empirically studied in management science-oriented studies, such as marketing and consumer psychology, which are integrated with research on behavioral economics. Certain concepts of economic psychology overlap with some ideas studied in these areas.

K. Takemura, *Foundations of Economic Psychology*,
https://doi.org/10.1007/978-981-13-9049-4_1

1.1 Economic Behavior as a Subject of Economic Psychology and Related Academic Societies

1.1.1 Economic Behavior

Economic behavior is a collective term for acts of judgment and decision making as well as behaviors displayed by humans in different socioeconomic situations. Economic behavior includes production (i.e., a series of behaviors wherein goods are made by using resources such as land) and consumption (i.e., a series of behaviors wherein people satisfy their desire with the consumed goods). In modern society, where the system of division of labor is established, some people are producers and the rest are consumers. Accordingly, as examples of economic behavior, consumers' judgment, decision making, and other behavioral aspects are emphasized in the research in this field.

Although production can be considered part of economic behavior, the latter is placed at the center of economic behavior research. Consumer behavior occupies an important position in the fields of economics, marketing, and industrial psychology. Consumer behavior is roughly divided into three aspects. The first aspect is the pre-purchase behavior. This consists of judgment, decision making, and behavior regarding the consumers' plans at the preparatory stage for purchasing goods and services. In other words, pre-purchase behavior mainly focuses on (1) allocating income to saving and consumption; (2) planning budgets, such as how to allocate income for consumption of which items; (3) forming product knowledge; and (4) forming knowledge about purchasing destinations and purchasing stores.

The second aspect is purchase behavior, which refers to the judgment, decision making, and behaviors directly related to the purchase of goods and services. In other words, purchase behavior includes (1) considering the types of products and services (decision of choice sets), (2) deciding purchasing destinations, (3) deciding purchasing stores, (4) deciding brands, (5) selecting a detailed model from the brand (selection of colors, design model numbers, optional functions, etc.), (6) deciding the quantity of the brand to be purchased, and (7) repetitive behavior of purchase, among others.

Finally, the third aspect of consumer behavior is post-purchase behavior. This kind of behavior refers to consumers' judgment, decision making, and behavior after purchasing goods and services. In other words, post-purchase behavior includes (1) how consumers receive services and how products are used, (2) what kind of experiences are provided as a result of the use behavior and what kind of evaluation judgment is made about the services and goods (evaluation judgment after use), (3) how the products are stored and maintained, (4) how the products are discarded, and/or (5) how the products are recycled, among others.

1.1.2 Academic Societies Related to Economic Psychology

Traditional economics, under the assumption of the so-called concept of "homo economicus," mainly explored economic phenomena by assuming the rationality of individuals. Recently, an area called experimental economics has emerged, with the aim of examining theoretical hypotheses on economic phenomena by conducting experiments. In addition, the term "behavioral economics" became popular after psychologist Daniel Kahneman received the Nobel Prize.

With respect to the psychological research of economic behavior, we can at least trace back to research on advertising psychology by W. D. Scott and D. Starch at the beginning of the twentieth century and the research on industrial psychology by M. Münsterburg. In addition, J. B. Watson, who initiated behaviorism at the beginning of the twentieth century, also succeeded in the advertising industry after entering the business world.

The International Association for Research in Economic Psychology (IAREP) is an international academic society of economic psychology founded in 1982. By referring to the website of this academic society (http://www.iarep.org), one can understand that it studies an area where economics and psychology intersect and is also related to research on business management, marketing, and consumer behavior. This academic society aims to study both the psychological mechanisms generating economic activities and the psychological effects of economic phenomena, with the results of the study being published through the *Journal of Economic Psychology*.

In economic psychology, the focus is consumer behavior, which is being investigated by researchers in various areas. Research on consumer psychology in recent years has not only focused on the mere application of cognitive psychology and social psychology but has also produced some unique theories. Indeed, the aforementioned consumer decision making research represented by Bettman et al. is based on their unique theory and methodology and is forming an individual area even in the psychonomic society in the USA, which deals with basic perception and cognition.

Representative periodical publications, such as academic journals related to economic psychology, are listed below.

(1) *Journal of Consumer Research*: This is a traditional academic journal in the consumption behavior research first published in 1974. Published by the University of Chicago Press, this journal carries interdisciplinary papers, such as those on psychology, marketing, economics, and sociology.
(2) *Journal of Consumer Psychology*: This is an official academic journal of the Society for Consumer Psychology.
(3) *Advances in Consumer Research*: This is a collection of papers in the annual conference of the society. This is published by the Association for Consumer Research (ARC).
(4) *Psychology and Marketing*: Published by Wiley Europe, this is an interdisciplinary academic journal on psychology and marketing that carries many papers related to consumer psychology.

(5) *Journal of Economic Psychology*: This is an academic journal of economic psychology and behavioral economics published by the International Association for Research in Economic Psychology.
(6) *Journal of Marketing Research*: This is an academic journal of marketing research published by the American Marketing Association (AMA). As a journal that is dedicated only to consumer behavior research, it carries experimental psychological papers on consumers.
(7) *Journal of Behavioral Decision making*: This is an academic journal concerning decision making and behavioral decision theory and carries a relatively large number of consumer decision making papers and is published by Wiley Europe.

Box 1: Herbert A. Simon

Born in 1916; deceased in 2001. He worked as a professor at the School of Computer Science and Psychology Department at Carnegie Mellon University and died in Pittsburgh in 2001. In 1978, he was awarded the Nobel Memorial Prize in Economic Sciences for his achievements in the research of human decision making process in organizations.

Photograph: Toshiya Tanaka, Kansai University

1.2 The Perspective of Economic Psychology

1.2.1 Micro and Macro in Economic Psychology

Economic psychology, like economics, has a micro-perspective, which views things through individual economic entities (e.g., the decision making behavior of one consumer), as well as a macro-perspective, which views things through the macro-level of economic phenomena (e.g., depression and unemployment). Economic psychology analyzes economic phenomena from the perspective of decision making in individual consumer behaviors, while microeconomics analyzes economic phenomena based on individual and household decision making activities and macroeconomics deals with macroeconomic phenomena. Economic psychology can analyze, from the macro-perspective, macro-phenomena such as product spreading among other popular phenomena.

Although both micro- and macro-viewpoints are important, considering the interaction between micro- and macro-phenomena is also necessary. For example, a popular phenomenon occurs and subsides suddenly, but because it can be thought of as an accumulation of individual social judgment and decision making, one can understand the cumulative popular phenomenon from the micro-perspective of individual decision making. The relationship between such micro- and macro-perspectives has not been fully explored so far, but it will make an important research topic in the future.

1.2.2 Process and Consequences

On the one hand, economic psychology analyzes the processes of economic behavioral phenomena. For example, when focusing on the decision making process as a micro-individual consumer, it is important to view the process by which a consumer finally decides to buy a product, including the process before the purchase, the reasons for selecting a brand, relevant information collection, and comparison with other brands. Moreover, analyzing the process of popularity and word of mouth is also important. On the other hand, apart from processes, economic psychology also focuses on the outcomes of economic behavior. For example, one must focus on results, such as preferred brands, the factors influencing the final purchase decision making, and the factors promoting a product‘s popularity, to further analyze economic behaviors. The analysis from this perspective rather emphasizes the aspect of “prediction.”

Economic psychology makes analyses from the perspectives of “process” and “result.” In other words, to successfully predict the results, one must identify the process. Moreover, explanations that cannot predict the results are not very useful. From these facts, it is understood that both process analysis and result analysis are necessary for research on economic psychology.

1.2.3 Explanation by Natural Language and Mathematical Explanation

In psychology, experimental data and survey data are often analyzed quantitatively, but in the theoretical analyses, except in research areas like visual perception and learning, most of the theoretical explanations are made using language (natural language). Meanwhile, in economics, except for areas like Marxian economics and economic thought, mathematics is typically used as the theoretical language. In economic psychology, an area of study where economics and psychology overlap, theoretical descriptions using mathematical expression and natural language are often used. To a certain extent, it can be expressed mathematically and it is actually described that way. Mathematical descriptions are possible not only for individual decision making and preference relationships but also for simple interactions that can be explained using game theory. In contrast, mathematically expressing popularity phenomena, consumer trends, and psychological processes within individuals is often difficult because these feature complicated interactions that require linguistic explanations.

In economic psychology, explanations by natural language and mathematical expressions are often used in combination and both are deemed useful in describing various phenomena. Mathematical expressions are suitable for explaining phenomena with a clear structure. For highly complicated phenomena with ambiguous structures, linguistic expression is more suitable. Therefore, to achieve balance, there is a need to acquire perspectives from both descriptions. Nevertheless, as mathematical expressions are considered scientific explanations, they can be better applied in other research areas. Moreover, because mathematical expressions are best used to make predictions, using these as foundation for developing a research track is an ideal option.

1.2.4 Qualitative Analysis and Quantitative Analysis

In psychology and economics, quantitative analysis of data is the mainstream approach, and data are often analyzed statistically. However, qualitative analysis is also an important method for economic psychology. Especially, it is extremely effective to use qualitative analysis methods for cases where phenomena are difficult to quantitatively describe or in studies wherein subjects are difficult to quantify. Economic phenomena that can be expressed in mathematical theoretical language can be easily analyzed quantitatively, but mathematical theories do not always correspond to quantitative analysis, and theoretical explanations by language do not always correspond to qualitative analysis either.

Economic psychology is a field of study that aims to describe human economic activities and explore human judgment and decision making in the economic situation mainly by conducting measurements based on experiments and surveys. Therefore, for this purpose, the measurement of the results of judgment and decision making must be theoretically guaranteed. People's judgment and decision making are qualitative in many cases, so from this qualitative relationship, researchers must have ideas of theories and measurement that guarantee quantification.

Various theories of measurement can be used, and these can be roughly divided into representational measurement and psychometric measurement based on the axiomatic measurement theory in mathematical psychology. The first type of measurement method, which is based on the axiomatic system, expresses an empirical relational system by a numerical relational system. This method basically attempts to construct a proportional scale and interval scale from ordinal scale judgment. Axiomatic measurement theory is a representative example of this approach. According to this theory, measurement is the process of associating an empirical relational system with a specific numerical relational system. First, the empirical relationship is a relationship between subjects observed empirically. For example, the preference relationship wherein a decision maker prefers B over option A is an empirical relationship. The approach based on this axiomatic measurement theory is mathematically the same as the axiomatic theory of utility, which expresses the preference relationship by the magnitude relationship of real numbers.

In comparison, the second type of measurement method is an extension of the traditional method of psychometrics, which presupposes the quantifiability of the economic phenomena of utility. Especially in economic psychology, conducting measurements based on the idea that psychological variables have random error terms is the mainstream approach. This includes quantitative methods based on psychophysical measurement and rating scales in general psychology. In addition, the willingness-to-pay (WTP) method that answers the payment intention amount and is used in the evaluation experiment, among others, can also be included in the psychometric measurement.

1.2.5 Abstraction and Materialization of Economic Behavior

The problems of abstraction and materialization can be encountered in the field of economic psychology. Academic researchers in economic psychology tend to prefer abstract discussion, whereas practitioners tend to ask for concrete discussion.

It seems that practitioners often criticize academic researchers who are seeking theories with too much abstraction. As for the abstraction of the theory of economic psychology, there is a desirable aspect, but at the same time, there is also a danger of seeing the reality by just the abstracted theoretical construct. Abstracted theory is necessary in viewing the reality, and theoretical abstraction is necessary to recognize the actual complexity.

The word "abstraction" is considered to be "A mental effect of understanding things or representations by extracting their certain aspects and properties. In that case, there arises behavior that naturally eliminates other aspects and properties, and this is also referred to as abstraction." Such a definition is considered natural from daily usage. Thus, abstraction has an aspect that selects remarkable attributes from a subject and ignores other aspects. However, this may be useful for simplifying our understanding of the subject and for viewing it further.

The significance of abstract theory cannot be doubted. However, it ignores many elements of concrete reality and has less information to offer compared with other similar theories. Subjects in social psychology have interactions in many ways even if they are abstracted and simplified, so their modelization makes quite a complicated construct. Here, we can understand the significance of viewing social psychological subjects through abstraction. In other words, the real economic situation and the interactions among individuals and organizations are more complicated than the models themselves. Thus, when viewing the reality, researchers must develop a much more complicated recognition framework. Abstract theories seem more useful than a highly complicated recognition framework. Even if one tries to look at the reality, which may be too complicated, one cannot even recognize its complexity. I believe that abstracted theories facilitate such a recognition.

1.3 The Goal of Economic Psychology and Its Approach

1.3.1 Elucidation of Economic Phenomena and Relationship with Behavioral Economics

Research on economic psychology analyzes human behaviors in economic situations. Thus, it has a strong interest in the theoretical aspect of the scientific elucidation of human decision making and the further explanation of various socioeconomic phenomena. At the same time, economic psychology not only has links with business administration, economics, and applied psychology, it can also contribute to the marketing activities of governments, public organizations, and companies and can be useful in improving the lives of ordinary citizens through consumer protection. The former goal is a rather theoretical one, whereas the latter focuses on the practical aspect. Both perspectives are necessary; practice without a theoretical perspective tends to result in wrong intuitions and some hindsight, whereas theory without a practical perspective can lead us to the direction of the study for the study itself, which is likely to accepted only in a narrow academic discipline. Hence, if one has studied a field thoroughly, one would come across many other studies and an association with practice. In the next part, I would like to explain the goals of the following items and describe their relationships with the adjacent study areas and practical problems.

Elucidating economic phenomena is a very important goal for economic psychology. Whether the elucidation of economic phenomena is done from a micro- or macro-perspective, it is important to observe economic phenomena and generate theoretical explanations for these. Regarding the theoretical explanation of economic phenomena, there is a considerable accumulation of such in traditional economics. As mentioned earlier, economics often uses mathematical models for theoretical statements and often assumes rationality on the part of the individual decision makers. Rationality, in a broad sense, is not preposterous decision making or behavior but is a property that allows an individual to make a convincing explanation when looking back on one's own behavior. Usually, however, it is understood to mean that one acquires information with as many options as possible, use all information, have a consistent preference relationship, and make the best decision from these options. Conversely, such an assumption is advantageous as it is compatible with a mathematical model and is easy to analyze. Moreover, given that the behavior of actual individuals and organizations features rationality to a certain extent (but not completely irrational), we cannot say that this assumption is completely wrong. However, assuming perfect rationality on the part of the decision makers in economic situations is an inappropriate explanatory theory in certain situations. For example, there are cases where personal decision making changes due to essentially irrelevant information or where economic panic phenomena occur due to the influence of others (e.g., the phenomenon of panic buying toilet paper), and it is often difficult to explain them from a full rationality model. Therefore, traditional economics alone is not sufficient for describing the wide range of economic phenomena that can be observed.

Meanwhile, psychology has a history of more than 100 years in terms of making accurate measurements; hence, there is a considerable accumulation of quantitative measurements of various phenomena, and many past studies have been conducted from the aspect of economic psychology. However, as psychology does not assume rationality very strongly as in traditional economics, there is no powerful theory at present. Therefore, in economic psychology, together with descriptive measurement and quantification of economic phenomena, theoretical explanations can become issues of further research.

The elucidation of economic phenomena is not only theoretically important but it also has a close relationship with economic policies and practice in society. In Japan, for example, some economists are appointed to become members of the advisory bodies of the government, as special advisors to the Cabinet Secretariat, and ministers; hence, economic research has a strong influence on economic policies. Other countries have the same circumstances. However, if politics and social policy are being carried out based on the theories with wrong assumptions, this can be quite problematic. In recent years, under these backgrounds, the areas of **behavioral economics** and **experimental economics** have emerged, and more people are thinking about economic theories based on the observation of people's actual behaviors. However, there are still many unknown occurrences of socioeconomic phenomena related to economic behavior. For example: How is the business condition related to consumer psychology? What influence does the consumption tax have on consumer demand? What kind of mathematical model is appropriate for approximating consumer decision making? There are still many important issues that have yet to be addressed. From these viewpoints, elucidating the actual behavior of consumers can continue to help clarify various socioeconomic phenomena. In fact, recent findings in the field of behavioral economics have had some degree of influence on economic policies. Thus, even in this sense, I believe research on economic psychology can develop further in the future.

1.3.2 Marketing

Economic behavior research provides practical viewpoints regarding the marketing activities carried out by producers (companies and public organizations) and has a theoretical contribution to marketing. Many people associate the word "marketing" with advertisements or efforts made to sell (products or services). Those activities, of course, comprise one aspect of marketing activities, but there is more to it. As marketing was originally part of commercial activities, it can be traced back to ancient Babylonia and ancient Japan. It was first developed in the USA, where mass production started at the beginning of the twentieth century. Thus, it would be a good idea to examine how it is used in the USA.

According to the 1985 definition of the American Marketing Association (AMA), a representative organization concerning marketing in America, marketing is defined as the "processes in which in order to create an exchange that satisfies the purpose of individuals and organizations, concept formation, price setting, promotion, and distribution regarding ideas, products, and services are planned and carried out." What is noteworthy in this definition is that marketing is not limited to corporate profit activities or promotional activities such as sales or advertisement. Therefore, the advertisement of public entities is considered to be marketing activities. In fact, schools, hospitals, and local governments are also consciously doing "marketing activities." In addition, we know that during the summer, electric power companies are promoting "energy conservation" as part of their public relations campaigns, which can also be classified as marketing activities. In 2007, the American Marketing Association extended its previous definition as follows: "Activities to create and transmit as well as to distribute and exchange valuable offerings for customers, clients, partners and society as a whole, and are a series of systems, and processes." In this way, marketing has become a fairly broad concept. This recent definition emphasizes society as a whole, focusing not only on the activities of mere commercial organizations but also those of public organizations and general individuals. As the term "social marketing" exists, we can easily see that marketing is also used for non-commercial purposes, such as communicating the risks of various hazardous substances or disseminating information on traffic safety issues to the public.

In marketing, the "4P" strategy has gained popularity in recent years. Consumer behavior research makes a very important contribution to the development of the 4P strategy. Let me explain this. The producer comprehensively develops a marketing strategy from the aspects of (1) **p**roduct, (2) **p**rice, (3) **p**lace or distribution, and (4) **p**romotion (4P) (Kiwata, Kakeda, & Mimura, 1989; Kotler & Armstrong, 1997; Sugimoto, 2012). Marketing practitioners are engaged in marketing activities by conducting marketing mix, which combines these factors. Figure 1.1 shows the factors of the 4P strategy and its specific implementation items (Kotler & Armstrong, 1997). In implementing the 4P strategy, utilizing the findings of consumer psychology is of utmost importance (Takemura, 2000).

First, in the product strategy, at the stage of providing products and services, it is necessary to know what kind of products and services consumers want and what kind

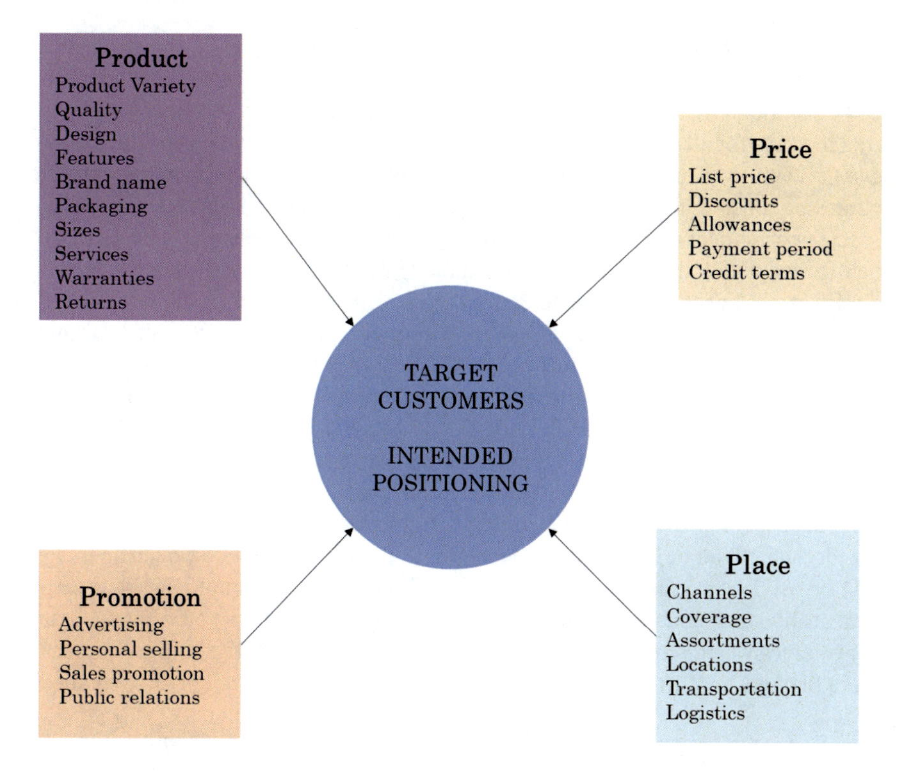

Fig. 1.1 Four Ps of the marketing mix. *Source* Kotler and Armstrong (1997)

of products they think are attractive. A product is preferred when it is recognized from the consumer's viewpoint that it has better features than other products. This is what is called "product differentiation" in marketing terms, and marketing activities will not succeed without it (see, e.g., Aoki & Onzo, 2004). For example, Tokyo Disneyland and Disney Sea are setting high entry fees compared with other amusement parks, but many people still visit and pay such high prices. In fact, these Disney resorts have welcomed nearly 30 million visitors annually in recent years. The reason why so many people visit these places despite the high entrance fees is that they achieved differentiation from other amusement parks. In other words, they have succeeded in distinguishing themselves from other amusement parks. They do so by incorporating unique amusement park features that change each year, employing staff providing high-quality service, and coming up with various schemes to produce extraordinary experiences. Hence, if companies succeed in product differentiation, consumers will spare no expense to patronize them even at high prices. Therefore, the important thing is to know what kind of product can differentiate itself from other products. Conversely, even if a technically high differentiation is made, if the product is not perceptually differentiated for consumers, the product strategy will not succeed. Here lies the importance of understanding the behavior of consumers.

Next, in pricing strategies, to improve sales, the act of setting prices must consider the consumers' purchase intention (see, e.g., Kojima, 1986; Ueda & Moriguchi, 2004). Although the cheaper price may be more attractive for some consumers, this is not the case for all consumers, and even if the product is similar, it does not always have the same cost rate. For example, prices of nutritional drinks, cosmetics, and so on, are set considerably higher compared to cost pries, but this strategy uses a key aspect of consumer psychology: The higher-priced product is assumed to be the better one in terms of quality. Generally, it is assumed in economics that demand decreases as price rises, but there are products for which demand grows as price rises (also known as "Giffen goods" in economics). Moreover, most of the refreshing drinks sold by vending machines are 120 yen or 150 yen; they are uniformly set regardless of cost. By analyzing consumer psychology and consumer behavior, effective price setting can thus be made.

In the distribution strategy, sellers consider what kind of distribution channel, including wholesale shops, retail shops, directly-managed stores, and mail-order sales, to use for providing products to consumers. At this point, they must also consider what kind of image consumers have of the product and how such an image must be used (see, e.g., Kobayashi & Minami, 2004). For example, in the case of products like toothbrushes, canned coffee, and seasonings, which you want to obtain immediately, one distribution strategy employed is to increase the number of stores where they are sold in order to increase opportunities for consumers to know these products.

In comparison, such a distribution strategy is not effective when selling a luxury product, such as Prada and Gucci handbags. What would happen to the image of these brands if anyone can easily buy them at convenience stores or supermarkets? Consumers highly appreciate these brands because they can only be bought at special luxury goods stores. Thus, when considering which distribution strategy to employ, sellers must be able to understand the psychology and behavior of the target consumers. In terms of distribution strategy, one measure to increase sales opportunities is by expanding distribution channels as much as possible, and another measure is to sell products only at specialty shops and direct sales stores, allowing the manufacturer to strictly exercise control through restricted distribution channels. Either measure can be adopted depending on how consumers perceive and recognize the product being sold.

Finally, in a promotion strategy that considers how to advertise and sell, sellers must also consider the psychological effects a strategy would have on the target consumers (see Ueda & Moriguchi, 2004). Specifically, advertising agencies analyze what kind of personality to use for advertisements, how much sales growth can be achieved with this celebrity, and how effective would the strategy be. The behavior analysis of consumers can also help in exploring how much attention consumers are paying to advertisements in stations and trains. In addition, detailed behavior observations of consumers at physical stores can reveal what kind of store information they pay attention to and which shelf they easily purchase products from. Such information help support future marketing activities at stores. These analyses also enable practitioners to consider effective promotion strategies based on the targets identified.

In recent years, it has been suggested that a marketing strategy that uses only the 4P strategy is no longer sufficient, paving the way for the emergence of 5P and 6P strategies (now including "**p**eople" and/or "**p**rocess"). When considering international marketing strategies, one must also pay due attention to people's consumption culture and their regionality. Findings in the field of consumer psychology are useful in considering such cultural and regional issues (Takemura, 2000).

Studies on marketing and consumer behavior research are closely related to the aforementioned practical goals. In consumer behavior research, the research approaches of various consumer psychologies and consumer behaviors are used, and the methodologies are diverse, including psychological experiments, surveys, interviews, mathematical psychology model configurations, and computer simulations. Research on consumer psychology is no longer just a mere application of cognitive psychology and social psychology as in the past but now produces unique theories. Indeed, the aforementioned consumer decision making research represented by Bettman et al. is based on their unique theory and methodology. This emerging field is also instrumental in forming a subfield in the psychonomic society in the USA, which deals with basic perceptions and cognition.

1.3.3 Public Policy for Consumers

Research on economic behavior is also useful in designing public policies to protect the rights of consumers (Takemura, 2000). Since the war, there have been cases such as the Morinaga arsenic milk case in 1955, the thalidomide case in 1962, and the drug-induced SMON disease, in which the defective products purchased by consumers caused serious health damages, accidents, and monetary damages due to so-called confidence tricks (Nishimura, 1999). Hence, effective policies are needed to prevent the occurrence of similar events in the future and to protect consumers' rights as well.

First, knowledge of economic behavior is necessary when conducting risk communication with consumers (Kikkawa, 1999, 2000). Even now, consumers are surrounded by various risks/hazards, such as accidents caused by defective cars or the improper use of products or food poisoning. When there is a possibility that consumers will be harmed when purchasing or using products, companies and public organizations (e.g., governments) should disseminate information on these risks and respond to consumer questions and complaints. Hence, they need to know what kind of knowledge consumers have, what kind of behavior they take, and what kind of information, when provided, would they understand well and are convinced by. For example, even if an organization describes hazards in detail in the instruction manual of a product, that alone does not guarantee the accurate communication of related risks. Consumers only read such information unwillingly or ignore it when they are provided too much information. The busier the consumer, the higher chances there are of them missing or ignoring information. Therefore, researchers and practitioners must grasp the psychology of consumers and think about communication through which consumers can be convinced. Further, in risk communication of food, misunderstanding of toxic substances can lead to a host of health-related damages. Thus, gaining a better understanding of consumers' cognition processes and behaviors is essential.

Next, findings of research on economic psychology and economic behavior are indispensable for consumer protection policies implemented by public organizations such as governments. For example, at consumer centers established for the purpose of consumer protection, counselors are assigned, negotiations with companies are conducted within the scope of law by the request of consumers, and aid for consumer damage and support for problem solving are provided (Nishimura, 1999). In order to consider the response of counselors to the damaged consumers, it is necessary to understand what kind of knowledge consumers have and what kind of behavior they take.

In addition, knowledge on consumer behavior is necessary in the enactment and implementation of laws for consumer protection. For example, when a consumer intends to take legal action against a company against a defective product-related accident, as recognized in the enactment of the so-called PL law, the consumer's burden of proof (i.e., the consumers bear a responsibility for proving the cause on which the action is based) is becoming more relaxed in recent years, shifting gradually to the company side. Such a revision of laws has been made because the authorities now realize the actual situation: The average consumers' product knowledge is considerably lower than that of companies. Even in trials, the characteristics of average consumers' knowledge and use behavior cause arguments, and sometimes the analysis of consumer behavior is carried out as evidence for that.

Finally, the findings of economic behavior research are necessary for improving consumer education on the part of the companies, the public organizations, and the consumers themselves (usually in the form of organizations promoting consumer rights). Consumer education from the companies includes the effective dissemination of product knowledge, good complaint management practices, and information exchange with consumer organizations. Especially since the enactment of the PL law, companies have been working diligently on consumer education as part of their management strategy. For example, regarding defective goods, some companies are studying responses, through which they can understand consumers' perceptions, allowing these firms to design effective management strategies. Consumer education carried out by public organizations includes such activities as the prevention of consumer damage and the provision of consumer aid and enlightenment.

In China, these types of consumer education have been partly integrated into the social education and school education through the former Ministry of Education. In the 1990s, the National Institute on Consumer Education was founded and jointly supervised by the Economic Planning Agency and Ministry of Education. In 1997, the Economic Planning Agency launched a dispatch system of consumer education experts, and since then, the Agency has been engaging in consumer education throughout the country (Nishimura, 1999). In implementing such consumer education activities, understanding the actual state of consumption behavior and clarifying the processes leading to damages are quite crucial. Consumer education includes enlightening activities, such as the dissemination of knowledge through product testing, as well as learning events related to environmental protection and consumer rights. Consumer organizations such as the Housewives Association, Japan Consumers Association, and Life Club Co-op are conducting such consumer education programs. Even in consumer education initiated by consumers themselves, there is a need to know what kind of knowledge consumers have, what points they misunderstand resulting in erroneous use behavior, and what kind of information they seek. In recent years, problems such as ore fraud and fraud of financial products are very much in the news. Hence, studies on economic psychology and economic behavior are indispensable in dealing with such problems. This kind of research is being carried out primarily by social psychologists.

1.3.4 Elucidation of Decision Making in Economic Behavior

Research on economic behavior also helps elucidate economic phenomena. Specifically, research on decision making in economic behavior is useful for the fields of psychology, economics, and neuroscience.

As an example, let us consider the concept of utility, which is often used as the basic concept for elucidating socioeconomic phenomena, and the system of utility theory based on this concept. Utility theory, as explained in a later chapter, is used in most theories in economics and is often employed in business administration and social engineering. In many cases, this also becomes the basis for the analysis of socioeconomic phenomena.

Let us consider decision making under certainty, wherein one selects a product: either brand a or brand b. In this case, utility means an actual value wherein the utility (u (brand a)) of brand a is higher than the utility (u (brand b)) of brand b, when and only when brand a is preferred to brand b. That is, u (brand a) $\geqq$ u (brand b) $\Leftrightarrow$ expresses the relationship wherein brand a is equal or preferred to brand b. This preference relationship is expressed by the utility function u. With regards utility theory based on the concept of utility, we can use some counterexample phenomena that cannot be justified from the perspective of consumer psychology. As described in a later chapter, human preferences do not necessarily satisfy transitivity (Tversky, 2004). Transitivity is the relationship between two options; for example, if you prefer oranges to bananas and prefer apples to oranges, then you prefer apples to bananas. This notion indicates the nature of a consistent preference relationship. There is also preference reversal, a phenomenon that deviates from the procedural invariance that preference is not reversed by the procedure of preference statement. This phenomenon cannot be explained simply by the idea of traditional utility theory (Takemura, 2014). In this phenomenon, for example, in the purchase decision making situation, a person prefers brand a when he/she is allowed to identify the evaluation value (the amount which he/she thinks is acceptable to pay, etc.) of both brands independently; yet, when he/she is made to actually choose either brand by comparing the two, he/she ends up choosing brand b. In economic psychology, decision makers think about the reasons why consumers sometimes cannot indicate preference-satisfying transitivity or cause preference reversal. In explaining such a phenomenon, researchers proposed consumer decision making theory, which does not necessarily satisfy transitivity or procedural invariance. This can influence studies that aim to elucidate socioeconomic phenomena, such as those on economics, business administration, social engineering.

The findings of economic psychology demonstrate that human decision making often does not follow utility theory. Such studies also show that explanations by alternative theories, such as prospect theory, also shown in a later chapter, can better explain economic phenomena. In this research on decision making, we focus on human judgment and decision making processes in an economic situation and try to understand them by using process-tracking techniques, namely the language protocol method and information monitoring method. The process-tracking techniques also include physiological and psychological measurement techniques, such as the measurement of eye movement and skin electrical activities. Recently, a method using functional brain imaging has also been proposed. Based on the development of neuroeconomics, which aims at exploring the neuroscientific basis of human decision making behavior, the noninvasive brain activity measurement methods (i.e., imaging methods for brain functions and the blood flow in the brain, such as functional magnetic resonance imaging (fMRI) and the positron emission tomography apparatus (PET)) have been developed and a system has been established, demonstrating the cooperation among psychologists, economists, and neuroscientists as they attempt to clarify the findings based on previously conducted behavioral experiments (Takemura, 2014). Especially, C. Camerer in the USA and E. Fehr in Switzerland have made a great contribution to the development of neuroeconomics. Since the beginning, economic psychology has mainly focused on personal analysis as the target for the decision making and behavioral phenomena analysis. However, at present, apart from personal analysis, this field has also targeted other research areas, such as the analysis of interactions among individuals, decision making in a group, decision making in an organization, social decision making, and crowd behavior, and so on, which social psychology has been targeting so far. In recent years, as a derived area of behavioral economics, behavioral game theory has also emerged, which describes decision making behaviors under human interactions using the method of experiment games. In these kinds of research, various games are played in experimental situations based on game theory.

1.4 Theoretical Framework of Economic Behavior Research

Based on the historical background of economic psychology, some classifications of the theoretical framework of economic behavior are presented.

1.4.1 Hermeneutic Paradigm

Shionoya (2009) stated that economics and psychology originally had a very close relationship. Economics was thought of as moral science and mental science, and it was also related to the hermeneutics of W. Dilthey. J. M. Keynes stated that moral science "includes all the individual sciences treating a human being as one with subjective ability, that is, as being who senses, thinks, and wills." In early classical economics, the method of introspection was used to obtain knowledge of mental phenomena. However, apart from psychology, economics has asserted its independence from the former since the beginning of the twentieth century.

Even in the Austrian school, exemplified by C. Menger, in the economics of the nineteenth century, they had close relationships with the Brentano school of psychology. Meanwhile, the tradition of "psychologism" claims that economics must find the basis of ultimate explanations in psychological facts (Shionoya, 2009). Menger and his colleagues thought of psychological processes such as necessity, desire, and satisfaction as the "essence" behind economic phenomena, and thought that these psychological experiences exist in their foundation (Shionoya, 2009).

Brentano created the branch of psychology called "descriptive psychology" or "act psychology," and formed the so-called Würzburg School, which conflicted with the notion of "experimental psychology" and "content psychology" represented by Wundt. Meanwhile, F Brentano emphasized the concept of "intentionality," whose abstract intentional behavior separated from the subject content is not inherent in human beings as a psychological phenomenon but a mental activity wherein mental phenomena affect outside subjects.

Brentano and a contemporary, Dilthey, founded "hermeneutics" as a methodology of mental science based on Brentano's psychology. Dilthey asserted that research on human behavior differed from research on natural phenomena, and while reliving it in light of the human spirit and its historical development, it should be understood and interpreted in light of a broader understanding of the world, including the self and others (Shionoya, 2009).

According to Shionoya (2009), such hermeneutic tradition is also inherited by the present cognitive psychology and behavioral economics. From this viewpoint, it is now possible to think that the position that emphasizes subjectivity in the current economic psychology belongs to this hermeneutic tradition.

Another research approach, sometimes called postmodern research in the area of marketing research, which has also inherited the hermeneutic tradition in a broad sense. This approach, developed by Hirschman and Holbrook (Hirschman & Holbrook, 1992; Holbrook & Hirschman, 1993), sought to fully deal with problems of consumer symbols and meanings as well as the social contexts and circumstances involved in them. This approach takes a position similar to social constructionism and hermeneutic research in the fields of social psychology and sociology. In this approach, however, conventional research has paid little attention to the interactions that occur among consumers. Moreover, there is an implicit assumption that the mental mechanisms, such as memory and thinking, have universality among cultures, situations and individuals, and this notion has been critically examined in this approach. This approach makes qualitative analysis the main research method. Although it cannot be said that sufficient results have been obtained with regards empirical research, this approach is accurate as to the problem identification of conventional research, and its future development is expected.

1.4.2 *Utility Theory Paradigm*

The tradition of utility theory can be traced back at least to the formulation of the mathematician D Bernoulli in the eighteenth century. He submitted the idea of expected utility to solve the St. Petersburg Paradox (to be discussed later) introduced by N. Bernoulli. The basis of this idea is that humans do not make decisions based on mathematical expected values but do so according to the expected values of utility expressed by the monotonic increase function of the result. The idea of this utility theory developed in the Austrian school represented by Menger, and many economists have also begun to utilize this approach. Menger's Austrian school, as Shionoya (2009) pointed out, sought the essence of economic behavior in the desire satisfaction by goods and attempted to explain economic behavior based on the principle of utility maximization and the law of diminishing marginal utility. The utility maximization theory is based on the hypothesis that economic behavior can be explained by the maximization of utility expressing a preference relationship and is based on the hypothesis that the increment of utility decreases in accordance with the increment of the monetary value (the slope of the utility function decreases).

Originally, utility theory was different from the hermeneutic paradigm, but a psychologism tendency was suggested by the fact that the subjectivity of utility was taken up (Shionoya, 2009). However, P. A. Samuelson, an economist, incorporated the hypothesis of revealed preference (i.e., the utility function expressing the preference relationship could be directly estimated from the actual economic behavior,) which distanced utility theory further from psychological ones. The hypothesis of revealed preference is based on human rationality, but it can correspond with behaviorism in psychology, which makes scientific statements only from observed ones.

Utility theory was further developed theoretically by von Neumann and Morgenstern (von Neumann & Morgenstern, 1944, 1947). They proved that utility functions exist based on objective probabilities if some axioms are satisfied and demonstrated that utility can be measured assuming the expected utility theory. Their expected utility theory does not necessarily assume a logarithmic utility function like Bernoulli's expected utility theory but formulates a utility function in a more abstract form. In addition, they not only axiomatized the expected utility theory but also laid the basis for interaction analysis by devising game theory that can treat interactions in economic behavior.

The theory that considers the expected value of utility is called the expected utility theory, and the one in which subjective probability is assumed is called the subjective expected utility theory. In recent years, it has developed into the system of nonlinear expected utility theory, as represented by modern prospect theory.

Moreover, in the measurement of actual economic behavior, a method called conjoint analysis is used. This analytical method, in multi-attribute decision making, is a quantitative psychological method of utility measurement satisfying the axiomatic system of the additive conjoint system, which in turn can express preference relationships among choices by the sum of the utility of each attribute. For conjoint analysis, a utility estimation does not make sense unless the preference relationship satisfies the axiomatic system of the additive conjoint, but quantitative analysis is often performed from priori assumptions that this axiom is satisfied. Marketing studies often employ this approach.

In the analysis of economic behavior, the discrete choice model assuming the utility theory is often used. This analysis was developed by the psychologist L. L. Thurstone and later by the economist L. D. McFadden, who improved this method and conducted extensive research. McFadden received the Nobel Prize in economics in 2000. In the discrete choice model, the utility of the choice is a metric model of decision making expressed by the sum of the deterministic utility for the choice and the probability term assuming a certain probability distribution. The discrete choice model is sometimes called a random utility model. The models assuming a normal distribution for the probability term and a Gumbel distribution are called the probit model and the logit model, respectively. The logit model that features two choices is called a binary logit model, and the one with three or more choices is called a multinomial logit model. This discrete choice model is also known as a disaggregated behavioral model. This is a method of estimating the parameter from the choice results of each individual, and not from the aggregated data of a group. This is sometimes used as another term for the discrete choice model.

The utility theory paradigm mathematically and quantitatively expresses people's decision making and preferences, and even now, economics in particular is following this tradition and it is often used in economic psychology as well.

1.4.3 Psychoanalytic Paradigm

After immigrating to New York in 1937, E. Dichter, influenced by S Freud, the founder of psychoanalysis, explored consumers' unconsciousness in "motivation research" and conducted research in the USA from the 1940s. Dichter tried to reveal the consumption motivation hidden behind behaviors related to purchasing products by conducting in-depth interviews with consumers. For example, he understood that a carpenter's tools were for the sake of satisfying masculinity and motivation for strength; meanwhile, red sports cars indicated motivation for eroticism (Lindquist & Sirgy, 2006).

Such a theoretical framework attempts to clarify the purchase motivation that consumers may be unaware of and understand why they purchase specific brands. This theoretical framework is heavily influenced by psychoanalysis, which attempts

Table 1.1 List presented in motivation research

List A	List B
Hamburger	Hamburger
Bread	Bread
Carrot	Carrot
Baking powder	Baking powder
Instant coffee	**Drip coffee**
Canned peach	Canned peach
Potato	Potato

Source Haire (1950)

to explain human motivation on a subconscious level. This theoretical framework was popular, especially in the 1950s. To date, many studies conducting marketing research still adopt this approach.

One of the famous studies under this framework is the purchase motivation research of Nescafe instant coffee conducted by Haire (1950). Nescafe's instant coffee sold to a certain extent at the beginning, but its sales gradually slowed down. Even though they asked consumers directly why they no longer bought instant coffee, the reason was not clear; moreover, they found that consumers could not distinguish between instant coffee and drip coffee in terms of taste. Then, Haire prepared shopping lists, in which only coffee was presented as different options (Table 1.1), and conducted a survey asking people what kind of woman they thought would buy these items. Such a survey is called a projection method, which aims to elucidate the unconscious desires and consciousness of consumers. From the results of the survey, Haire found that those who received the list with drip coffee thought of the purchaser positively. On the other hand, those who received the list with instant coffee assumed common consumer images, such as lazy people and a lack of planning skills. Based on the results of this survey, Haire concluded that through marketing and advertisements, Nestlé succeeded in conveying the notion that drinking Nescafe is rather active and necessary for family members.

In motivation research, such projection methods and interview methods are often used. Currently, they are used in the practical sphere as part of research on consumer insights.

Box 2: Sigmund Freud

Born in 1856; deceased in 1939. He received a medical degree in 1881 at the University of Vienna. He is the founder of psychoanalysis, who created a new approach to the understanding of the human unconsciousness. His thought influenced the psychoanalytic approach of consumer research such as motivation research proposed by Ernest Dichter.

Photograph by Robert Goldstein:

https://www.flickr.com/photos/dream9/44646019415/in/photolist-2b2dw8r-8BgLAS-oR6pcZ-91Fj2x-esEk8-8BgLLY-foyQ1-byht52-dDu9Pi-dDu9XH-4J2Z8F-98USDS-8aF3Fu-onSLGH-GJ4EXa-bNxVZR-74yRpK-e6WFu9-4JyzG-4PXpsz-6sZzX-erK8U-esE3v-8aF221-4JyCz-erGKe-hwB63k-aDHBca-erHHb-4JyFm-4JyGU-erKBY-ZjmhpV-erK8T-8aBMeg-84BRpP-bvzUsp-erKBX-97hp8H-8BgLaU-5GVeYZ-itwgXB-4JyHg-XTyDTs-erJPt-esEka-ZiYu2X-E47G6S-erJPu-4JyAf

1.4.4 Macroeconomic Psychology Paradigm

This paradigm can be attributed to G. Katona, who was born in Hungary and studied basic psychology under the influence of Gestalt psychology. He later went to the USA to begin his research on the psychological factors of the problem of inflation during the Second World War. In his works, he studied the psychological factors of macroeconomic phenomena such as inflation through extensive research. Katona tried to clarify macroeconomic phenomena psychologically without assuming the rational "homo economicus" in traditional economics theory. Regarding "mass consumption society," he pointed out the importance of general affluence, which allows discretionary purchase (i.e., consumer rights that affect the economy and consumer psychology) and tried to elucidate the characteristics of consumer behavior from the psychological aspect. He established research methods of economic society by elucidating consumers' motivation, attitudes, expectations, and so on, which are factors that have a significant effect on macroeconomic phenomena.

Katona also pointed out that the economic behavior of individuals does not exist independently, but imitation behavior, occur due to the simultaneous transmission of information by the mass media; moreover, economic behaviors correlate with one another, so macroeconomic phenomena cannot be predicted or explained by the quantitative analysis assuming independence of ordinary individuals (Maeda, 1999).

With respect to economic psychology, Katona stated that "The fundamental need for psychology in economic research is to discover factors behind economic processes, that is, factors responsible for economic activities, decision making, and choices, and to analyze them… 'Economics without psychology' did not succeed in explaining important economic processes, and 'psychology without economics' cannot explain an aspect of the most common aspects of human behavior" (Katona, 1975, p. 9). Furthermore, Katona indicated that the socialization of the economy is a long-term process of the education of children by family members, teachers, and friends, which would take a decade to produce changes in economic behavior. He also predicted the effect this learning process would have on various economic phenomena. Furthermore, Katona believed that consumer optimism and pessimistic indicators against income changes can explain consumer spending and savings in the aggregate level, adding that the combination of these two factors shows a higher predictive value than that of only income data or expectation data (Katona, 1975).

Research on such a macroeconomic psychology paradigm was energetically done at the University of Michigan, where Katona belonged, although this trend had rather been inherited especially in Europe. Vigorous studies are being conducted by F. van Raiij, G. Antonides, H. Montgomery, and T. Gärling. For example, Gärling, Kirchler, Lewis, and van Raaij (2009) explained the psychology of people and what kind of economic behavior may occur related to the financial crisis of this century from the economic psychology perspective; they also attempted to explain macroeconomic phenomena. A survey-based research from a macroeconomic psychological perspective was also attempted by Japanese psychologists such as H. Akuto and S. Kojima in the 1960s. Currently, such research is being conducted by Japanese economists such as Y. Tsutsui, F. Otake, S. Ikeda, T. Ida, M. Ogakim and Y. Ukai.

Box 3: George Katona

Born in 1901; deceased in 1981. He graduated with a doctorate in Experimental Psychology from the University of Göttingen, Germany in 1921, and then moved to USA in 1933. He was working on the application of economic psychology to macroeconomics, developing measures of consumer expectations. Katona wrote many books and papers for macroeconomic aspects of economic psychology.

Photograph: Institute for Social Research University of Michigan

https://isr.umich.edu/fellowships-awards/george-katona-economic-behavior-research-award-fund/

1.4.5 Sociopsychological Paradigm

This paradigm overlaps with the macroeconomic psychology paradigm of Katona and colleagues. Rather than explaining macroeconomic phenomena, it is used as a general term for research approaches applying the concept of "attitude" in the prediction of personal economic behavior. The attitude is a concept that is commonly used in social psychology, and is defined as the "Psychological or neurophysiological prepared state organized through experience, which directs or changes the behavior of the organism itself for all of the subjects and situations the organism is involved" (Allport, 1935). Studies on predicting consumer behavior using the questionnaire method based on the concept of this attitude were particularly active in the 1970s.

Fishbein and Ajzen (1975) proposed a reasoned action theory to explain the behavioral relationship from the attitude. This theory is also used extensively in consumer behavior, and because this theory assumes the attitudes of various attributes, it is considered to be the study of a multi-attribute attitude model. As shown in Fig. 1.2, researchers stated that it is not the attitude but the intention to perform a behavior that directly defines that behavior, and that the intention is the subjective norm of taking the attitude and its behavior. Here, the subjective norm is one that is related to the degree by which others (parents, friends, a spouse, etc.) are expecting you to act upon such a behavior. The weight of how much the subjective norm and attitude contribute to intention and ultimately lead to behavior is estimated by the statistical analysis of actual research and experimental data.

Ajzen (1991) further developed the reasoned action theory and proposed a planned behavior theory. As shown in Fig. 1.3, it is intention that directly defines behavior, but in addition to the attitude and subjective norm, the sense of behavioral control

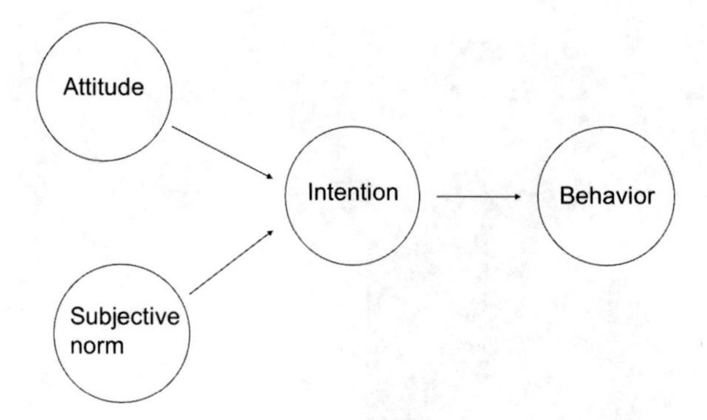

Fig. 1.2 Path diagram of reasoned action theory (Fishbein and Ajzen 1975)

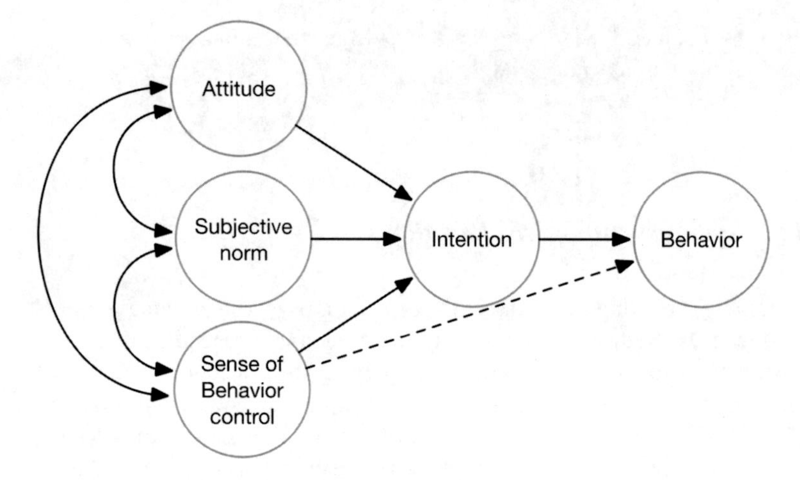

Fig. 1.3 Planned behavior theory. *Source* Ajzen (1991)

contributes to the intention. The sense of behavioral control is related to carrying out a behavior and is a cognition of how easy it is for the actor to perform such behavior. These three factors are interrelated and affect intention, as shown below. Moreover, the current version of the model was shown in Fig. 1.4 (Ajzen, 2019).

Sociopsychological studies based on such attitudes are still extensively carried out. Especially, in the field of consumer behavior, studies using the construct level theory of social psychology are actively carried out (Abe, 2009). Construct level theory, developed originally in the area of social psychology, posits that the spiritual representation is divided into construct levels with higher-order and lower-order ones according to the psychological distances that people feel about the subject and event being studied (Trope & Liberman, 2010). In addition, the high and low classification of construct levels can lead to differences in choice criteria. For example, in evaluating two radios with a clock, in which one radio has good sound quality but the accuracy of the clock is inferior, and the other radio has bad sound quality but is accurate, the

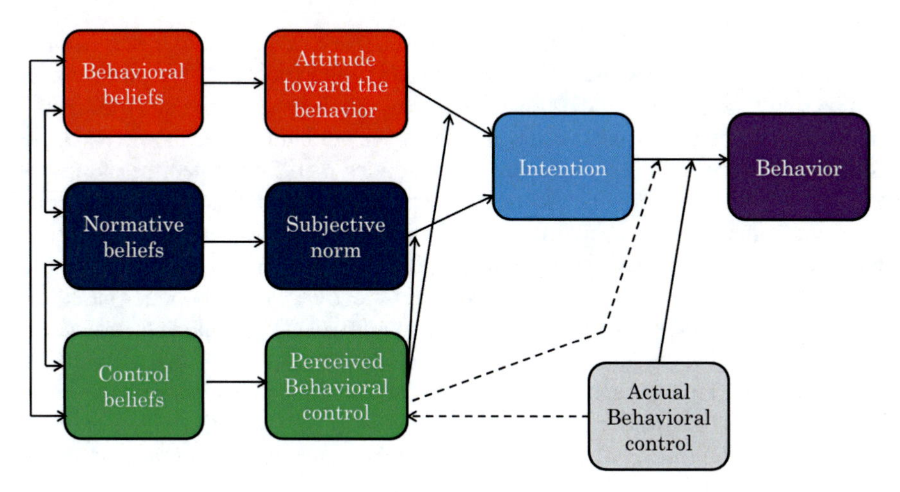

Fig. 1.4 Theory of planned behavior diagram. *Source* Ajzen (2019)

former is highly evaluated in the far future choice, whereas the latter's evaluation becomes higher on the condition of the near future choice (Abe, 2009). If construct level theory is correct, the purchase decision making is made depending on the secondary features of the product immediately before purchasing, but when it is long before purchasing, the purchase decision making is made by prioritizing the basic and essential features of the product. This idea emphasizes the essential features of the product in the advertisements, such as television and/or newspaper advertisements, which are effective at a stage considerably earlier than the actual purchase. Doing so also emphasizes the secondary features of the product in the POP and leaflets, which are effective immediately before the purchase.

1.4.6 Behavioral Decision Theory Paradigm

In research on decision making, studies on normative theory traditionally precede others, and in a way of comparing the theory there with actual human decision making behavior, research on behavioral decision theory—a descriptive theoretical research—has also been conducted (Kobashi, 1988). W. Edwards is the founder of behavioral decision theory. He began conducting psychological research on decision making from 1948 and already wrote a review paper titled "Behavioral Decision Theory" in 1961 (Edwards, 1961). In this behavioral decision theory research, Edwards conducted psychological experiments on how much the expected utility principle in economics applies.

The original behavioral decision theory research has traditionally been conducted in psychology. Especially, from a methodological viewpoint, the research has been done by mathematical psychologists and experimental psychologists as well as by cognitive psychologists and social psychologists, when dividing it by research subject area. These studies have a close relationship with expected utility theory, so it can be said that they are studies on economic psychology and have no strong relationship with actual consumer behavior. Under such circumstances, however, H. A. Simon, who received the Nobel Prize in economics in 1978, refuted the old traditional economics based on the theories and experiments on human decision making behavior from the 1950s. Since the 1980s, D. Kahneman has conducted research that has challenged the expected utility principle, Bayesian inference, problems of uncertainty, and the hypotheses of the universality of preferences in economics based on the psychological experiments. Kahneman proposed a decision making theory based on psychological value functions, which he called prospect theory, together with his collaborator A. Tversky, and tried to explain the economic phenomena.

In consumer behavior research, such as those on marketing, studies on consumer information processing by Bettman (1979) gained momentum, and those that focused on the consumers' information processing and decision making serve as the paradigm. In this approach, as shown in Fig. 1.5, consumers' decision making process is conceptually assumed. This model is proposed by Blackwell, Miniard, and Engel (2001). Basically, it assumes a time process: desire recognition > information search > option evaluation > purchase > Result (satisfaction, dissatisfaction, disposal). Consumer information processing theory clarifies what kind of factor and store environment consumer decision strategies and heuristics are easy to use and explains the kind of information processing the act of purchase decision making goes through.

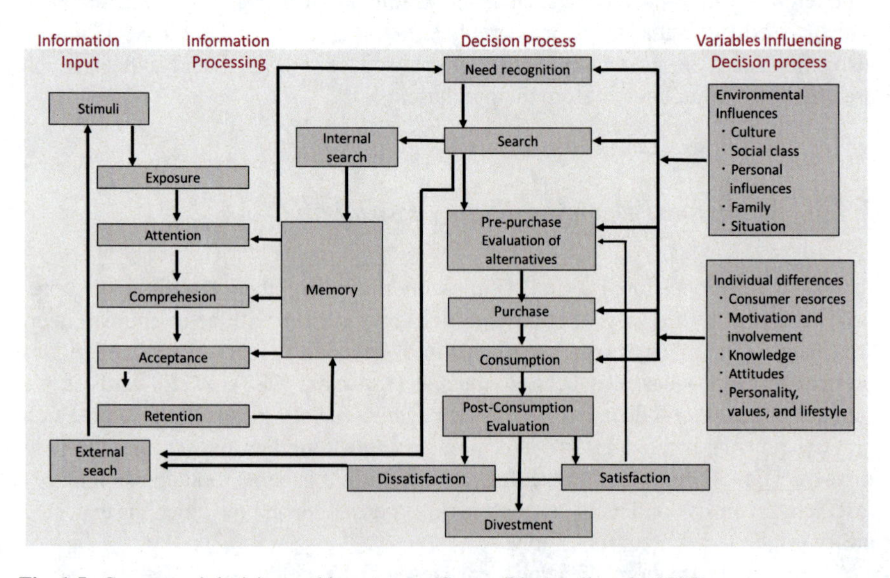

Fig. 1.5 Consumers' decision making process. *Source* Blackwell et al. (1995)

The reason why this approach focuses on the information processing process of consumers is that the way of information processing may affect the results of decision making. Hence, it is sometimes impossible to predict the behavior of consumers without knowing the actual information processing.

A method called process-tracking technique is used to examine the information processing of consumers. This technique features the verbal protocol method to encourage participants talk about what they think and the method of monitoring information acquisition to analyze the order of information acquisition. The verbal protocol method identifies how to decide and what decision making strategies to employ by recording speech and verbal reports in the decision making process. In addition to the decision making process, there are cases where verbal reporting is made after a decision is made. In addition, the method of monitoring information acquisition presents problems to the decision making process, recording the search process of information in the process, and identifying how to make decisions and strategies from the search process. There is also a method that uses a device to measure eye movements.

As described above, the process-tracking technique also includes physiological psychological measurement techniques such as eye movement measurement and the measurement of skin electrical activities, and recently, a method using functional brain imaging has also appeared. Based on the development of neuroeconomics, which aims at exploring the neuroscientific basis of human decision making behavior, noninvasive brain activity measurement methods have been developed. These include imaging methods concerning brain functions and the blood flow in the brain, such as functional magnetic resonance imaging (fMRI) and the positron emission tomography apparatus (PET). Furthermore, a system has been established that allows psychologists and economists to cooperate with neuroscientists in clarifying the findings having been handled only by behavioral experiments.

1.4.7 Behavioral Economics Paradigm

Behavioral economics is a study that aims to describe human economic and to explore human judgment and decision making in the economic situation, mainly by conducting measurements based on experiments and surveys. In this respect, it is common to describe the decision making process with the behavioral decision theory paradigm shown earlier. In addition, as the research by Kahneman and others is also called behavioral economics, it is common in this field as well. However, in the case of a behavioral economics paradigm, not only decision making phenomena but also human interactions and macro-collective behaviors are further explored.

The findings of behavioral economics so far have indicated that both human judgment and decision making are not necessarily rational and have clarified their psychological factors, environmental factors, and processes related to them. Some readers may think that human judgment and decision making are not reasonable, which is consistent with the commonsense of the world, and that the findings do not have any novelty. However, traditional economics has thus far constructed theories by implicitly assuming that humans are the rational homo economicus and always adopt the best choice; moreover, economic policies are often considered based on those theories. Hence, it has been suggested that theories of traditional economics and economic policies based on these must be revised based on some of the findings of behavioral economics, which have had a great impact. Eventually, Simon, Kahneman, and Smith were awarded the Nobel Prize in economics for their research achievements related to behavioral economics. Research on economic theories that consider the actual human judgment and the characteristics of decision making have also been carried out relatively recently, and many other attempts are still in progress.

Behavioral economics has been regarded as an area of economics, but based on the studies of Simon and Kahneman et al.., it has a close relationship with psychology. Indeed, the relationship between behavioral economics and psychology has a very long history.

The early version of "behavioral economics," which already existed in the area of psychology, originated from the research called behavior analysis, in the 1930s by American psychologist B. F. Skinner. Skinner tried to explain the behaviors of humans and other animals with the concept of operant conditioning, which he developed from the studies of condition reflexology by I. Pavlov and the trial-and-error learning theory by E. Thorndike. Operant conditioning is the process of learning in which the frequency of the voluntary occurrence of the following behavior changes in accordance with the change of the environment immediately after the behavior has occurred. In his theory, Skinner tried to explain many learning behaviors through operant conditioning. Behavioral economics research in the 1970s began to link behavior analysis with economics. In 1980, Hursh concretely investigated behavioral experimental data on animals by linking them with economic concepts, such as closed/open economic environments, price elasticity, or substitution and supplementarity (Hursh, 1980). Behavioral economics, in this vein, attempted to explain animal behavior by using concepts assumed in traditional economics, such as the assumption of the maximization of utility, price elasticity, substitution, and supplementarity.

In comparison, the current version of behavioral economics can be traced back to the research on the behavioral decision theory described earlier. Conventionally, in decision making research in psychology, studies on normative theory traditionally preceded others. In order to compare the theory with actual human decision making behavior, research on behavioral decision theory, a descriptive kind of theoretical research, has been conducted. Behavioral economics derived from the behavioral decision theory examined phenomena where human behaviors are not consistent with normative theories, as in utility theory and the reasons behind such phenomena. Conversely, behavioral economics derived from behavioral analysis applied theories of traditional economics to animals, such as pigeons and rats, rather than to humans. However, in recent years, both approaches have become very close. In particular, there is a tendency to conduct a unified research applicable to both humans and the other animals on delayed value discounting and choice behavior research under risk, among other topics.

The relationship between behavioral economics and psychology dates back to the nineteenth century when interpreted in a broad sense. In a book published in 1860, G. T. Fechner, famous for the classical psychology theory, proposed a psychophysical method and developed a constant measurement method and a scale construction method to specify the functional relationship with psychological quantity produced through stimulus intensity and judgment. He then derived the theory of sensory quantity expressed by logarithmic functions; this research used expected utility theory proposed by mathematical scientist D. Bernoulli in the eighteenth century. Thus, psychology and economics have had a lengthy relationship. In addition, Shionoya (2009) described in detail the close relationship between psychology and economics, not only in the mathematical scientific aspect but also in the hermeneutic terms. He further stated that using psychological methods in economics did not seem to be unique in the nineteenth century.

Since the beginning, the field of behavioral economics has mainly focused on personal analysis as the target for the decision making and behavioral phenomena analysis. However, at present, in addition to this personal analysis, the research areas like the analysis of interactions between individuals, decision making in a group, decision making in an organization, social decision making, and crowd behavior, are targeted. These topics are also being investigated in social psychology research. Hence, we can say that there is also an emerging specialization, called behavioral game theory, which describes decision making behaviors under human interaction using the method of experimental games, in which various games are played in experimental situations based on game theory.

Summary

- Economic psychology studies the psychological aspects behind various economic phenomena and explores the mechanisms of people's judgment, decision making, and behaviors in different economic situations.
- Economic behavior is directly involved in activities through which humans acquire goods and services and consume and dispose of in their socioeconomic lives.
- Economic psychology research is useful for elucidating various economic phenomena, marketing activities of organizations like companies, public policies like consumer protection and education, and consumers' decision making behaviors.
- Economic psychology research features theoretical frameworks, such as the hermeneutic paradigm, utility theory paradigm, psychoanalytic paradigm, macroeconomic psychological paradigm, social psychological paradigm, behavioral decision theory paradigm, and behavioral economics paradigm, and all of these are closely related to one another.

Reading Guide for More Advanced Learning

- Antonides, G., & van Raaij, W. F. (1999). *Consumer behavior: A European perspective*. Chichester, UK: Wiley.

This book is a textbook of European consumer research and covers economic psychology, sociology, and historical studies.

- Kirchler, E., & Hoelzl, E. (2018). *Economic psychology: An introduction*. Cambridge, UK: Cambridge University Press.

Among books of economic psychology, this book features the general items of economic psychological research, and the reader can easily understand how economic psychological research is conducted and how it is related to other psychology-related sciences.

- Ranyard, R. (Ed.) (2018). *Economic psychology*. Chichester, UK: Wiley.

This book concretely explains how economic behavior research is carried out and how they make contributions to the society. It is particularly useful when you want to have a concrete image of economic psychological research.

- Leiser, D., & Shemesh, Y. (2018). *How we misunderstand economics and why its matters: Psychology of bias, distortion, and conspiracy*. London, UK: Routledge.

This book is explaining why ordinary people misunderstand micro- and macroeconomical phenomenon. The author gave insightful discussions on relationships between psychology and economics.

References

Abe, S. (2009). Kaishaku reberu riron to shōhisha kodo kenkyū. *Ryūtū jōhō* [*The Journal of marketing and distribution*], *41*, 6–11 (in Japanese).

Ajzen, I. (1991). The theory of planned behavior. *Organizational Behavior and Human Decision Processes, 50,* 179–211.

Ajzen, I. (2019). *Theory of planned behavior diagram [Figure]*. Retrieved from https://people.umass.edu/aizen/tpb.diag.html.

Allport, G. W. (1935). Attitudes. In C. Murchison, et al. (Eds.), *A handbook of social psychology*. Worcester, MA: Clark University Press.

Aoki, Y., & Onzo, N. (2004). *seihin burando senryaku* [*Product and brand strategy*]. Tokyo, JP: Yuhikaku Publishing.

Bettman, J. (1979). *An information processing theory of consumer choice*. Reading, MA: Addison-Wesley.

Blackwell, R. D., Miniard, P. W., & Engel, J. F. (1995). *Consumer behavior* (8th ed.). Fort Worth, TX: The Dryden Press.

Blackwell, R. D., Miniard, P. W., & Engel, J. F. (2001). *Consumer behavior* (9th ed.). Fort Worth, TX: The Dryden Press.

Edwards, W. (1961). Behavioral decision theory. *Annual Review of Psychology, 12,* 473–498.

Fishbein, M., & Ajzen, I. (1975). *Belief, attitude, and behavior: An introduction to theory and research.* Reading, MA: Addison-Weseley.

Gärling, T., Kirchler, E., Lewis, A., & van Raaij, F. (2009). Psychology, financial decision making, and financial crises. *Psychological Science in the Public Interest, 10,* 1–47.

Haire, M. (1950). Projective techniques in marketing research. *Journal of Marketing Research, 3,* 649–652.

Hirschman, E. C., & Holbrook, M. B. (1992). *Postmodern consumer research: The study of consumption as text.* Newbury Park, CA: Sage Publications.

Holbrook, M. B., & Hirschman, E. C. (1993). *The semiotics of consumption: Interpreting symbolic consumer behavior in popular culture and art.* New York, NY: Mouton de Gruyter.

Hursh, S. R. (1980). Economic concepts for the analysis of behavior. *Journal of the Experimental Analysis of Behavior, 34*(2), 219–238.

Katona, G. (1975). *Psychological economics.* New York, NY: Elsevier.

Kikkawa, T. (1999). *Risuku komyunikēshon: Sōgo rikai to yoriyoi ishi kettei o mezashite* [*Risk communication: Toward mutual understanding and better decision*]. Tokyo, JP: Fukumura Shuppan Inc.

Kikkawa, T. (2000). Kigyō no risuku komyunikēshon to shōhi kōdō [Firm risk communication and consumer behavior]. In K. Takemura (Ed.), *Shōhi kōdō no shakai shinrigaku: shōhisuru ningen no kokoro to kōdō* [*Social psychology of consumer behavior: Human mind and action in consumption*] (pp. 118–128). Tokyo, JP: Fukumura Shuppan.

Kiwata, Y., Kakeda, Y., & Mimura, Y. (1989). *Tekisutobukku gendai māketinguron* [*Textbook of contemporary marketing theory*]. Tokyo, JP: Yuhikaku Publishing.

Kobashi, Y. (1988). *Ninchi kagaku sensho 18: Kettei wo shien suru* [*Selected works in cognitive science 18: Support decisions*]. Tokyo, JP: University of Tokyo Press.

Kobayashi, T., & Minami, T. (2004). *ryūtsū eigyō senryaku* [*Distribution channel and sales strategy*]. Tokyo, JP: Yuhikaku Publishing.

Kojima, S. (1986). *Kakaku no shinri: Shōhisya ha nani wo kōnyū kettei no monosashi ni surunoka* [*Price psychology: What measure do consumers use to make purchasing decisions?*]. Tokyo, JP: Diamond Inc.

Kotler, P., & Armstrong, G. (1997). *Marketing: An instruction* (4th ed.). Upper Saddle River, NJ: Prentice Hall.

Lindquist, J. D., & Sirgy, M. J. (2006). *Shopper, Buyer, and consumer behavior: Theory, marketing applicability, and public policy implications* (3rd ed.). Cincinnati, OH: Atomic Dog Publishing.

Maeda, Y. (1999). Keizai ishiki, keizaiteki taido, keizaiteki sinnenkenkyū no keihu (1): G. Katona no makuroteki shōhisha taido kenkyū. [Main lines of research on economic attitudes, beliefs and consciousness (1): G. Katona's consumer attitudes studies]. *Memoirs of the Nara University, 27,* 107–118.

Nishimura, T. (1999). *Nihon no shōhisha kyōiku: Sono seisei to hatten* [*Consumer education in Japan: Its origin and development*]. Tokyo, JP: Yuhikaku Publishing.

Shionoya, Y. (2009). *keizai tetsugaku genri: Kaishakugakuteki sekkin* [*Principles of Economic Philosophy: The hermeneutical approach*]. Tokyo, JP: University of Tokyo Press.

Sugimoto, T. (Ed.). (2012). *Shin shōhisha rikai no tame no shinrigaku* [*New psychology for understanding consumer*]. Tokyo, JP: Fukumura Shuppan.

Takemura, K. (2000). Vagueness in human judgment and decision making. In Z. Q. Liu & S. Miyamoto (Eds.), *Soft computing for human centered machines* (pp. 249–281). Tokyo, JP: Springer.

Takemura, K. (2014). *Behavioral decision theory: Psychological and mathematical representations of human choice behavior.* Tokyo, JP: Springer.

Trope, Y., & Liberman, N. (2010). Construal-level theory of psychological distance. *Psychological Review, 117,* 440–463.

Tversky, A. (2004). *Preference, belief, and similarity.* Cambridge, MA: MIT Press.

Ueda, T., & Moriguchi, T. (Eds.). (2004). *Kakaku puromōshon senryaku* [*Pricing and promotion strategy*]. Tokyo, JP: Yuhikaku Publishing.

von Neumann, J., & Morgenstern, O. (1944/1947). *Theory and games and economic behavior* (1st and 2nd ed.). Princeton, NJ: Princeton University Press.

Chapter 2
Rational Choice and Revealed Preference: Theoretical Representation of Preference Relations Leading to the Best Choice

Keywords Rational choice · Best option · Revealed preference · Indifference curve · Marginal substitution rate

Traditional economics has relied on the assumption of "rational homo economicus." This chapter investigates the rationality of decision making and focuses on the concept of rational choice. We utilize set theory to conduct a rigorous discussion. The concept of rational choice is based on the premise that at least the best option can be chosen and the options can be ordered in descending order of preference. Furthermore, we present the idea of revealed preference, which can deduce preference relations that produce rational choices from actual economic behavior. We elucidate the concept of utility, which is inferred from the idea of revealed preference, and briefly present the concept of traditional ordinal utility theory. Numerous economic theories have utilized the ordinal utility theory. However, some counterexample phenomena exist, which cannot be justified from the perspective of behavioral decision theory. This chapter presents examples that do not satisfy transitivity, which is the premise of the ordinal utility concept. Transitivity indicates the relation between two options. For example, if a shopper prefers oranges to bananas and apples to oranges, then the transitivity assumption requires him/her to prefer apples to bananas. This idea reflects the nature of a consistent preference relation.

2.1 Framework to Describe Decision Making

2.1.1 Relationship Recognition and Set Theory

People in ancient India, ancient China, and ancient Greece considered human preference judgment. Basically, judgment begins with the recognition of relations.

K. Takemura, *Foundations of Economic Psychology*,
https://doi.org/10.1007/978-981-13-9049-4_2

Consider a situation. An apple and an orange are placed on a plate. Two boys, Taro and Jiro, appear; Taro eats the apple, and Jiro eats the orange. A set {apple, orange} of fruits is placed on the plate, and a set {Taro, Jiro} of persons exists. The possibilities of who eats what are the sets of combinations between the sets of the plate and the person {(apple, Taro), (apple, Jiro), (orange, Taro), (orange, Jiro)}. This implies that possibilities exist where Taro eats the apple, Jiro eats the apple, Taro eats the orange, and Jiro eats the orange. In set theory, these combinations of two sets are called a direct product, which indicates the possibilities that can occur.

Let a part (called a subset) of this direct product be denoted by $R = \{(\text{apple, Taro}), (\text{orange, Jiro})\}$. This set expresses the relationship that Taro eats the apple and Jiro eats oranges. In this manner, we can express the relationship between Taro and Jiro and the apple and orange on the plate using subset R of the direct product showing possibilities. Set theory is used in mathematics. However, with the use of abstract theory, the relation that forms the basis of judgment can be expressed. When describing the relationship somewhat more abstractly from another perspective, consider that an ordered pair (x, y) belongs to R such that x and y are in the relationship with R. Accordingly, we can express it by writing as xRy and state that "x is more or at least equally prefered to y with relation R."

In the aforementioned example, the relationship between two sets of people and fruits can be called a binary relation. Furthermore, for example, considering a plate with different fruits, we can express the fruits Taro and Jiro ate among those on the second plate using the subset of the direct product. In this case, the relationship can be expressed using the three-term relation. In general, relationships can be expressed by the n-term relation.

2.1.2 Ordering and Comparative Judgment

Accordingly, we have shown that relationships can be expressed using set theory. However, how can we describe the ordering of subjects? Ordering judgment is a situation of binary judgment wherein when two subjects are compared, you judge which is higher or lower, larger or smaller, or whether subjects are different or the same. For example, when comparing the length of two line segments, you judge which segment is longer or shorter. Although such binary judgment may not look like ordering, comparative judgment is actually being performed in this case. This is the basis of judgment of ordering. A slightly different example is a case where you compare two subjects and choose one as per your preference. From the word "ordering," you may imagine a situation where you order many subjects, but from a basic point of view, ordering can be considered to occur from the comparison of two subjects. We regard such comparative judgment between two simple subjects as judgment of ordering.

Such judgment of ordering can be thought of as a result of comparative judgment among elements of a collection (set) of a subject. We can express the result of comparative judgment with the subset by considering the direct product of the subject set. In the above example, if you consider a set X with two elements X = {apple, orange} and make the direct product, then you have $X \times X$ = {(apple, apple), (apple, orange), (orange, apple), (orange, orange)}. Then, for example, if R = {(apple, orange)}, then the result of comparative judgment—namely, that the apple is preferred to the orange—may be expressed.

2.1.3 *Various Forms of Comparative Judgments*

Comparative judgments have various forms. In ordinary psychology, cases for judgment exist where on being shown two options, you are asked to "choose either one." This approach is called a forced choice method, and you may choose neither option in some cases. Even if you make a comparative judgment, ordering the subjects appropriately is not always possible. We list the standards that divide the judgment properties as follows:

(1) Completeness (comparability): In decision making, either xRy or yRx is true. For example, if a set of fruits is X and xRy is defined as the relation such that y is preferred to x or they are indifferent, then we can deduce that banana is preferred to strawberry or they are indifferent and that strawberry is preferred to banana or they are indifferent. It is not comparable and does not satisfy completeness in case "you do not know which one you prefer or whether they are indifferent."
(2) Reflectivity: In a comparative judgment of the same subject, a relation such as xRx may exist. For example, when a set of fruits is X and R makes the relation of the same preference, it is selfevident that the banana and banana, or the strawberry and strawberry are prefered equally, thereby they satisfy reflectivity.
(3) Symmetry: When making a comparative judgment, when the order relation of the subjects is reversed, if xRz, then the same relation as zRx can be obtained. For example, when a set of fruits is X and R makes the relation of the same preference, if banana is preferred as much as strawberry, then strawberry is also preferred as much as banana, thereby satisfying the symmetry.
(4) Antisymmetry: When making a comparative judgment, if the order relation of the subjects is reversed, the same relationship is obtained. If xRz and zRx, then $x = z$ is always obtained; thus, the antisymmetry is satisfied. For example, when the set of real numbers is X and the relation of equal magnitude is R, this relationship is satisfied. However, when the set of fruits is X and R makes the relation of the same preference, even if banana is preferred as much as strawberry, it does not make the strawberry and banana equal. Thus, they do not satisfy antisymmetry.
(5) Transitivity: A transitive relationship states that for elements x, y, z, if xRy and yRz, then xRz holds true. For example, if a set of fruits is X and xRy is defined as a relation such that y is preferred to x or they are indifferent, then transitivity is satisfied. This means that if a relation exists where banana is preferred to orange or they are indifferent, banana is preferred to strawberry or they are indifferent, and strawberry is preferred to orange or they are indifferent, then

transitivity is satisfied. Alternatively, considering the relation xRy such that the weight of object x is heavier than that of y, if x is heavier than y and y is heavier than z, then x is heavier than z. This idea holds true as long as the balance is functioning. In addition, when the transitivity does not hold true, as shown in Chap. 1, the relation is a three-way standoff. For example, if the relation of paper–rock–scissors is shown by $\succ$, then the relation among paper, rock, and scissors is such that rock $\succ$ scissors and scissors $\succ$ paper, but because rock $\succ$ paper does not hold true, $\succ$ does not satisfy transitivity.

2.2 Best Option and Selection Function

2.2.1 *Best Option*

In economic behavior, people are considered to act reasonably. Consider, for example, a case where only two brand options exist in consumer decision making. For example, decision making is needed when choosing either brand A or brand B. Which option is the best is unknown unless you examine the content of the options, but judging from the formal features is easy. For example, if you choose brand A, then you can say A is formally the best. In such a case, if brands A and B are comparable, then this result will be acceptable.

However, when three or more brand options are available, the situation becomes somewhat complicated. For example, when brands A, B, and C are presented, choosing the best brand among them becomes rather complicated. If brand A is preferred to B, brand B is preferred to C, and brand A is preferred to C, then brand A will be the best option. However, if brand C is preferred to brand A, then the preference circles around A to B to C to A and the best option cannot be decided. If the preference order is circling, then you will not be able to choose the best option.

Here, let us define the best option for decision makers using symbols from set theory. Let x be an element of X, where X denotes the set of options. For example, if $X = \{\text{brand } A, \text{brand B}, \text{brand } C\}$, then its elements are brand A, brand B, and brand C. If $X = \{x, y, z\}$, then its elements are options x, y, z. If you prefer x at least as much as the other elements in X, then x is called the best option. $C(X, R)$ as the best option is defined as "an element of X that satisfies the relationship R such that x is at least as good as y for y in any X." Formally, this can be expressed as follows:

$$C(X, R) = \{x \in X | \text{ satisfies the relationship } R \text{ such that } x \text{ is at least as good as } y \text{ for } y \text{ in any } X.\}$$

Furthermore, if xRy is expressed as "the relationship R such that x is at least as good as y," then

$$C(X, R) = \{x \in X | \text{ also for } y \text{ in any } X, xRy\}$$

can be obtained. The above formula is called "$C(X, R)$ is the set of elements of X that satisfies xRy for y in any X." The fact that $C(X, R)$ is not an empty set implies that the best option exists.

Next, another relation is defined from the relation R. xRy has been understood as "x is at least as good as y," and the following relations are defined below.

If it is not xRy or yRx, then it is xPy.

If it is xRy and yRx, then it is xIy.

If x is at least as good as y and if it is not true that y is at least as good as x, then xPy and x are preferred to y. If x is at least as good as y and y is at least as good as x, then x is indifferent from y.

2.2.2 *Conditions for Guaranteeing Preference Relations and the Best Option*

Usually, in economic behavior, consumers are often assumed to be able to compare two options. This idea shows that completeness (connectivity) holds true. Thus, when at least either xRy or yRx holds true, for example, if the set of brands is X and when xRy is defined as the relation such that y is preferred to x or they are indifferent, then we can deduce that brand x is preferred to y or they are indifferent and the brand y is preferred to z or they are indifferent. It is not comparable and does not satisfy completeness in case you do not know which one you prefer or whether they are indifferent.

That is, completeness is expressed as follows:

Completeness: $\forall x, y \in X, xRy \vee yRx$.

Accordingly, the relation ($\forall x, y \in X$), xRy, or yRx holds for any elements x and y in the option set X. Here, $\vee$ is a logical symbol, indicating that at least either one holds true. The following reflective properties can be derived further from the completeness property. Reflectivity: $\forall x \in X, xRx$

Next, we explain acyclicity, which can be defined as follows: for any selection subjects $x_1, x_2, \ldots, x_k$, if $x_1Px_2, x_2Px_3, \ldots, x_{k-1}Px_k$ hold true, then x_kPx_1 does not hold true. An example where acyclicity does not hold true is the three-way standoff relation. For example, if the relation of paper–rock–scissors is shown by P, then the relation among paper, rock, and scissors is ${}_{\text{rock}}P_{\text{scissors}}$ and ${}_{\text{scissors}}P_{\text{paper}}$; however, acyclicity is not satisfied because ${}_{\text{rock}}P_{\text{paper}}$ does not hold true.

A theorem suggests that in the selective set of finite elements, if the preference relation satisfies the properties of completeness and acyclicity, then the best option exists. Below, we show the theorem on the best option and rational choice based on interpretations by Feldman and Serrano (2005).

2.2.2.1 Theorem on the Best Option

Let X be a finite set to be selected. If the relation R is complete and acyclic, then $C(R, X)$ is not empty. That is, under this condition, the best option exists.

Proof If you consider any element from X and it is judged as the best, then this proof will end. Given that X has a finite number of elements and the relation R has completeness, either the existence of the best option can be proved from the finite choices or the idea that the choice lasts forever holds true. Given that the elements of X are finite, if the choice lasts forever, because also for selection subjects $x_1, x_2, \ldots, x_k$, if $x_1Px_2, x_2Px_3, \ldots, x_{k-1}Px_k$ hold true, then circling such as x_kPx_1 occurs. This finding is contrary to the assumption. Therefore, under this condition, the best option will be available. Thus, the proof is completed.

The following theorem focuses on the existence of the best option.

2.2.2.2 Theorem on the Necessary and Sufficient Conditions for the Existence of the Best Option

Let R be complete. Only when R is acyclic, the selection subject X with finite elements has $C(X, R)$ that is not empty. That is, under R satisfying completeness, the necessary and sufficient condition for the selection function to be the best option is that R is acyclic.

Proof In this proof, the idea that when R is complete and acyclic, it has a nonempty $C(X, R)$ is the same as mentioned earlier. We now demonstrate that under R satisfying completeness, having nonempty $C(X, R)$ indicates that R is acyclic. Suppose R is not acyclic. Then, $x_1, x_2, x_3, \ldots, x_k$ exist, which become $x_1Px_2, x_2Px_3, \ldots, x_{k-1}Px_k$, and x_kPx_1. If set X is $\{x_1, x_2, x_3, \ldots, x_k\}$, then $C(X, R)$ is empty. However, this contraposition result is that when R is complete, if it has a nonempty $C(X, R)$, then R is acyclic. Thus, the proof is completed.

Box 1 Amartya Kumar Sen

was born in 1933. He received Ph.D. at Cambridge University in 1959. He was awarded the Nobel Prize in economics in 1998 for his work of welfare economics. He has also contributed to the decision theory, social choice theory, and philosophical study of human value and decision making. He is currently working as a Professor at Harvard University.

Photograph by public.resource.org:

https://www.flickr.com/photos/publicresourceorg/26917255153/in/photolist-c9CdcL-bUKj54-c9C8TA-c9C4q7-GfdmNU-c9Ca7Y-231M5Xk-ajS14E-EHXwHF-c9C6dY-cDrnFW-b6fNAe-H1zYb6

2.3 Criteria of Rationality and Weak Order

2.3.1 Two Criteria of Rationality

We revealed that acyclicity and completeness are necessary to make the best decision. However, even if you have selected the best option, it is unclear as to whether it is rational. Next, let us consider the following two criteria for rationality (Sen, 1970):

Property α: If an element x belonging to X_1, which is a subset of the selected set X_2, is the best element of X_2, then x is the best element in all X_1.

That is, $\forall x \in X_1 \subset X_2 \rightarrow [x \in C(X_2, R) \rightarrow x \in C(X_1, R)]$.

This property β is also called the condition of independence from unrelated options (Sen, 1970).

Property β: If elements x, y belonging to X_1, which is a subset of the selected set X_2, are the best elements in X_1 and x is the best in X_2, then y is also the best element in X_2.

That is, $\forall x, y, [x, y \in C(X_1, R) \& X_1 \subset X_2] \rightarrow [x \in C(X_2, R) \Leftrightarrow y \in C(X_2, R)]$

Property β is such that if both x and y are the best options in X_1, which is a subset of X_2, if one x is the best option in X_2, then the other y is also a subset of X_2; if y is the best option in X_2, then x is also the best option in X_2. This property suggests that, for example, if brand A, the best option in Japanese stationery, is the best stationery in the world, then brand B, the best type of the same Japanese stationery, must also be the best stationery in the world.

All the selection functions $C(X, R)$ generated by R satisfy property α but do not necessarily satisfy property β (Sen, 1970). If x belongs to $C(X, R)$ for all y in X, then it becomes xRy. Thus, property α is satisfied. However, for example, considering the following binary relation with three options, property β is not satisfied. That is, when $X = \{x, y, z\}$, consider the preference relation of xIy, xPz, and zPy. Then, given that $\{x, y\} = C(\{x, y\}, R)$ and $\{x\} = C(\{x, y, z\}, R)$ are obtained, this relation obviously does not satisfy property β.

2.3.2 *Rational Choice and Weak Order*

An important theorem about rational choice has been presented (Sen, 1970).

2.3.2.1 Weak Order Theorem on Rational Choice

A necessary and sufficient condition for the selection function $C(X, R)$ derived from the binary relation R to satisfy the property β is that R is in weak order.

The weak order is a relation that satisfies completeness and transitivity. Completeness has been explained, and transitivity is the relation as described below, that is,

Transitivity: $\forall x, y, z \in A, xRy \& yRz \rightarrow xRy$.

That is, if xRy and yRz hold true for any elements x, y, z ($\forall x, y, z \in A$) of A, then xRz holds true. For example, when A is the set of options of product brands in the same way above, if you interpret xRy and if a relation exists such that brand x is preferred to brand y or they are indifferent, brand y is preferred to z or they are indifferent, and brand x is preferred to z or they are indifferent, then transitivity is satisfied.

If transitivity does not hold true, then a three-way standoff relation exists. Acyclicity and transitivity are slightly different concepts. If transitivity holds true, then the relation is acyclic, but the converse does not necessarily hold true. For instance, when $X = \{x, y, z\}$, let us consider the preference relation of xIy, yIz, and xPz. This idea obviously satisfies acyclicity, but it does not satisfy transitivity.

Proof We prove this theorem using the proof by contradiction. First, assuming that it is not in weak order, we need to prove that the selection function $C\ (X, R)$ derived from the binary relation R does not satisfy property β. Second, assuming that R does not satisfy property β, we need to prove that it is not in weak order.

First, R is assumed to not be in weak order. In this case, either completeness is not satisfied or transitivity does not hold true. When completeness is not satisfied, given that the selection function cannot be generated, property β is not satisfied. Next, transitivity is assumed to not be satisfied. Then, options x, y, z that become xPy, yIz, and zRx exist. Even though $\{y, z\} = C(\{y, z\}, R)$ and $\{y\} = C(\{x, y, z\}, R)$ result, $\{x\} = C(\{x, y, z\}, R)$ does not hold true. Thus, property β is not satisfied.

Next, R is assumed to not satisfy property β. Then, $x, y \in C(X_1, R)$ results when $X_1 \subset X_2$ and x, y, which result in $x \in C(X_2, R)$ and $\neg[y \in C(X_2, R)]$ ($\neg$ means negation in parentheses). Obviously, z can be present in X_2 to be zPy and xRz. Given that $x, y \in C(X_1, R)$ hold true, it becomes xIy; from the assumption of transitivity, $zPy \,\&\, yIx \rightarrow zPx$ result. However, this finding is inconsistent with the assumed xRz. Therefore, R does not satisfy transitivity and is not in weak order. Thus, the proof is completed.

In this manner, we found that the preference relation that satisfies the rationality criteria α and β can only be in weak order. Therefore, weak order is considered to be a formal criterion that must be satisfied in rational choice, especially in traditional economics wherein we assume weak order for the decision making of an economic entity.

The fact that transitivity and completeness hold true indicates that the best decision making that satisfies properties α and β can be made in daily living. If transitivity does not hold true, then preference circles similar to the way paper–rock–scissors does, thereby preventing you from making the best decision. In addition, comparing things is impossible, and you cannot select the best decision in the first place.

2.4 Criteria of Rationality and Utility Maximization

2.4.1 Expression Theorem of Weak Order

So far, we have shown that this preference relation of weak order leads to rational selective functions, actually equivalent to utility maximization as explained in the following paragraphs. We have described the preference relation using the symbol of R; here, we use the symbol $\succcurlyeq$ to express the relation of weak order. Let $x \succcurlyeq y$ be used with the same meaning as xRy (however, R is in weak order here).

Utility is a subjective value or desirability for adopting an option in daily usage; in decision making theory, it is often operationally considered as a real-valued function that expresses a preference relation (sometimes referred to as a utility function). We consider utility by using a real number value because mathematical analysis of the phenomena of decision making makes predicting and explaining phenomena easier.

Let us give a simple example of utility. We consider decision making under certainty that you select either brand A or brand B. In this case, the utility means a real number value such that the utility (u[brand A]) of brand A is higher than the utility (u[brand B]) of brand B when and only when brand A is preferred to brand b (brand $A \succcurlyeq$ brand B). That is, when the relation of u(brand A) $\geq u$(brand B) $\Leftrightarrow$ brand $A \succcurlyeq$ brand B holds true, the preference relation $\succcurlyeq$ is expressed by the utility function u. In particular, the utility that preserves only the order of preference is called ordinal utility. The ordinal utility does not lose its essential meaning even if the monotonically increasing transformation of its utility function is applied, and it corresponds to the ordinal scale used in psychology and statistics. For example, if u is an ordinal utility and when u(brand A) = 5 and u(brand B) = 2 are applied as well as in the function φ, which monotonically increases and transforms its value, when φ (u[brand A]) = 8 and φ (u[brand B]) = 3 are applied, then preference relations are preserved.

Let us express the ordinal utility slightly more formally. We assume that the set of options X is finite and the preference structure $\langle X, \succcurlyeq \rangle$ is in weak order. The preference structure is a set that arranges the set of options and some preference relations $\succcurlyeq$ together. The following theorems hold true for the weak order that satisfies these two properties (Krantz, Luce, Suppes, & Tversky, 1971).

2.4.1.1 Theorem on Weak Order (Cases of Finite Sets)

If the preference structure $\langle X, \succcurlyeq \rangle$ on the finite set X is in weak order, only then does a real-valued function (ordinal utility function) u: $X \rightarrow Re$ on X exist

$$\forall x, y \in X, x \succcurlyeq y \Leftrightarrow u(x) \geq u(y)$$

In other words, this theorem means that if preference is in weak order, then the preference relation is expressed with the function that has real number values that preserve its preference. That is, we can consider the qualitative preference relation in weak order through quantification by ordinal utility. Although we are considering this theorem by finite sets, it holds true in countable infinite sets and in uncountable infinite sets if a certain condition is added (Krantz et al., 1971).

2.4.2 *Uniqueness Theorem on Weak Order*

For ordinal utility, the following theorem holds true:

2.4.2.1 Theorem on Uniqueness of Weak Order (Cases of Finite Sets)

When the preference structure $\langle X, \succcurlyeq \rangle$ on the finite set A is in weak order and $\langle X, \succcurlyeq \rangle$ is expressed with $\langle Re, \geq \rangle$ through the real-valued function u: $X \to Re$ on A shown in the above theorem and the structure $\langle \langle X, \succcurlyeq \rangle, \langle Re, \geq \rangle, u \rangle$ is an ordinal scale.

This theorem assumes finite sets, but it also holds true for countable and uncountable infinite sets. In this sense, if the preference satisfies the weak order, then the superiority or equality for any pair of options can be clearly stated and at least one best option can be selected. This preference is equal to utility maximization.

2.4.3 Utility Maximization

Therefore, the following can be stated:

2.4.3.1 Theorem on Utility Maximization

The selection that satisfies the properties α and β of the rational criteria equals the maximization of the ordinal utility derived from the preference relation in weak order.

The assumption of "rational homo economicus" is often said to maximize utility, which is the maximization of ordinal utility, and it does not necessarily assume that the utility can be added or multiplied in the cardinal meaning (meaning by the interval scale).

2.5 Utility Function and Indifference Curve

2.5.1 Indifference Curve

Consider a consumption plan. For example, consider goods constituting two items: rice (x) and wheat (y). Assume that these goods are measured in kg, and consider that these goods are expressed by any vectors in the form, such as 5 kg of rice and 1 kg of wheat. This expression is made abstract as (x_1, y_1), (x_2, y_2). Given two consumption vectors (x_1, y_1), (x_2, y_2), we consider that consumers make a comparative judgment based on their preferences. Assuming that the preference structure $\langle X, \succcurlyeq \rangle$ is in weak order, the relation where y is preferred to x is expressed as $x \succ y$. Accordingly,

$$x \succ y \Leftrightarrow x \succcurlyeq y \,\&\, \neg(y \succcurlyeq x)$$

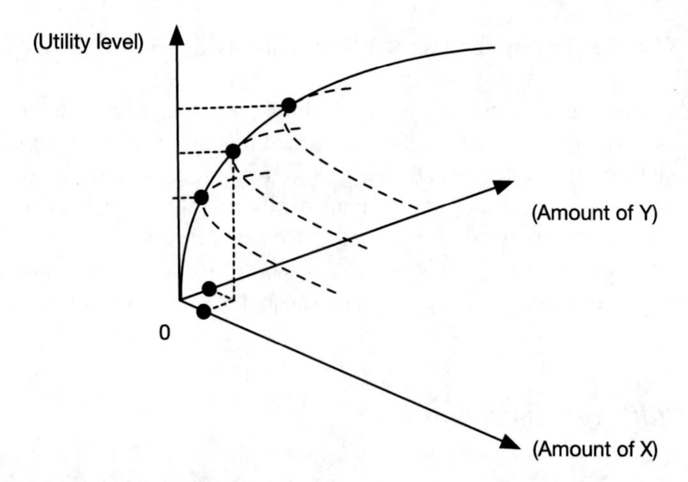

Fig. 2.1 Utility function of two goods

can be obtained. This concept expresses that x is preferred to y in a stronger sense. Furthermore, $x \succcurlyeq y$ indicates that x is preferred in a weaker sense.

Further, when a preference with the same degree is expressed as $x \sim y$,

$$x \sim y \Leftrightarrow x \succcurlyeq y \,\&\, y \succcurlyeq x$$

can be obtained and x and y are indifferent.

A set of consumption vectors indifferent from a consumption vector $x_i \in X$,

$$Ix = \{y \in X | y \sim x\}$$

is called an indifference set, and a set of consumption vectors that prefers a consumption vector in a weak sense

$$Rx = \{y \in X | y \succcurlyeq x\}$$

is called an upper contour set.

With regard to the weight of rice (first good) and wheat (second good), when the consumption vectors' consumption amount of the first good and the second good are on both axes, curves that draw consumption vectors indifferent from the consumption vectors are indifference curves. This concept may be easier to understand if it is considered from the utility function shown above. All the consumption vectors that make the utility function of a consumption vector $u(x, y) = k$ are connected to form indifference curves. As shown in Fig. 2.1, the utility is higher in the upper part of the figure. Two-dimensional indifference curves are drawn in Fig. 2.2, where the amounts of x and y of the same utility height are connected and projected onto a two-dimensional plane.

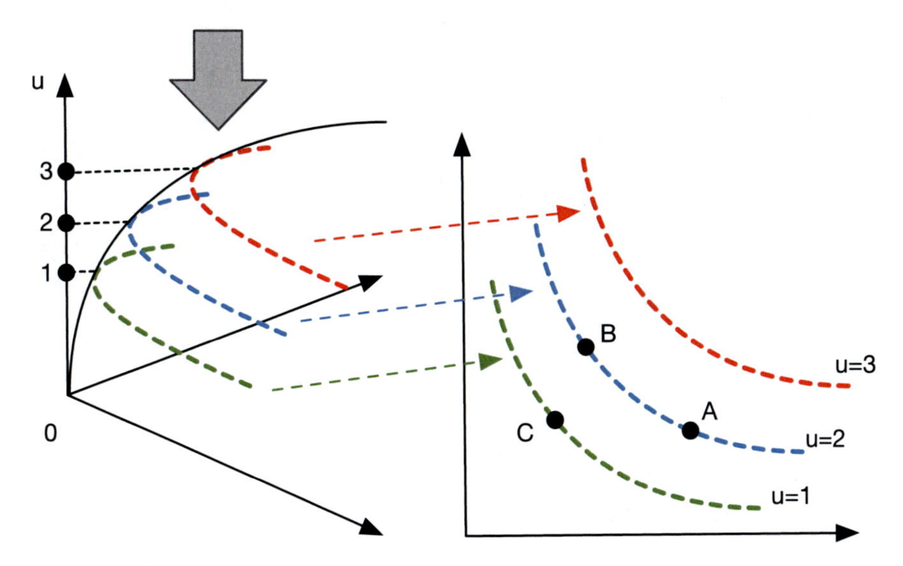

Fig. 2.2 Indifference curves from the utility function of two goods. According to Fig. 2.2, the consumption vectors *A* and *B* are indifferent, indicating strong preference for consumption vector *C*

2.5.2 *Perfect Substitutes*

When a consumer can substitute a good with another good with a certain exchange ratio, these two goods are called perfect substitutes. For example, when a hamburger of company *A* can be substituted with a hamburger of company *B* with a one-to-one exchange ratio, indifference curves, such as those shown in Fig. 2.3, can be drawn. In this case,

$$u(x, y) = x + y$$

expresses the utility curve. In the case of perfect substitutes, preferences are generally indicated by the utility curve as follows:

$$u(x, y) = ax + by$$

The slope of the indifference curve is $-a/b$. The slope of the indifference curve may be referred to as the marginal rate of substitution. It shows the exchange ratio when the consumer substitutes the first good with the second good.

Let us explain this marginal rate of substitution from the viewpoint of marginal utility. The marginal utility of a good is the rate of change in utility ΔUx linked to the minute change Δx in the good while the change in the quantity of other goods is fixed.

If the marginal utility for the first good is MUx,

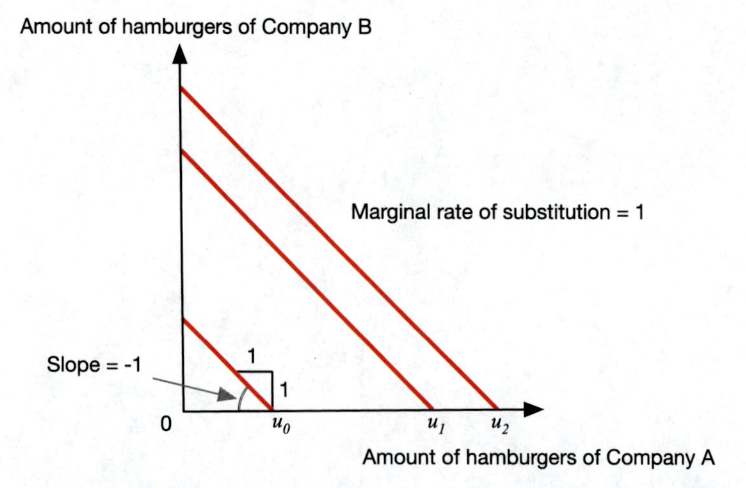

Fig. 2.3 Indifference curves for perfect substitutes

$$\mathrm{MU}_x = \frac{\Delta U}{\Delta x} = \frac{u(x + \Delta x, y) - u(x, y)}{\Delta x}$$

If the marginal utility for the second good is MU*y*,

$$\mathrm{MU}_y = \frac{\Delta U}{\Delta y} = \frac{u(x, y + \Delta y) - u(x, y)}{\Delta y}$$

can be obtained. If the utility function is partially differentiable, then MU*x* is a value obtained by partially differentiating the utility function with *x* and MU*y* is a value obtained by partially differentiating the utility function with *y*.

If the marginal rate of substitution is expressed with marginal utility,

$$\mathrm{MRS} = \frac{\Delta y}{\Delta x} = -\frac{\mathrm{MU}_x}{\mathrm{MU}_y}$$

is obtained.

2.5.3 *Perfect Completeness*

Goods that are consumed together at a certain fixed ratio indicate perfect completeness. For example, a pair of shoes and a pair of glasses are goods that make little sense if both objects in the pair are not provided. Another example is a pair of shoes. Consider that a shoe from a pair is the first good and a shoe from the same pair is the

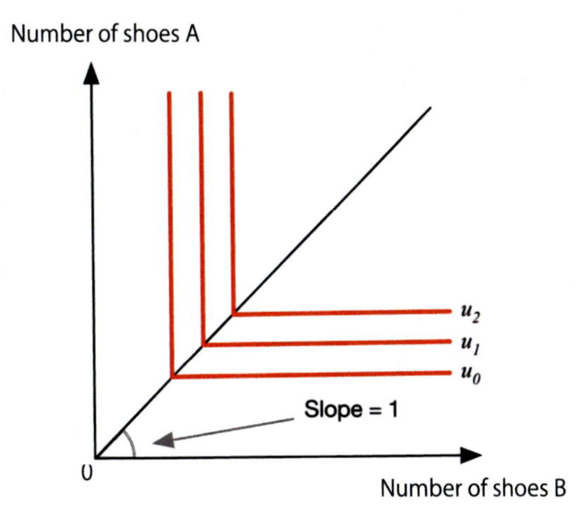

Fig. 2.4 Indifference curves for perfect completeness

second good. In such a case, indifferent curves, as shown in Fig. 2.4, are drawn. In the case of perfect completeness, as shown in Fig. 2.4,

$$(x, y) = \min(x, y)$$

expresses the utility function because utility increases only when both pairs have the same number of pieces. In the case of perfect substitutes, preferences are generally indicated by the utility curve as follows:

$$(x, y) = \min(ax, by)$$

Here, a and b are positive numbers that indicate the ratio at which each good is consumed.

2.5.4 Indifference Curves of Noneconomic Goods

In general, goods satisfy the monotonic property that utility becomes higher as the quantity increases. However, noneconomic goods (called "bads") or negative goods are not preferred as the quantity increases. Environmental wastes and pollutants are examples of such goods. Figure 2.5 shows an example of noneconomic goods for one good x, and Fig. 2.6 shows the indifference curves of noneconomic goods for both goods x, y. Figure 2.5 shows the increase in the utility level in the upper left direction, and Fig. 2.6 shows the increase in the utility level in the bottom right direction.

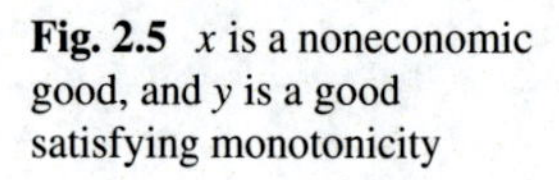

Fig. 2.5 x is a noneconomic good, and y is a good satisfying monotonicity

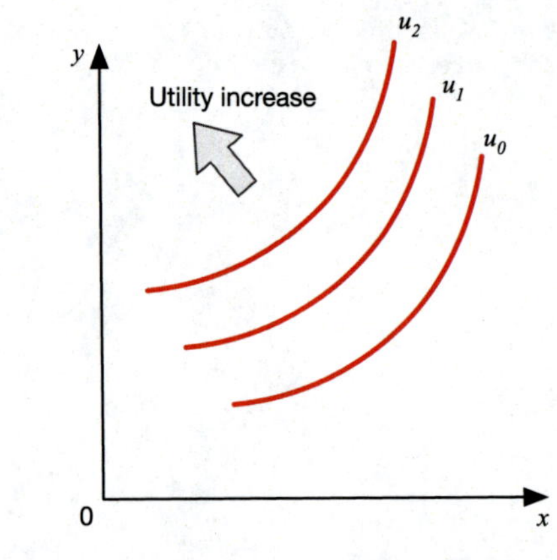

Fig. 2.6 Both x and y are noneconomic goods

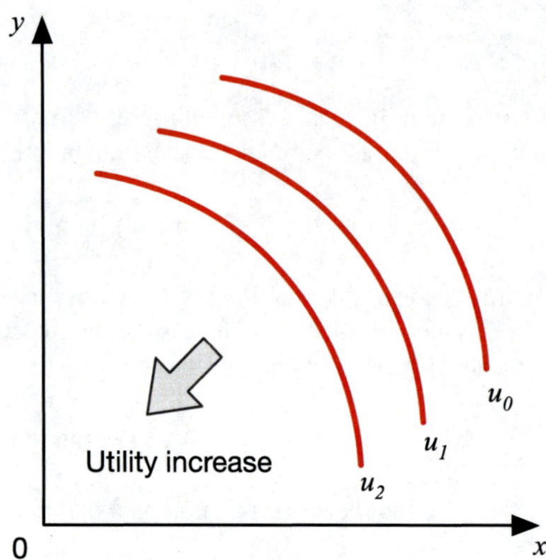

2.5.5 *Indifference Curves of Neutral Goods*

Goods wherein utility for consumers does not increase even when the amount increases are called neutral goods. At a hamburger shop, no matter how much the amount of potato chips increases, if your utility does not increase, then the potato chips are neutral goods. Figure 2.7 shows the indifference curves in the case where one good x is a neutral good and the other good y satisfies normal monotonicity.

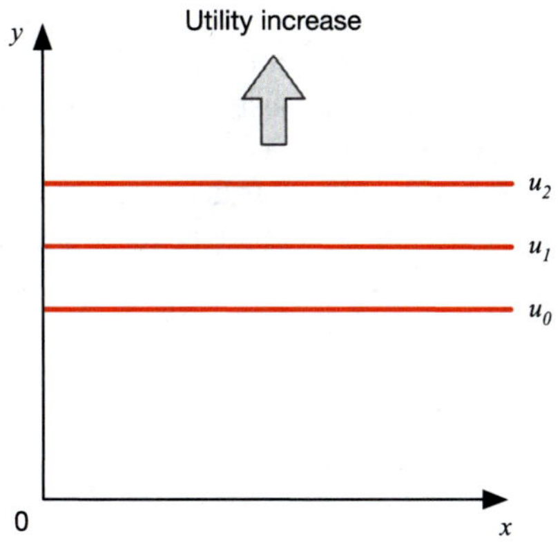

Fig. 2.7 x is a neutral good, and y is a normal good that satisfies monotonicity

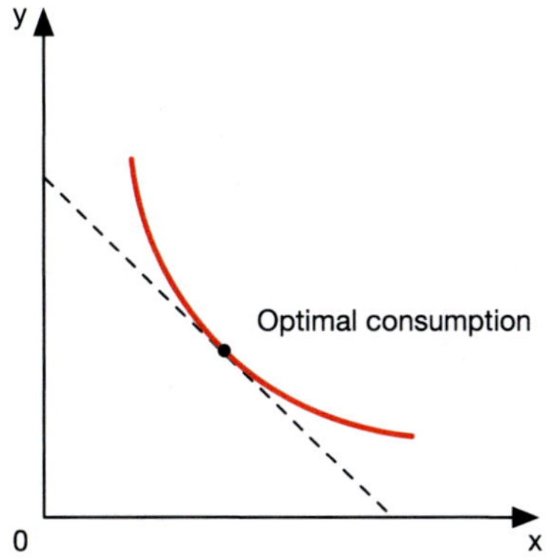

Fig. 2.8 Consumption vector realizing utility maximization under certain budget constraints

2.6 Revealed Preference

2.6.1 What is Revealed Preference?

Determining consumer preferences is a problem when you want to know these preferences. This condition also leads to the issue of inferring the utility function. Preference may be asked directly by using the questionnaire method, as used in social psychology. However, no guarantee exists that consumers will answer honestly. A methodology that obtains information about consumer preference, assuming rationality of consumers, is called revealed preference theory.

As shown in Fig. 2.8, assume that an indifference curve of goods x and y exists. When the price per unit of a good x is p_1 and the price per unit of y is p_2, the budget m is defined as

$$m = p_1 x + p_2 y.$$

The straight line going down to the right in Fig. 2.8 is the budget constraint line. If the indifference curve is convex, as shown in the figure, then the combination of goods with the highest utility is the point where the indifference curve connects the budget constraint line. The consumption vector of the black · mark is the optimal consumption plan. Furthermore, the utility value of the consumption vector is low at the lower left of the budget constraint line.

The principle of revealed preference can be shown as follows when considering the problem of budget constraint.

2.6.2 Principle of Revealed Preference

Suppose that a consumption vector (x_1, y_1) is selected under the price (p_1, p_2). Suppose that (x_2, y_2) is another consumption vector satisfying $p_1x_1 + p_2y_1 \geq p_1x_2 + p_2y_2$. At this time, if the consumer intends to select the best consumption vector, then (x_1, y_1) is strongly preferred to (x_2, y_2).

When an inequality $p_1x_1 + p_2y_1 \geq p_1x_2 + p_2y_2$ concerning the budget constraint holds true and (x_1, y_1) and (x_2, y_2) are different consumption vectors, then (x_1, y_1) is directly and revealingly preferred to (x_2, y_2). Furthermore, when three consumption vectors (x_1, y_1), (x_2, y_2), and (x_3, y_3) exist, for example, even if (x_1, y_1) is not directly and revealingly preferred to (x_3, y_3), (x_1, y_1) is directly and revealingly preferred to (x_2, y_2), (x_2, y_2) is directly and revealingly preferred to (x_3, y_3), and the transitivity of preference holds true, then (x_1, y_1) is indirectly and revealingly preferred to (x_3, y_3).

2.6.3 Weak Axiom of Revealed Preference

Suppose that a consumption vector (x_1, y_1) is selected under the price (p_1, p_2). Suppose that (x_2, y_2) is another consumption vector satisfying $p_1x_1 + p_2y_1 \geq p_1x_2 + p_2y_2$. At this time, when (x_1, y_1) is directly and revealingly preferred to (x_2, y_2) and these consumption vectors are not identical, (x_2, y_2) is not directly and revealingly preferred to (x_1, y_1).

To express this axiom in another way, when a consumption vector (x_2, y_2) can be purchased, if (x_1, y_1) is purchased when (x_2, y_2) is purchased, then (x_1, y_1) is not available for purchase.

The weak axiom of revealed preference holds true when a consumption vector is defined as $x\ (p, m)$, the price vector with an arbitrary budget m is p (expressed as (p, m)), a consumption vector is defined as $x(p', m')$ when the price vector with another arbitrary budget m' is p' (expressed as (p', m')), and the following relationship is satisfied. That is, If $px\left(p', m'\right) \leq m, x\left(p', m'\right) \neq x(p, m)$, then$p'x(p, m) > m'$.

Therefore, if the weak axiom of revealed preference is satisfied $px\left(p', m'\right) > m$, or $p'x(p, m) > m'$ holds true, and $px\left(p', m'\right) \leq m$ and $p'x(p, m) \leq m'$ are never satisfied at the same time. Therefore, in the example in Fig. 2.9, $px\left(p', m'\right) > m$ and

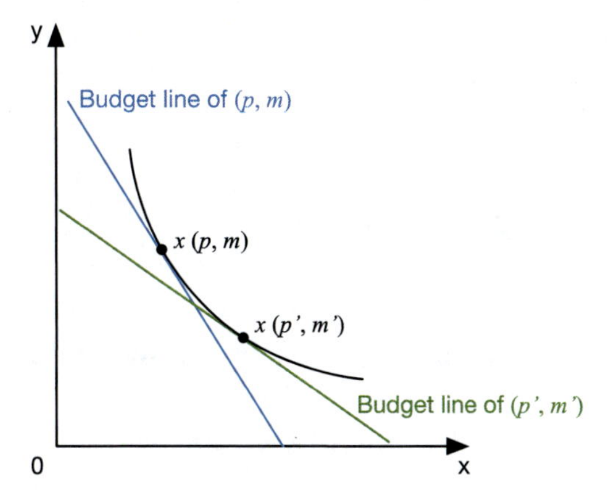

Fig. 2.9 Examples where the weak axiom of revealed preference is satisfied

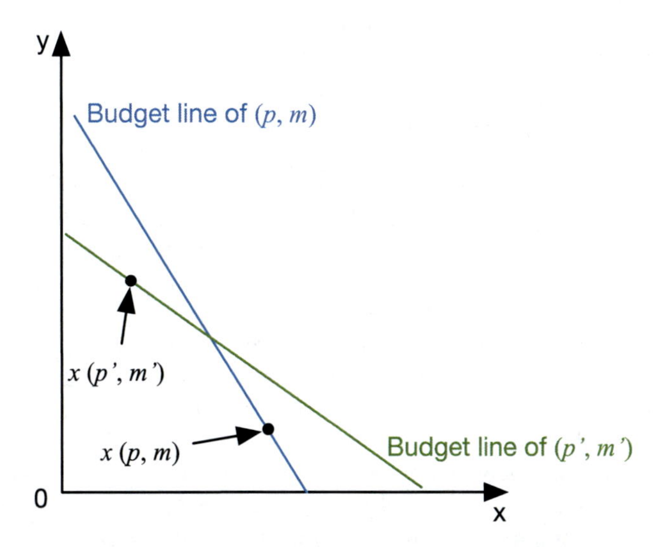

Fig. 2.10 Cases where the weak axiom of revealed preference is not satisfied

$p'x(p, m) > m'$ hold true, and the weak axiom of revealed preference is satisfied. In the example of Fig. 2.10, $px(p', m') \leq m$ and $p'x(p, m) \leq m'$ are satisfied at the same time. Thus, the weak axiom of revealed preference is not satisfied.

2.6.4 Strong Axiom of Revealed Preference

Suppose that a consumption vector (x_1, y_1) is selected under the price (p_1, p_2). Suppose that (x_2, y_2) is another consumption vector satisfying $p_1x_1 + p_2y_1 \geq p_1x_2 +$

$p_2 y_2$. At this time, (x_1, y_1) is directly or indirectly revealingly preferred to (x_2, y_2), and if these consumption vectors are not identical, then (x_2, y_2) is not directly or indirectly revealingly preferred to (x_1, y_1).

Consumers are maximizing utility when the strong axiom of this revealing preference is satisfied. On the basis of the axiom of revealed preference, we can infer preference from the actual purchasing patterns of consumers.

2.7 Empirical Testing of Rational Preference Relations

2.7.1 Empirical Examination of Weak Order

Does the property of weak order assumed in rational decision making, strong axiom of revealed preference, and ordinal utility theory hold true in actual preference judgment and decision making?

Tversky (1969) experimentally examined whether the transitivity assumed in weak order is satisfied in decision making. This examination is also an empirical study of acyclicity, with more relaxed conditions, rather than that of transitivity.

He asked subjects which gamble they preferred by showing them two cards of pie charts, such as that shown in Fig. 2.11. At this time, the statement of indifferent preference relations was not permitted. Thus, they had to indicate which one they preferred. Therefore, this reflects a relation of strong preference $x \succ y$, that is, $x \succcurlyeq y \,\&\, \mathrm{not}(y \succcurlyeq x)$ ($\succcurlyeq$ indicates weak order). On the card, the amount of prize money was written on the pie chart, and the percentage of the black-painted area of the fan shape of a circle area was expressed as the winning rate. In the experiments, several patterns were prepared, and in a typical pattern, the subjects were asked which one they preferred by combining five cards, as shown in Table 2.1. From a down to e, the winning rate is higher and the amount of prize money is lower. In the case of comparative judgment such as a and b as well as b and c, the slight difference in the winning rate was ignored and the one with the larger prize amount was selected. However, in the case of the combinations of the winning rate with larger differences such as a and e, e with a higher winning rate was selected. This finding shows the relations of $a \succ b$, $b \succ c$, $c \succ d$, $d \succ e$, and $e \succ a$ and does not satisfy transitivity. This result indicates that the condition of acyclicity with the relaxed condition of transitivity is also not satisfied.

Tversky (1969) also presented to participants the percentile ranking scores of the ratings of the intelligence, emotional stability, and sociality of five university applicants, as shown in Table 2.2. With intelligence given the highest scores, he had the subjects answer which applicant they think should be enrolled in the university by pair comparison. In the case of comparative judgment such as a and b as well as b and c, the slight difference in the intelligence rating was ignored and the one with the higher ratings in other dimensions was selected. However, in the case of combinations with larger differences in the intelligence rating such as a and e, e

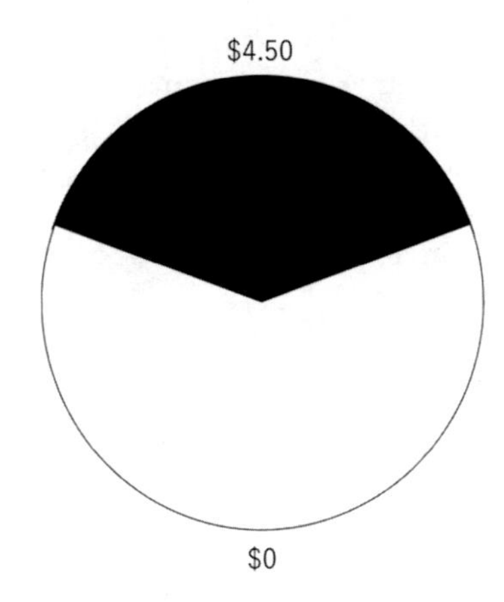

Fig. 2.11 Example of a gambling card used in the experiment. Drawn from Tversky (1969)

Table 2.1 Experimental subjects to examine transitivity

Gamble	Winning rate	Prize money	Expected value
A	7/24	$5.00	$1.46
B	8/24	$4.75	$1.58
C	9/24	$4.50	$1.69
D	10/24	$4.25	$1.77
E	11/24	$4.00	$1.83

Source Tversky (1969)

Table 2.2 Experimental subjects to examine transitivity

Applicant	Intelligence	Emotional stability	Sociality
A	69 points	84 points	75 points
B	72 points	78 points	65 points
C	75 points	72 points	55 points
D	78 points	66 points	45 points
E	81 points	60 points	35 points

Source Tversky (1969)

with the higher intelligence rating had a tendency to be selected. This finding also shows the relations of $a \succ b$, $b \succ c$, $c \succ d$, $d \succ e$, and $e \succ a$ and does not satisfy transitivity.

2.7.2 Nontransitivity and Thresholds

Nontransitivity can be explained by thinking that preference becomes indifferent within a certain threshold. Indifference can be defined as a binary relation I on X when assuming a strong preference relation P on set X (a relation where you can state which is preferred). That is, for $x, y \in S$,

When $xIy \Leftrightarrow$ not$[xP'y]$ & not$[xPy]$,
x and y are indifferent.

In this way, in judgment and decision making, a relation should exist where you cannot state which is preferred. Therefore, when considering the preference relation R on the set X, a real function with thresholds can be assumed as follows, where indifference occurs with respect to a certain degree of difference, that is, for any $x, y \in X$,

$$xRy \Leftrightarrow v(x) > v(y) + \delta(x, y)$$

Here, v is a utility function, and δ is a function of thresholds taking positive values and is assumed to vary depending on the subjects x, y. For simplicity, assuming that the threshold is constant within situation X, the following is obviously gained from the above equation, that is, for any $x, y \in S$,

$$xRy \Leftrightarrow v(x) > v(y) + \delta$$

However, δ is a positive constant.

The necessary and sufficient conditions that satisfy this formula, according to the theorem of Scott and Suppes (1958), are understood to be semiorder with the preference structure (X, R) below, that is, for any elements w, x, y, z,

(i) not$[xPx]$
(ii) wRx and $yRz \Rightarrow [wRz$ or $yRx]$
(iii) wRx and $yRw \Rightarrow [zRw$ or $yRz]$.

These conditions are necessary and sufficient conditions to satisfy $xRy \Leftrightarrow v(x) > v(y) + \delta$.

2.7.3 *Decision Making Model to Explain Nontransitivity*

Tversky (1969) proposed a mathematical model called additive difference model to explain human preference that does not satisfy such transitivity. In this model, first, consider a set of options comprising multi-attributes, as shown in Table 2.2, such as option set $A = A_1 \times A_2 \times \cdots \times A_n$. Assume that each option consists of multi-attribute values, such as $x = (x_1, x_2, \ldots, x_n)$ and $y = (y_1, y_2, \ldots, y_n)$. The additive difference model is expressed as follows, with ui as a real-valued function and φi as an increasing function.

$$x \succsim y \Leftrightarrow \sum_{i=1}^{n} \phi_i[u_i(x_i) - u_i(y_i)] \geq 0$$

However, for any attribute i, $\phi_i(-\delta_i) = -\phi_i(\delta_i)$, $\delta_i = u_i(x_i) - u_i(y_i)$

Here, assuming that $\phi_i(\delta_i) = t_i(\delta_i)$ and $t_i > 0$,

$$\sum_{i=1}^{n} \phi_i[u_i(x_i) - u_i(y_i)] = \sum_{i=1}^{n} t_i u_i(x_i) - \sum_{i=1}^{n} t_i u_i(y_i)$$

is obtained. By setting $v_i(x_i) = tu_i(x_i)$

$$x \succsim y \Leftrightarrow \sum_{i=1}^{n} v_i(x_i) \geq \sum_{i=1}^{n} v_i(y_i)$$

is obtained, and an additive utility model result is obtained. When ϕ_i can be supposed to be linear in this manner, nontransitivity cannot be explained. However, in the case of a step function in which ϕ_i has a threshold (e.g., if $\varepsilon \geq \delta$, $\phi_i(\delta_i) = 0$), this additive difference model can explain nontransitivity.

Nakamura (1992) conducted an experimental study on conditions that deviate from transitivity. He indicated that (1) in the case of preference judgment by a single attribute, judgment is relatively clear even if the utility difference is small; (2) in the case where the utility with two or more attributes allows a trade-off and the difference is about the same, the judgment becomes ambiguous; and (3) in the case where the utility can be regarded as equivalent for a certain attribute and the utility cannot be regarded as equivalent for some other attribute, the effect of the attribute regarded as equivalent is almost neglected. To explain the nontransitivity of human preference, he proposed a preference model called the additive fuzzy utility difference structure model, which assumes utility as a set with ambiguous boundaries called a fuzzy set.

Thus, although we showed that transitivity in weak order does not necessarily hold true, empirically, comparability will not always hold true either. For example, we consider that when insufficient knowledge about brands of goods exists, always showing preference relations that satisfy comparability is difficult. In the study by Tversky (1969), subjects were forced to select one of the alternatives, but in fact, a situation where selecting one of the alternatives is difficult appears to exist. Takemura (2007, 2012) expanded the model of Nakamura (1992) by proposing a model with a fuzzy set as the weight function of utility. He expressed the preference relation approximately when comparability and transitivity are not satisfied and actually conducted surveys in product selection (Fig. 2.12).

Box 2 Robert Duncan Luce

was born in 1925 and deceased in 2012. He received Ph.D. in mathematics at Massachusetts Institute of Technology in 1950. He finished his position of Distinguished Research Professor of cognitive science at the University of California, Irvine. He contributed much to the axiomatic approach of psychological measurement, utility theory, global psychophysics, and mathematical behavioral sciences.

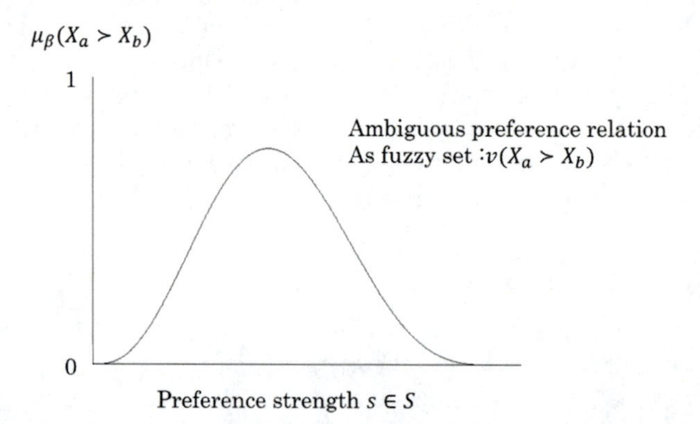

Fig. 2.12 Expression by the fuzzy set of comparative judgment when comparability does not hold true. *Source* Takemura (2007, 2012)

Summary

- Traditional economics relies on the assumption of "rational homo economicus," and a rational choice is based on the premise that at least the best option can be selected and the options can be ordered in descending order of preference.
- The preference relation in weak order in which completeness and transitivity hold true leads to rational choice.
- The preference relation in weak order can be expressed by the magnitude relation of ordinal utility.
- When the preference relation is in weak order, making the best choice is equivalent to maximizing the ordinal utility.
- A methodology that obtains information about preference, assuming rationality of preference, is called revealed preference theory.
- Ordinal utility theory is also assumed in many economic theories. However, some counterexample phenomena cannot be justified from the perspective of behavioral decision theory.
- Decision making models with thresholds such as those used in psychology do not satisfy transitivity, cannot be explained by normal utility theory, and require another decision making model.

Reading Guide for More Advanced Learning

- Krantz, D. H., Luce, R. D., Suppes, P., & Tversky, A. (1971). *Foundations of measurement Vol. 1: Additive and polynomial representations*. New York, NY: Academic Press.

The axiomatic theory base about the utility theory and measurement theory is explained in detail. This book is somewhat difficult to understand but is very important as a specialized book in this area.

- Varian, H. R. (2014). *Intermediate Microeconomics: A modern approach* (9th ed.). New York, NY: W. W. Norton & Company.

This introductory book on microeconomics explains concepts such as ordinal utility and revealed preference in an easy-to-understand manner and provides an elaborate treatment of various concepts of microeconomics.

- Takemura, K. (2014). *Behavioral decision theory: Psychological and mathematical representations of human choice behavior*. Tokyo: Springer.

The concept of decision making utility theory is explained clearly while comparing the findings of behavioral experiments.

References

Feldman, M. A., & Serrano, R. (2005). *Welfare Economics and social choice theory*. New York, NY: Springer.

Krantz, D. H., Luce, R. D., Suppes, P., & Tversky, A. (1971). *Foundations of measurement, Vol. 1: Additive and polynomial representations*. New York, NY: Academic Press.

Nakamura, K. (1992). On the nature of intransitivity in human preferential judgments. In V. Novak, J. Ramik, M. Mares, M. Cherny, & J. Nekola (Eds.), *Fuzzy approach to reasoning and decision making* (pp. 147–162). Dordrecht, The Netherlands: Kluwer.

Scott, D., & Suppes, P. (1958). Foundational aspects of theories of measurement. *The Journal of Symbolic Logic, 23,* 113–128.

Sen, K. A. (1970). *Collective choice and social welfare*. London, UK: Oliver and Boyd.

Takemura, K. (2007). Ambiguous comparative judgment: Fuzzy set model and data analysis. *Japanese Psychological Research, 49,* 148–156.

Takemura, K. (2012). Ambiguity and social judgment: fuzzy set model and data analysis. In E. P. Dadios (Ed.), *Fuzzy logic-algorithms, techniques and implementations* (pp. 1–22). Shinjuku City, Japan: IntechOpen Access Publisher.

Tversky, A. (1969). Intransitivity of preferences. *Psychological Review, 76,* 31–48.

Chapter 3
Expected Utility Theory and Economic Behavior: Predicting Decision Making Based on the Expected Value of Utility

Keywords Expected value · Expected utility theory · St. Petersburg paradox · Ellsberg's paradox · Allais paradox

Although economic behavior is often exhibited under conditions of uncertainty, we can decide future behavior to a certain extent by assigning a probability value to the uncertainty of the future. A situation is "under risk" when the probability of the future is known, and it is "under uncertainty" when the probability of the result is unknown. In this chapter, we examine the problem of decision making on the basis of the expected value in decision making under risk and propose expected utility theory as its solution. Expected utility theory has a long history that dates back to the 18th century and has a close relationship with economic psychology. In this chapter, we explain the basic idea of the original expected utility theory, explain the theories (e.g., the axiomatic system of expected utility theory), and introduce the economic psychological study of utility measurement on the basis of expected utility theory.

3.1 Decision Making Problems Based on Expected Values and Expected Utility Theory

3.1.1 St. Petersburg Paradox

Expected utility theory, which deals with decision making under risk, can be traced back to the formulation of mathematician D Bernoulli in the 18th century. He proposed the idea of expected utility to solve the St. Petersburg Paradox that had been introduced by N Bernoulli earlier. The name of this paradox comes from the fact that D Bernoulli solved this problem when he was staying in St. Petersburg, Russia.

K. Takemura, *Foundations of Economic Psychology*,
https://doi.org/10.1007/978-981-13-9049-4_3

The paradox is described (Tamura Nakamura & Fujita, 1997) as follows: "Keep tossing a gold coin with a probability of 1/2 on each side until the coin comes up with heads, and you get 2^n dollar when the coin has showed heads for the first time in the nth time. How much can you pay for participating in this game?" (For the sake of simplicity, the unit of the monetary amount here has been adjusted to dollar here).

Assuming that the number of trials is infinite, when participating in this game, expected value (EV) is expressed as follows:

$$\begin{aligned} \text{EV} &= \sum_{i=1}^{\infty} 2^{-i} \cdot 2^{i} \\ &= (1/2) \times 2 + (1/4) \times 4 + (1/8) \times 8 + \ldots \\ &= 1 + 1 + 1 + \ldots = \infty. \end{aligned}$$

Then, if the expected value is taken as the criterion for judgment, the finite participation expenses, whatever they are, will be exceeded. As this is contrary to people's intuition, it is called a "paradox." Generally, when considering decision making in a situation under risk, it is often considered to be based on the expected value. For example, the smoking risk is assessed according to the extent which the expected value of a lifetime decreases because of the act of smoking cigarettes. Moreover, ideas that are based on the expected value are also common in relation to radiation risk. However, this paradox shows that decision making that is based on expected values only leads to unacceptable decision making that is intuitively difficult to understand.

3.1.2 Expected Utility Theory by Logarithmic Functions

Bernoulli considered the expected utility (EU), which is the expected value of the logarithmic function utility $u(x) = \log_e x$ (see Fig. 3.1), as described by the following:

$$\begin{aligned} \text{EU} &= \sum_{i=1}^{\infty} 2^{-i} \log_e(2^i) = \log_e 2 \sum_{i=1}^{\infty} i/2^i \\ &= \log_e 2(1/2 + 2/4 + 3/8 + 4/16 + \cdots) \\ &= \log_e 4. \end{aligned}$$

Then, the expected utility becomes $\log_e 4$. However, in seeking the inverse function of the original utility function, we can see that it is equal to the utility value under certainty with a very small and finite amount of money, such as 4 dollars. He clarified this and asserted that the paradox could be solved by considering the expected value of the utility function of such a logarithmic function.

However, in this paradox, even if the banker has as large a budget of 1 trillion dollar that exceeds to the Japanese national budget in 2019 by about 900 billion dollars (101 trillion yen) is only between the 39th and the 40th power of 2 dollars,

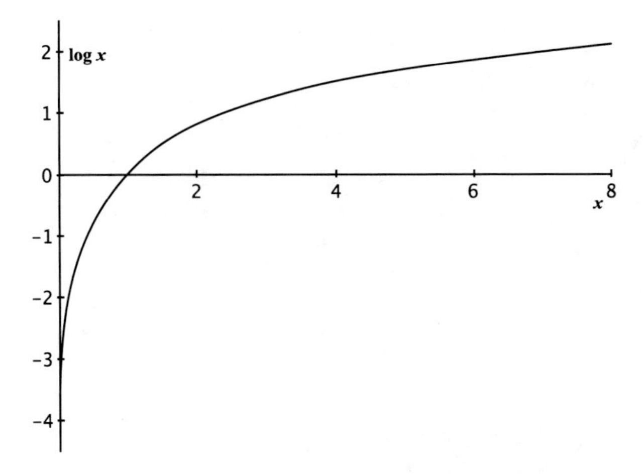

Fig. 3.1 Logarithmic function assumed by Bernoulli as a Utility function

the expected value of the game is only between 39 dollars and 40 dollars. Thus, even on considering the case of a banker who has as much money as the Japanese or the US national budgets, the expected value is considerably lower, which does not lead to a paradox. Likewise, even in the case of the USA, which has a national budget ranging from about 3 trillion dollars to 5 trillion dollars, the expected value is between 41 dollars and 43 dollars, and when it is at 50 dollars, you need much more than 1000 trillion dollars. Furthermore, in the case of the St. Petersburg problem, if it is assumed that the prize money increases with the nth power of the nth power of 2 dollars for the first coin with heads in the nth time, even if the logarithmic utility function is considered, the expected utility diverges infinitely and renders the paradox inexplicable. Such a problem of the St. Petersburg paradox has not been elucidated as yet; however, the idea of expected utility theory is already widely accepted even today.

Box 1 Daniel Bernoulli

Born in 1700; deceased in 1782. He is well known by his work on the applications of mathematics to mechanics and the work in probability and statistics. Bernoulli also provided a solution to the St. Petersburg paradox as the basis of the expected utility theory. His theory describes that decision makers do not always maximize their expected value of monetary gain, but rather maximize the expected utility.

Photograph. Wikipedia

3.1.3 Expected Utility Theory and Subjective Expected Utility Theory

Bernoulli showed that decision making under risk can be explained by the idea of expected utility, namely the expected value of utility. For example, if you consider the utility of going out with a raincoat and an umbrella when heavy rain is imminent, we consider this problem as follows:

EU (going out with a raincoat) $= p_1$ (heavy rain falls) $\cdot\, u_1$ (going out with a raincoat in heavy rain) $+\, p_2$ (no heavy rain) $\cdot\, u_2$ (going out with a raincoat when heavy rain does not fall).

Here, p_1 and p_2 are probabilities, and $p_1 + p_2 = 1$ according to the axiom of probability. Hence, in the decision making under risk, the theory that considers the expected value of utility is called expected utility theory. In particular, the one assuming subjective probability is called the subjective expected utility theory. The subjective probability has the property of probability, in which subjectivity is accepted. This was first introduced by FP Ramsey, B de Finetti, LJ Savage, and others in the 20th century.

3.2 Expected Utility and Risk Attitudes

3.2.1 *Logarithmic Utility Function and Risk Avoidance*

The utility function expressed by the logarithmic function shown in Fig. 3.1 shows the property of "diminishing marginal utility," which states that the larger the amount of money involved, the lesser the rate of utility increase that can be expected. In addition, this function means risk aversive decision making. In risk aversive decision making, even for options with the same expected value, reliable options are strongly preferred to risky options. For example, suppose there are two options: option A, through which 200 dollars can be surely obtained, and a gamble Option B, wherein 300 dollars is obtained with a probability of 1/2 and 100 dollars with a probability of 1/2; options A and B have the same expected value. However, in the case of Option A, the expected utility is expressed as

$$\begin{aligned} \mathrm{EU(A)} &= \log_e 200 \risingdotseq 5.30 \\ \mathrm{EU(B)} &= (1/2)\log_e 300 + (1/2)\log_e 100 \risingdotseq 5.15 \end{aligned}$$

This means that EU (A) > EU (B) is gained.

Option A has no risk and Option B has a risk with a gamble, thus indicating risk avoidance from the expected utility of the logarithmic function.

3.2.2 *Risk Attitudes*

Risk attitudes can be defined by expected utility for gambling when the results are stochastically generated. Specifically, in comparing options that produce reliable results (options without risk) with those that produce stochastic results (options with risk), when the expected utility of a risk-free option is equal to that of a risky option, then such an option is said to be risk neutral. Furthermore, when the expected utility of a risk-free option is higher than that of a risky option, this is considered risk aversive. The logarithmic utility function shown above is risk aversive, and utility functions, such as $u(x) = x^{0.5}(x \geq 0)$ also indicate risk aversive decision making.

In addition, when the expected utility of a risk-free option is lower than that of a risky option, it is said to be risk-seeking. For example, utility functions, such as $u(x) = x^2(x \geq 0)$, also indicate risk-seeking decision making. Given that people follow expected utility theory, risk neutral decision makers make decisions according to the expected values, risk aversive decision makers select risk-free options even though the expected values are the same, and risk-seeking decision makers select risky options (gambles) if the expected values are the same. Therefore, risk aversive decision makers dislike gambling, such as lotteries and horse races, whereas risk-seeking decision makers prefer gambling.

3.2.3 *Risk Attitudes and the Type of Utility Functions*

The following theorem holds for risk attitudes.

3.2.3.1 Theorem on Risk Attitudes

First, let X be the set of results, R be the set of real numbers, and let the real-valued function $u : X \to R$ be the utility function.

(i) If the decision maker is risk aversive, only then can the utility function u become a concave function (strict concave function). The concave function (strict concave function) means that for any t of any different two points x, y and in the open section (0, 1),

$$u(tx + (1 - t)y) > tu(x) + (1 - t)u(y)$$

holds.

(ii) If the decision maker is risk neutral, only then can u become a linear function. That is, for any t of any different two points x, y and in the open section (0, 1),

$$u(tx + (1 - t)y) = tu(x) + (1 - t)u(y)$$

holds.

(iii) If the decision maker is risk-seeking, only then can u become a convex function (strict convex function). The concave function (strict concave function) means that for any t of any different two points x, y and in the open section (0, 1),

$$u(tx + (1 - t)y) < tu(x) + (1 - t)u(y)$$

holds.

3.2.3.2 Proof of the Theorems Above

Prove these theorems.

First, we prove (i). Consider a gamble wherein the result x occurs at the probability p and the result y occurs at the probability $(1 - p)$. For risk aversive decision makers, by the definition of risk avoidance,

$$u(tx + (1 - t)y) > tu(x) + (1 - t)u(y)$$

holds. On the one hand, given that this expression itself is the definition of u being a concave function, it is proven that u is a concave function in risk aversive decision making. On the other hand, if u is a concave function, for all $x_1, \ldots, x_n \in X$ and $a_1, \ldots, a_n \in (0, 1)$;

however, $\sum_{i=1}^{n} \alpha_i = 1$,

$$u\left(\sum_{i=1}^{n} \alpha_i x_i\right) > \sum_{i=1}^{n} \alpha_i u(x_i)$$

holds. Then, the probabilities $p_1, \ldots, p_n \in (0, 1)$; however, by $\sum_{i=1}^{n} p_i = 1$, for a gamble that $x_1, \ldots, x_n \in X$ occurs,

$$u\left(\sum_{i=1}^{n} p_i x_i\right) > \sum_{i=1}^{n} p_i u(x_i)$$

is gained. Therefore, if the utility function u is a concave function, and this can lead to risk aversive decision making. Hence, theorem (i) is proven.

Next, we prove (ii). Consider a gamble wherein the result x occurs at the probability p and the result y occurs at the probability $(1-p)$. For risk neutral decision makers, by the definition,

$$u(tx + (1-t)y) = tu(x) + (1-t)u(y)$$

holds. On the one hand, given that this expression shows that u is a linear function, it is proven that u is a linear function in risk neutral decision making. On the other hand, if u is a linear function, for all $x_1, \ldots, x_n \in X$ and $a_1, \ldots, a_n \in (0, 1)$; however, $\sum_{i=1}^{n} \alpha_i = 1$,

$$u\left(\sum_{i=1}^{n} \alpha_i x_i\right) = \sum_{i=1}^{n} \alpha_i u(x_i)$$

holds. Then, probabilities $p_1, \ldots, p_n \in (0, 1)$; however, by $\sum_{i=1}^{n} p_i = 1$, for a gamble that $x_1, \ldots, x_n \in X$ occurs,

$$u\left(\sum_{i=1}^{n} p_i x_i\right) = \sum_{i=1}^{n} p_i u(x_i)$$

is gained. Therefore, if the utility function u is a linear function, this leads to risk neutral decision making. Hence, theorem (ii) is proven.

Finally, we prove (iii). Consider a gamble wherein the result x occurs at the probability p and the result y occurs at the probability $(1-p)$. For risk aversive decision makers, by the definition of risk-seeking,

$$u(tx + (1-t)y) < tu(x) + (1-t)u(y)$$

holds. On the one hand, given that this expression itself is the definition of u being a convex function, it is proven that u is a convex function in risk-seeking decision making. On the other hand, if u is a convex function, for all $x_1, \ldots, x_n \in X$ and $a_1, \ldots, a_n \in (0, 1)$; however, $\sum_{i=1}^{n} \alpha_i = 1$,

$$u\left(\sum_{i=1}^{n} \alpha_i x_i\right) < \sum_{i=1}^{n} \alpha_i u(x_i)$$

holds. Then, the probabilities $p_1, \ldots, p_n \in (0, 1)$; however, by

$$\sum_{i=1}^{n} p_i = 1$$

for a gamble that $x_1, \ldots, x_n \in X$ occurs,

$$u\left(\sum_{i=1}^{n} p_i x_i\right) < \sum_{i=1}^{n} p_i u(x_i)$$

is gained. Therefore, if the utility function u is a convex function, this can lead to risk-seeking decision making. Hence, theorem (iii) is proven.

3.3 The Axiomatic System of Expected Utility Theory

3.3.1 Decision Making Under Risk and Expected Utility Theory

von Neumann and Morgenstern (1944, 1947) proved the existence of utility functions based on objective probabilities if the axioms above are satisfied. They also demonstrated that utility can be measured when assuming expected utility theory. Their expected utility theory did not necessarily assume a logarithmic utility function like Bernoulli's expected utility theory; instead, they formulated a utility function in a more abstract form.

In expected utility theory proposed by von Neumann and Morgenstern, expected utility is expressed as follows (Tamura et al., 1997). First, let the set of options be $A = \{a_l, a_m, \ldots\}$, and suppose that when the decision maker chooses the option $a_i \in A$, the probability with which the result x_i is gained is p_i. In addition, when the decision maker chooses $a_m \in A$, the probability with which the result x_i is gained is q_i ..., and then we let the set of all possible results be $X = \{x_1, x_2, \ldots\}$. At this time, suppose $p_i \geq 0, q_i \geq 0$, for all i, and suppose that $\sum_i p_i = \sum_i q_i = \cdots = 1$ is satisfied. In addition, when the utility function on X is $u : X \rightarrow R$, the expected utility when adopting the options $a_1, a_m, \ldots$ is $Ea_1 = \sum_i p_i u(x_i)$, $Ea_m =$

Table 3.1 Examples of results according to options and conditions

A / Θ	θ_1: 1, 2, 3	θ_2: 4, 5	Θ_3: 6
a_1: Gamble 1	x_1: 10,000 dollars	x_2: 0 dollar	x_1: 10,000 dollars
a_2: Gamble 2	x_1: 10,000 dollars	x_2: 0 dollar	x_3: 20,000 dollars
a_3: Gamble 3	x_3: 20,000 dollars	x_3: 20,000 dollars	x_1: 10,000 dollars

$\sum_i q_i u(x_i), \ldots$. In this expected utility theory, it is assumed that decision makers adopt options that maximize the expected utility from option set A. Furthermore, this utility function does not lose its essential meaning even if a positive linear transformation is performed and that it has the property of radix utility (interval scale).

Before explaining the axioms of expected utility theory, let us organize the structure of decision making under risk. First, let the set of finite options be A, after which we organize the elements into contradictory options $a_1, \ldots, a_i, \ldots, a_l$ (l is the number of options) of one another, then they can be described as a set $A = \{a_1, \ldots, a_i, \ldots, a_l\}$. Next, consider a set $X = \{x_1, \ldots, x_j, \ldots, x_m\}$ occurring as a result of adopting these options. For example, the elements of X are as follows: x_1 = you can get 10,000 dollars, x_2 = you can get nothing, x_3 = you can get 20,000 dollars, and so on. When a certain option α_i is adopted, a certain result x_j appears, but α_i and x_j do not necessarily have a one-to-one correspondence. The resulting x_j caused by adopting the option α_i can be considered to be dependent on at least some conditions $\Theta = \{\theta_1, \ldots, \theta_k, \ldots, \theta_n\}$, and in decision making under risk, it means that the probability distribution of Θ is known.

For example, consider the following conditions when you throw a die:

$\theta_1 = 1$, 2 or 3 comes up,
$\theta_2 = 4$ or 5 comes up, or
$\theta_3 = 6$ comes up.

Then, as shown in Table 3.1, suppose that the prize amount is decided depending on the upcoming spot of the die thrown. As can be seen from Table 3.1, the result is decided by the function (mapping) from the adopted option and condition to the result, that is, $f : A \times \Theta \rightarrow X$. However, $A \times \Theta = \{(a_i, \theta_k) | a_i \in A, \theta_k \in \Theta\}$. Considering the probabilities, then $p(\theta_1)$ of $\theta_1 = 1/2$, $p(\theta_2)$ of $\theta_2 = 1/3$, and $p(\theta_3)$ of $\theta_3 = 1/6$. Notably, these may be considered from the viewpoint of a frequency theory or subjective probability. Then, the probabilities on the result X can be decided for each option $a_j \in A$, as shown in Table 3.2. For example, p_{33} in Table 3.2 is the probability of a result (x_3) that 20,000 dollars is obtained when a gamble 3(α_3) is chosen. From Table 3.1, as this result occurs when the conditions θ_1 and θ_2 occur, the probability p_{33} is $p_{33} = 5/6$, as shown in Table 3.2, from $p(\theta_1) + p(\theta_2) = 1/2 + 1/3 = 5/6$.

Table 3.2 Examples of the probability distribution of results in decision making under risk

A / X	x_1: 10,000 dollars	x_2: 0 dollar	x_3: 20,000 dollars
a_1: Gamble 1	p_{11}: 2/3	p_{12}: 1/3	p_{13}: 0
a_2: Gamble 2	p_{21}: 1/2	p_{22}: 1/3	p_{23}: 1/6
a_3: Gamble 3	p_{31}: 1/6	p_{32}: 0	p_{33}: 5/6

From this, the decision making problem under the risk of selecting which of the options $\alpha_j \in$ A can be replaced by the problem of selecting the right option from the following probability distributions on *X*:

$p_1 = [p_{11}, p_{12}, \ldots, p_{1m}]$,
$p_2 = [p_{21}, p_{22}, \ldots, p_{2m}]$, and
$p_l = [p_{l1}, p_{l2}, \ldots, p_{lm}]$.

This means that expressing decision making under risk is possible by the preference relationship $\succ$ on the set of probabilities $P = \{p_1, p_2, \ldots, p_l\}$ on *X*.

3.3.2 Probability Measurement and Gambles

In order to further consider decision making under risk, in accordance with the explanation by Tamura et al. (1997), let us first think about the definition of probability and redefine a "gamble."

First, consider the set *X* of results. The subset *E* ($E \subset X$) of this set *X* is the elements of $2^{\times}$ of the power set of *X* ($E \in 2^{\times}$). Here, the power set of *X* is the set of all the subsets of the set *X*, which is expressed by $2^{\times}$. Notably, the elements of the power set are sets themselves. For example, when $X = \{x_1, x_2, x_3\}$, $2^{\times}$ is a set of eight elements as follows (where φ is an empty set):

$$2^{\times} = \{\varphi, \{x_1\}, \{x_2\}, \{x_3\}, \{x_1, x_2\}, \{x_1, x_3\}, \{x_2, x_3\}, \{x_1, x_2, x_3\}\}.$$

Here, consider a finite additive probability measure *p* on $2^{\times}$. For example, a finite additive probability measure is a "probability," such as $p\ (\{x_1\}) = 0.2$. Hence, the finite additive probability measure *p* on $2^{\times}$ is a set function that satisfies, for all E_i, $E_j \in 2^{\times}$, the following:

$$\begin{aligned} & p(X) = 1, \\ & p(E_i) \geqq 0, \\ & E_i \cap E_j = \varphi \\ & \Rightarrow p(E_i \cup E_j) = p(E_i) + p(E_j). \end{aligned}$$

That is, it has the following properties: (1) the overall probability of the result set X is 1, (2) the probability of any subset E_i of X is 0 or more, and (3) if the product set of any subset of X, $E_i \cap E_j$ is the empty set (i.e., if there is no intersection of E_i and E_j), the probability of the union of E_i and E_j (i.e., the set of E_i and E_j summed) is equal to $p(E_i) + p(E_j)$.

Next, consider the convex set P_X of a finite additive probability measure on $2^{\times}$ (hereinafter referred to as a "probability measure" for simplicity). That P_X is a convex set means that if $0 \leqq \lambda \leqq 1$ and any p and q are elements of P_X ($p, q \in P_X$), then $\lambda p + (1 - \lambda)q$ is also an element of P_X ($(\lambda p + (1 - \lambda)q) \in P_X$). Hence, even if the probabilities of any two results are mixed, it makes an element of P_X.

Here, when $E_i \in 2^{\times}$ is a finite set, the probability measure that $p(E_i) = 1$ is said to be simple. From the example in Table 3.2, this simple probability measure can be interpreted as a gamble or lottery. Therefore, the fact that P_X is a convex set means that it is now possible to interpret that the composite lottery and composite gamble combining the lottery and gamble with a certain probability λ and $(1 - \lambda)$ are also elements of P_X.

3.3.3 *The Axiomatic System of Expected Utility Theory*

Let us continue to explain the axiomatic system of expected utility theory based on the descriptions of Tamura et al. (1997). First, as P_X can be interpreted as a set of options, considering the binary relationship on P_X, for all $p, q \in P_X$,

$$p \succ q \Leftrightarrow \Phi(p, q) > 0,$$

then, a real-valued function Φ on $P_X \times P_X$ that satisfies the above can be assumed. Here, $\succ$ is a strong preference relationship (i.e., $\forall p, q \in P_X, p \succcurlyeq q$ & not $(q \succcurlyeq p)$, and $\succcurlyeq$ is a weak preference relationship). On the basis of this real-valued function Φ, expected utility theory of von Neumann and Morgenstern (1944, 1947) is explained from the following linear utility model.

The linear utility model is a linear functional U on P_X that makes $\Phi(p, q) = U(p) - U(q)$ for all $p, q \in P_X$. A linear functional can be defined as follows. When P_X is a linear space on R, when the mapping U: $P_X \rightarrow R$ has the following two properties (linearity), that is,

(1) $\forall p, q \in P_X, U(p+q) = U(p) + U(q)$ and
(2) $\forall a \in R, \forall p \in P_X, U(ap) = aU(p)$

hold, and U is said to be a linear function in P_X. To say that U is linear means that, for all $p, q \in P_X$ and for all $0 < \lambda < 1$,

$$U(\lambda p + (1 - \lambda)q) = \lambda U(p) + (1 - \lambda)U(q) \tag{5}$$

holds.

From the definition of the linearity of U, because U has uniqueness even if it is multiplied by a positive constant (i.e., it is a proportional scale), this means that U has uniqueness in the range of positive linear transformations (i.e., it is an interval scale). This is because if $U' = \alpha U + \beta$ $(\alpha > 0)$, $\alpha > (p, q) = U'(p) - U'(q)$ is gained.

A linear utility model based on the utility $U(p_i)$ of a simple probability measure p_i that causes m pieces results $x_j \in X$ of a gamble $\alpha_i \in A$ with the probability $p_{ij}\left(\sum_{j=1}^{m} p_{ij} = 1\right)$, respectively, can be thought of as seeking the expected values of $U(x_j)$. This is because $U(p_i) = \sum_{j=1}^{m} p_{ij} U(x_j)$ is obtained from the linearity of U, and $U(p_i)$ is seeking the expected value of $U(x_j)$. In this sense, this linear utility model U can be considered as an expected utility model. Moreover, expected utility theory proposed by von Neumann and Morgenstern (1944, 1947) features the act of seeking expected utility by the linear utility model U.

There are several necessary and sufficient conditions for establishing expected utility theory proposed by von Neumann and Morgenstern. They also submitted an axiomatic system showing the necessary and sufficient conditions. However, given that the axiomatic system of Jensen (1967) is often cited in general, this is the one shown below. In addition, the following axiomatic system holds for all $p, q \in P_X$ and all $0 < \lambda < 1$ defined above. Here, the expression of the axiomatic system is according to Tamura et al. (1997) and Gilboa (2009).

Axiom A1 (Order Axiom)

$\succsim$ on P_x is for weak order. That is, for the preference relationship $\succsim$, this is equal for the following to hold:

(1) Transitivity $\forall\, p, q, r \in P_X, p \succsim q \;\&\; q \succsim r \Rightarrow p \succsim r$;
(2) Completeness $\forall\, p, q \in P_X, p \succsim q \vee q \succsim p$.

Axiom A2 (Independence Axiom)

If $\forall\, p, q, r \in P_X, \lambda \in (0, 1), p \succ q, \lambda p + (1 - \lambda) r \succ \lambda q + (1 - \lambda) r$ i is given. In addition, $p > q$ is $p \succsim q$ and not $(q \succsim p)$.

Axiom A3 (Continuity Axiom)

If $p \succ q$ and $q \succ r$, there exists some $\alpha, \beta \in (0, 1)$, and $\alpha p + (1 - \alpha) r \succ q$ and $q \succ \beta p + (1 - \beta) r$.

The Theorem of Expected Utility of von Neumann and Morgenstern

When axioms A1, A2, and A3 hold, only then can there be a linear functional U on P_X and for all $p, q \in P_X$,

$$p \succsim q \Leftrightarrow U(p) \geqq U(q)$$

holds. Furthermore, U has uniqueness in the range of positive linear transformations (U is an interval scale).

The independence axiom of axiom A2 is a necessary and sufficient condition for U being linear, and the continuity axiom of axiom A3 is the axiom necessary for U to become a mapping to the set of real numbers of P_X. In particular, the independence axiom is an important property in expected utility theory, but it can be interpreted that deviation from this axiom can cause Ellsberg's paradox. The independence

Table 3.3 Example of composite gambles

A / X	x_1: 10,000 dollars	x_2: 0 dollar	x_3: 20,000 dollars
a'_1: Gamble 1′	p_{11}: 5/12	p_{12}: 1/6	p_{13}: 5/12
a'_2 : Gamble 2′	p_{21}: 1/3	p_{22}: 1/6	p_{23}: 1/2

axiom means that when preference relationships of certain two options (gambles) are determined, then the preference relationships of those options are preserved. This holds even in the case of combining different gambles, whose results are equivalent and probabilities gaining each result are equal with those options, respectively. For example, in the example of gambles in Table 3.2, assume that gamble 2 is preferred to gamble 1. Combining gamble 1 with gamble 3, and gamble 2 with gamble 3 with a probability of 0.5 to make up composite gambles, then, gamble 1′ and gamble 2′ in Table 3.3 are obtained. The independence axiom requests that gamble 2′ is preferred to gamble 1′ if gamble 2 is preferred to gamble 1.

Box 2 John von Neumann

Born in 1903; deceased in 1957. Von Neumann received his diploma in chemical engineering from the Technische Hochschule in Zürich. He also received his doctorate in mathematics from the University of Budapest in 1926. He is well known by his work on mathematics, physics, and computer science. He is also a pioneer of game theory and axiomatic utility theory. He wrote Theory of Games and Economic Behavior (1944) with Princeton economist Oskar Morgenstern.

Photograph: Wikpedia

3.4 Counter-Examples of Expected Utility Theory—The Allais Paradox and Ellsberg's Paradox

Does such utility theory reflect actual people's decision making? The phenomena called the Allais paradox (see Fig. 3.2) and Ellsberg's paradox (see Fig. 3.3) are counter-examples of expected utility theory and deviate from the independence axiom of this theory described earlier. These phenomena demonstrate that expected utility theory does not fully reflect real decision making (Slovic & Tversky, 1974).

3.4.1 Allais Paradox

Allais (1953) gave a counter-example of the expected utility theory (Takemura, 2014). Consider the following decision making problem. Problem 1 involves the selection of options A and B, as shown in Fig. 3.2. By selecting Option A, you will definitely obtain one million dollars. Option B is an option that gives you 5 million dollars

〔Problem 1〕

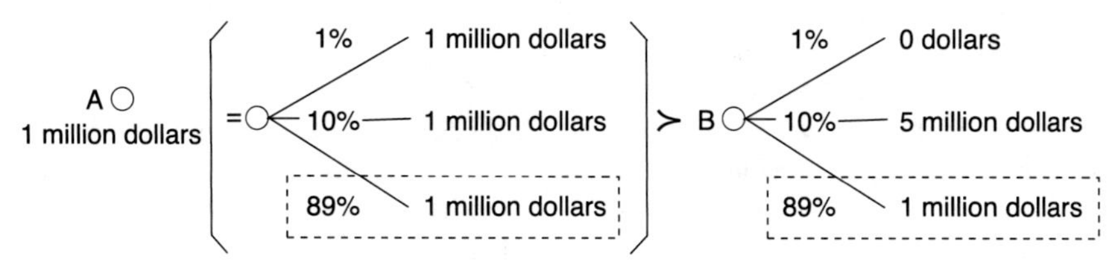

If A ≻ B, then
0.01 u(1 million dollars)+0.10 u(1 million dollars)>0.01 u(0 dollar)+0.10 u(5 million dollars) ……①

〔Problem 2〕

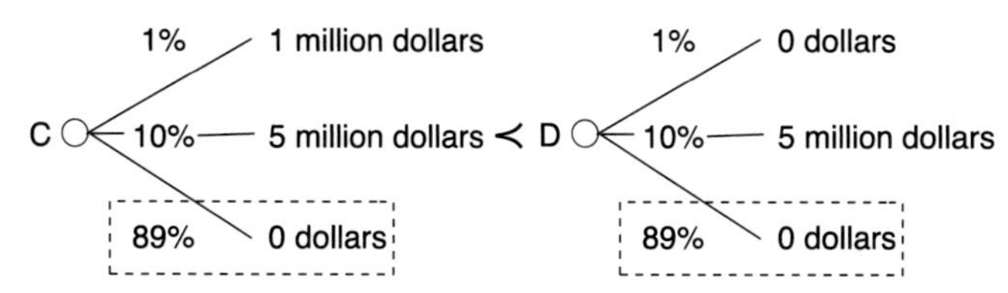

If C ≺ D, then
0.01 u(1 million dollars)+0.10 u(1 million dollars)<0.01 u(0 dollar)+0.10 u(5 million dollars) ……②

There is a contradiction between ① and ②

Fig. 3.2 Allais paradox. *Source* Takemura (1996)

〔Problem 1〕

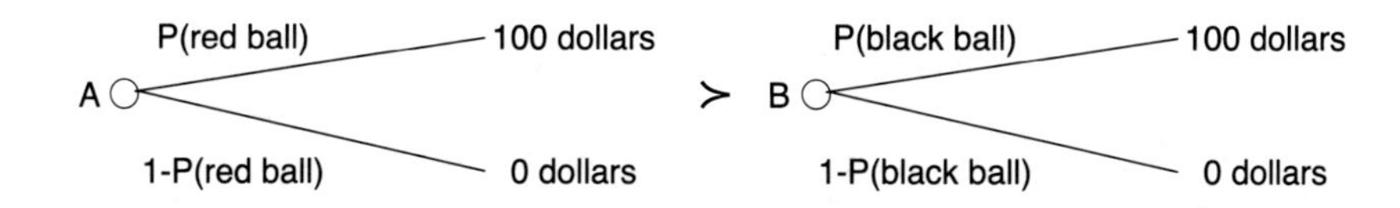

〔Problem 2〕

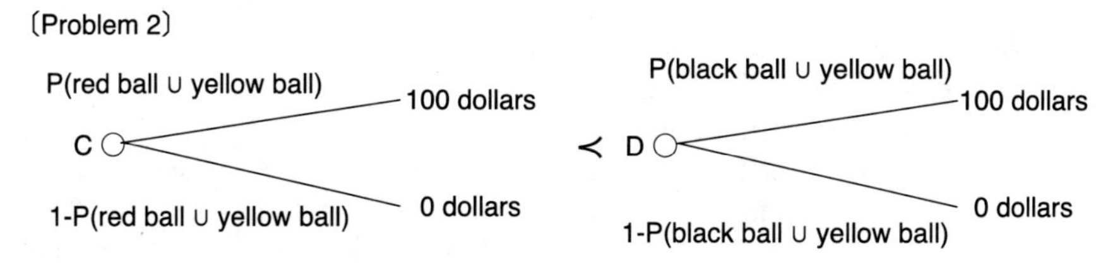

Fig. 3.3 Ellsberg's paradox. *Source* Takemura (2014)

with 10% probability, 1 million dollars with 89% probability, and 0 dollars (no prize money) with 1% probability. When comparing A and B, many people will prefer Option A as it guarantees to bring them the prize money. Next, in Problem 2, we consider two options: option C, which gives you 1 million dollars with a probability of 11%, and option D, which gives you 5 million dollars with a probability of 10%. In this case, many people would prefer Option D to C. However, this result obviously contradicts expected utility theory. For the reasons that, first, as the parts surrounded by the dashed rectangle in the figure is common in each problem, considering them in the preference by the independence axiom of expected utility theory is not necessary. Moreover, the parts not surrounded by the dashed rectangle is the same as A in Problem 1 and C in Problem 2, B in Problem 1 and D in Problem 2 (see Fig. 3.2). In psychological experiments, this is known as the Allais paradox, and is demonstrated by a number of subjects (Slovic & Tversky, 1974; Tversky & Kahneman, 1992). From a psychological perspective, this occurs from the certainty effect of preferring a certain gain to an uncertain gain.

3.4.2 Ellsberg's Paradox

Ellsberg (1961) expressed preference on ambiguity when the probability distribution of results is unknown by a concrete example, and gave a counter-example of expected utility theory (Takemura 2014). Following the paradox he presented, consider the following situation (see Fig. 3.3). There are a total of 90 balls in a pot, of which you know red balls are 30 and black and yellow balls are collectively 60, but you do not know the true composition ratio. In taking out one ball from this pot, consider the following decision making problem. In Problem 1, as shown in Fig. 3.3, option A is a bet that if a red ball (r) comes out, you get 100 dollars; otherwise, you get nothing. Another option B is a bet that if a black ball (b) comes out, you get 100 dollars; otherwise, you get nothing. Comparing both options, many people would prefer Option A to B (A $\succ$ B). Next, in Problem 2, as shown in Fig. 3.3, option C is a bet that if a red ball or a yellow ball (*r* or *y*) comes out, you get 100 dollars; otherwise, you get nothing. Another option D is a bet that if a black ball or a yellow ball (*b* or *y*) comes out, you get 100 dollars; otherwise, you get nothing. In this case, many people would prefer Option D to C (D $\succ$ C).

However, the results of this preference clearly contradict expected utility theory, assuming the additivity of the probability that the probability of the sum event of contradictory events is equal to the sum of the probabilities of each event. That is, the preference (A $\succ$ B) in Problem 1 means that the probability $P(r)$ of taking out a red ball is higher than the probability $P(b)$ of taking out a black ball ($P(r) > P(b)$), and the preference (D $\succ$ C) in Problem 2 means that the probability ($P(r \cup y)$) of taking out a red ball or a yellow ball is lower than the probability ($P(b \cup y)$) of taking out a black ball or a yellow ball ($P(r \cup y)$) ($P(r \cup y) < P(b \cup y)$). Given that r and y and b and y are two events that contradict each other, assuming the additivity of the probability, $P(r \cup y) = P(r) + P(y)$, $P(b \cup y) = P(b) + P(y)$ is given. From this, the preference (D ≻ C) in Problem 2 means $P(b) > P(r)$, which clearly contradicts the conclusion $P(r) > P(b)$ from the preference in Problem 1. This Ellsberg's paradox can be interpreted as indicating a deviation from the independence axiom in expected utility theory. As a psychological cause of Ellsberg's paradox, the ambiguity aversion that decision makers try to avoid ambiguity is being considered. In other words, because of this property, people tend to dislike ambiguity and avoid choosing options with ambiguity when the probability of the result is unknown. In recent years, various explanations have been proposed as to why such an aversion occurs, and various empirical studies have been carried out to understand such a phenomenon.

Yates and Zukowski (1976) compared people's decision making under risk; decision making under ambiguity, in which the second-order probability distribution (the probabilities for probabilities) is uniform; and decision making under ambiguity, in which the second-order probability distribution is unknown. They recognized ambiguity aversion irrespective of whether the second-order probability distribution is known or unknown. They also found no difference in the selective tendency between the decision making scenarios under two ambiguities, but in the willingness-to-pay (WTP) method, the one with the second-order probability distribution being uniform is more highly evaluated. Meanwhile, considering a gamble with two results of winning and losing, Curley and Yates (1985) further examined the influence of the median of possible values of the winning probability and the range of possible values of the probability. They found a tendency that ambiguity aversion tends to become stronger as the median becomes higher, and explained that ambiguity aversion is less likely to be observed when the median is low. They also found no difference in preference due to the difference in the range (degree of ambiguity). In Japan, Shigemasu (1988) conducted an additional examination of Ellsberg's two pots problem as an example of the Bayesian rationality in the 1980s, and found the tendency of ambiguity aversion.

Keren and Gerritsen (1999) carried out several experiments and examined the robustness of ambiguity aversion, recognizing the ambiguity aversion commonly both in the gain region and in the loss region. They observed no change in the strength of ambiguity aversion depending on the degree of ambiguity. In a series of studies examining sex differences on risk attitudes and decision strategies, among others, which are related to decision making, sex differences are not observed in ambiguity aversion (Powell & Ansic, 1997).

Research findings that generally accept ambiguity aversion continued at first, but in recent years, it has become apparent that ambiguity aversion has become less frequent. Fox and Tversky (1995) contend that ambiguous options must be presented together with risky options in order for ambiguity aversion to occur. They demonstrated that ambiguity aversion occurs when both ambiguous and risky options are presented comparatively in such problems as two colors, three colors, the problems of Ellsberg's, and problems using real events; however, ambiguity aversion does not occur when options are presented individually. Heath and Tversky (1991) also found a tendency, in which, for a gamble with the same probability as the confidence of the correct answer to a quiz, answering the quiz is selected even if it is ambiguous when the confidence is high. This indicates that, when there exists a high sense of capability, ambiguity preference occurs rather than ambiguity aversion. Meanwhile, Masuda (2010) found a tendency to judge a task less ambiguous when putting elements that can be controlled by subjects themselves in the task. Furthermore, Masuda, Sakagami, and Hirota (2002) examined the influence of selection opportunities on ambiguity aversion, and found a preference for ambiguity preference when there is competition, and that ambiguity aversion is strengthened when there is no freedom of choice. In this way, some psychological factors and factors of circumstances may have an influence on whether ambiguity is preferred or averted.

3.4.3 *Paradox of Independence Axiom and Expected Utility Theory*

The paradoxes of Allais and Ellsberg can be explained by deviating from the independence axiom in expected utility theory. Given that the Allais paradox is the paradox of decision making under risk, it is the case where the probability distribution of the natural condition is known, and since the Ellsberg's paradox is the case where only the natural condition is generally known, it becomes a problem under uncertainty.

In the case of decision making under risk, the independence axiom requests that, for any probability distribution p, t, or r, if $p > t$, the convex combination ($\lambda p + (1 - \lambda)r$) of probability distributions p and r and the preference relationship with $\lambda t + (1 - \lambda)r$, which is the convex combination of t and r, are also the same. This means that, for all probability distributions $p, t, r \in P_X$ and for all probabilities $0 < \lambda < 1$,

$$p \succ t \Rightarrow \lambda p + (1 - \lambda)r \succ \lambda t + (1 - \lambda)r$$

holds. Therefore, the independence axiom does not hold, and there exist a certain probability distribution $p, t, r \in P_X$ and a certain probability $0 < \alpha < 1$, and despite $p > t$, $\alpha p + (1 - \alpha)r \preccurlyeq \alpha t + (1 - \alpha)r$ holds (Tamura et al., 1997).

In the case of the Allais paradox, in Problem 1, by selecting option A you will definitely get 1 million dollars, and by option B, you will select an option that will bring you 5 million dollars with 10% probability, 1 million dollars with 89% probability, and 0 dollar (no prize money) with 1% probability. Given that option A can be broken down into 1 million dollars with 10% probability, 1 million dollars with 89% probability, and 1 million dollars with 1% probability, what is common to A and B is that you can get at least 1 million dollars with 89% probability. Here, when option A is expressed by p and option B by q, and if the probability that 5 million dollars cannot be obtained with a probability of 10/11 is expressed by an option t of 1/11, we arrive at the expressions

$$p = 0.11p + 0.89p,$$
$$q = 0.11t + 0.89p.$$

Therefore, from independence axiom, if $p \succ t$, $p \succ q$ is given.

Furthermore, in Problem 2, there are two options: option C, which gives you 1 million dollars with a probability of 11%, and option D, which gives you 5 million dollars with a probability of 10%. What is common to C and D is that there is a probability of at least 89% that you can get nothing. Option C is expressed by r, option D by s, and when the option that definitely gives you nothing is expressed by t', we arrive at the expressions

$$r = 0.11p + 0.89t',$$
$$s = 0.11t + 0.89t'.$$

Therefore, from the independence axiom, if $p \succ t$, $r \succ s$ is given.

To summarize the above, from the independence axiom, it is requested that $p \succ$ q and $r \succ s$ if $p \succ t$ and q $\succ p$ and $s \succ r$ if $t \succ p$. However, in the actual selection, subjects have expressed $p \succ q$ and $s \succ r$ (Slovic & Tversky, 1974) , so it is understood that the independence axiom is not satisfied.

In addition, the independence axioms under uncertainty are as follows (Tamura et al., 1997). Let X be a set of results, Θ a set of natural conditions, $A \subseteq \Theta$ an event, and let the two options be f: $\Theta \to X$, g: $\Theta \to X$. The independence axiom requests that for any $\theta \notin A$, if $f(\theta) = g(\theta)$, the preference relationship between f and g does not depend on A's complementary event A^c. Therefore, the fact that independence does not hold is as follows. This means that if some options f, g, f', g' are $\theta \in A$ for an event A, $f\ (\theta) = f'(\theta)$, $g(\theta) = g'(\theta)$, and if $\Theta \notin A$ when $f\ (\theta) = g(\theta)$ and $f'(\theta) = g'(\theta)$, it becomes $g' \succsim f'$ despite being $f \succ g$.

Let us explain this with the Ellsberg's paradox. There are a total of 90 balls in a pot, of which you know red balls are 30 and black and yellow balls are collectively 60, but you do not know the exact composition ratio. In this decision making problem under uncertainty, we assume that the decision maker constructs some subjective probability p. In Problem 1, option A is a bet that if a red ball (r) comes out, you get 100 dollars; otherwise, a black ball or a yellow ball gets you nothing. Another option B is a bet that if a black ball (b) comes out, you get 100 dollars; otherwise, you get nothing. If options A and B are expressed as f and g, respectively, then we obtain

$$\text{Expected utility of } f = p(r)u(100) + p(b \cup y)u(0),$$
$$\text{Expected utility of } g = p(b)u(100) + p(r \cup y)u(0).$$

In Ellsberg's paradox, given that $f \succ g$,

$$f \succ g \Leftrightarrow \text{expected utilityof } f > \text{expected utility of } g$$
$$\Leftrightarrow p(r)u(100) + p(b \cup y)u(0) > p(b)u(100) + p(r \cup y)u(0).$$

Moreover, given that p is a probability, additivity holds for contradictory events, and as $u(100) > u(0)$ can be assumed,

Table 3.4 Ellsberg's paradox and natural conditions

Option	Natural condition		
	Red (r) (dollars)	Black (b)	Yellow (y)
f	100	0	0
g	0	100	0
f'	100	0	100
g'	0	100	100

$$\begin{aligned} f \succ g &\Leftrightarrow p(r)u(100) + p(b)u(0) + p(y)u(0) > p(b)u(100) + p(r)u(0) + p(y)u(0) \\ &\Leftrightarrow p(r)u(100) + p(b)u(0) - p(b)u(100) - p(r)u(0) > 0 \\ &\Leftrightarrow (p(r) - p(b))(u(100) - u(0)) > 0 \\ &\Leftrightarrow p(r) > p(b) \end{aligned}$$

holds.

Likewise, in Problem 2, option C is a bet that if a red ball or a yellow ball (r or y) comes out, you get 100 dollars; otherwise, you get nothing. Another option D is a bet that if a black ball or a yellow ball (b or y) comes out, you get 100 dollars; otherwise, you get nothing. When option C is expressed by f', and option D by g', since the preference becomes $g' \succ f'$,

$$g' \succ f' \Leftrightarrow p(b) > p(r)$$

must hold. This is clearly inconsistent with $p(r) > p(b)$, which indicates that $f \succ g$ and $g' \succ f'$ do not hold at the same time. It also demonstrates that, in expected utility theory, it is impossible to explain Ellsberg's paradox regardless of how the subjective probability is set.

As can be clearly seen from Table 3.4, Ellsberg's paradox does not satisfy the independence axiom in decision making under uncertainty. That is, the fact that independence does not hold means that if some options f, g, f', g' are $\theta \in A$ (if θ is red or black) for a certain event A (red or black), $f(\theta) = f'(\theta)$, $g(\theta) = g'(\theta)$, and if $\theta \notin$ A (if θ is yellow), when $f(\theta) = g(\theta)$ and $f'(\theta) = g'(\theta)$, it becomes $g' \succsim f'$ **despite being** $f \succ g$. Ellsberg's paradox shows $g' \succ f'$ **despite being** $f \succ g$, so it does not satisfy the independence axiom.

There are some decision making phenomena that do not satisfy such independence axioms; hence, expected utility theory seems to have some problem as a descriptive theory. In order to explain these phenomena, nonlinear expected utility theory and prospect theory, which are introduced in the next chapter, can be used.

Box 3 Maurice Félix Charles Allais

Born in 1911; deceased in 2010. He was graduated from the École Polytechnique in 1933 and also studied at the École nationale supérieure des mines de

Paris. He received the Nobel Prize in economics in 1988 for his pioneering contributions to the theory of markets and efficient utilization of resources. He contributed to not only economics but also physics in the fields of gravitation, special relativity, and electromagnetism.

Photograph: Wikipedia

Summary

- Decision making based on expected values can lead to results that are contradictory to the intuition of most people, as demonstrated by the St. Petersburg paradox.
- As a theory to rationally explain the St. Petersburg paradox, Bernoulli's expected utility theory examines the expected value of utility.
- Expected utility and risk attitudes are related. When the utility function concave is downward, it indicates risk avoidance. When the utility function convex is downward, it indicates risk-seeking. Meanwhile, the linear utility function indicates risk neutrality.
- Expected utility theory has been axiomized by von Neumann and Morgenstern, and the axiomatic system expressing that theory has been found. This axiomatic system is composed of the axioms of weak order, independence, and continuity.
- According to the axiomatic system of expected utility theory, utility is an interval scale.

- As phenomena contrary to independence axioms of expected utility theory, the Allais paradox and Ellsberg's paradox, along with many psychological studies showing these phenomena, have been presented and discussed.

Reading Guide for More Advanced Learning

- Fishburn, P. C. (1988). *Nonlinear preference and utility theory (John Hopkins series in the mathematical sciences).* Baltimore, MD: John Hopkins University Press.

In this work, the axiomatic theory base about expected utility theory and nonlinear expected utility theory is explained in detail. In addition, the problem of the counter-example of expected utility theory is explained in detail in relation to the axioms. This book is somewhat difficult to understand but is very important as a specialized book in this area.

- Gilboa, I. (2009). *Theory of decision under uncertainty (Econometric Society Monographs).* Cambridge, UK: Cambridge University Press.

In this work, the epistemological base and the axiomatic theory base about expected utility theory and nonlinear expected utility theory are explained in detail. This book is somewhat difficult to understand but is very important as a specialized book in the area of theoretical research on decision making under risk and uncertainty.

- Dhami, S. (2017). *The foundations of behavioral economic analysis.* Oxford, UK: Oxford University Press,

Expected utility theory and its empirical research, the relationship between the research on the Allais and Ellsberg's paradoxes and utility theory are all explained clearly while comparing the findings of past behavioral experiments.

References

Allais, M. (1953). Le comportement de l'homme rationnel devant le risque: Critique des postulats et axiomes de l' ecole americaine. *Econometrica, 21,* 503–546.

Curley, S. P., & Yates, J. F. (1985). The center and range of the probability interval as factors affecting ambiguity preference. *Organizational Behavior and Human Decision Processes, 36,* 273–287.

Ellsberg, D. (1961). Risk, ambiguity, and the Savage axiom. *Quarterly Journal of Economics, 75,* 643–669.

Fox, C. R., & Tversky, A. (1995). Ambiguity aversion and comparative ignorance. *The Quarterly Journal of Economics, 110,* 585–603.

Gilboa, I. (2009). *Theory of decision under uncertainty*. Cambridge, UK: Cambridge University Press.

Heath, C., & Tversky, A. (1991). Preference and belief: Ambiguity and competence in choice under uncertainty. *Journal of Risk and Uncertainty, 4,* 5–28.

Jensen, N. E. (1967). An introduction to Bernoullian utility theory. I. Utility functions. *Swedish Journal of Economics, 69,* 163–183.

Keren, G., & Gerritsen, L. E. M. (1999). On the robustness and possible accounts of ambiguity aversion. *Acta Psychologica, 103,* 149–172.
Masuda, S. (2010). Aimai na jōhō to seigyo gensō ga wariai no suitei ni oyobosu eikyō [Effects of ambiguous information and illusion of control on making proportion estimates]. *Nihon Kansei Kougakukai Ronbunshi [Transactions of Japan Society of Kansei Engineering], 9*(4), 232–240.
Masuda, S., Sakagami, T., & Hirota, S. (2002). Sentaku no kikai ga aimaisei kihi ni ataeru eikyō: Kotonaru shurui no aimaisei deno kentō [The effect of available choices on ambiguity aversion; An examination of different types of ambiguity]. *Shinrigaku Kenkyū [The Japanese Journal of Psychology], 73*(1), 34–41.
Powell, M., & Ansic, D. (1997). Gender differences in risk behavior in financial decision-making: An experimental analysis. *Journal of Economic Psychology, 18,* 605–628.
Shigemasu, K. (1988). Aimaisa no ninchi ni okeru gōrisei [Rationality in probability judgement]. *Kōdō keiryōgaku [The Japanese Journal of Behaviormetrics], 16*(1), 39–48.
Slovic, P., & Tversky, A. (1974). Who accepts savage's axiom? *Behavioral Science, 19*, 368–373.
Takemura, K. (1996). Ishikettei to Sono Shien [Decision-making and support for decision-making]. In S. Ichikawa (Ed.), *Ninchi Shinrigaku 4kan Shiko [Cognitive psychology vol. 4 thoughts]* (pp. 81–105). Tokyo, JP: University of Tokyo Press.
Takemura, K. (2014). *Behavioral decision theory: Psychological and mathematical representations of human choice behavior.* New York, NY: Springer.
Tamura, H., Nakamura, Y., & Fujita, S. (1997). *Kōyō bunseki no sūri to ōyō [Mathematical principles and application of utility analysis].* Tokyo, JP: Corona Publishing.
Tversky, A., & Kahneman, D. (1992). Advances in prospect theory: Cumulative representation of uncertainty. *Journal of Risk and Uncertainty, 5*, 297–323.
von Neumann, J., & Morgenstern, O. (1944/1947). *Theory and games and economic behavior.* Princeton, NJ: Princeton University Press.
Yates, J. F., & Zukowski, L. G. (1976). Characterization of ambiguity in decision making. *Behavioral Science, 21,* 19–25.

Chapter 4
Nonlinear Utility Theory and Prospect Theory: Eliminating the Paradoxes of Linear Expected Utility Theory

Keywords Nonlinear expected utility theory · Non-additive probability · Choquet integral · Prospect theory

As introduced inthe previous chapter, expected utility theory has counter-examples called the Allais paradox (Allais, 1953) and Ellsberg's paradox (Ellsberg, 1961), and we know that these counter-examples are related to independence axioms. In recent years, these paradoxes have been explained by nonlinear utility theory (Edwards, 1992; Fishburn, 1988), which does not assume this independence axiom, and by the theory system called "generalized expected utility theory" (Quiggin, 1993). Among them, we introduce utility theory on the basis of the expected utility of Choquet. Furthermore, prospect theory proposed by Kahneman and Tversky (Kahneman & Tversky, 1979; Tversky & Kahneman, 1992) was revised in 1992 and came to be calledcumulative prospect theory. This theory is partly based on utility theory from the Choquet integrals and is also a nonlinear utility theory that synthesizes the past findings of behavioral decision theory. This chapter explains the basic assumptions of prospect theory, the phenomena that prospect theory can explain, and cumulative prospect theory by using the Choquet integrals assumed in nonlinear utility theory. The related empirical studies are also explained.

4.1 Non-additive Probability and Nonlinear Utility Theory

4.1.1 Non-additive Probability and Paradox

As explained Chap. 3 in, both the Allais and Ellsberg's paradoxes arise from the fact that independence axioms do not hold empirically. Psychologically, the Allais paradox can be explained by the certainty effect, which prefers certainty, and Ellsberg's paradox can be explained by the ambiguity aversive effect (Takemura, 2014). Vari-

K. Takemura, *Foundations of Economic Psychology*,
https://doi.org/10.1007/978-981-13-9049-4_4

ous theoretical frameworks describing such paradoxes have been proposed (Camerer, Lowenstein, & Rabin, 2004; Takemura, 2000; Tamura, Nakamura, & Fujita, 1997; Takemura, 2011, 2014). The first representative explanation is the one proposed through nonlinear utility theory with relaxed independence axioms. This theory system generalizes expected utility theory (Starmer, 2000; Tamura et al., 1997). In the field of economics, this theory system is called nonlinear utility theory (Edwards, 1992; Fishburn, 1988) or generalized expected utility theory (Quiggin, 1993). Nevertheless, it is mathematically almost the same as the theory system of the fuzzy integral by fuzzy measure theory in the field of engineering (Sugano & Murofushi, 1993). The second representative explanation is the one by a description model with psychological assumptions (Einhorn & Hogarth, 1985, 1986; Payne, Bettman, & Johnson, 1993). These models are not as axiomatic as nonlinear utility theory; however, they are able to explain various experimental results well with psychological assumptions. Prospect theory, which is described later, is regarded as one integrating the first and second theoretical frameworks.

The third representative explanation is the one expressing the ambiguity and inconsistency of preference by fuzzy sets with ambiguous boundaries (Nakamura, 1992; Seo, 1994; Takemura, 2000). In this approach, the situation in which the probability distribution of the result cannot be defined is expressed by the spreading of the membership function. In this chapter, we present the explanations in the system of nonlinear utility theory, which is the first representative explanation theory framework.

In the nonlinear utility theory system, we consider non-additive probability weighting functions that transform the probability that additivity does not hold even when probability information is given in decision making under risk as in the case of the Allais paradox. In the case of Ellsberg's paradox, formulation is performed as non-additive probabilities, wherein additivity does not hold true for measuring subjective beliefs for natural conditions.

The non-additive probability is sometimes referred to as "capacity" because it has been originally used in the field of physics. In the field of fuzzy engineering, it is called fuzzy measure. Although the names are different, the mathematical definitions are the same. The non-additive probability is the set function $\pi : 2^{\Theta} \rightarrow [0, 1]$ from the aggregate consisting of the subsets of the non-empty set Θ, which satisfies the following conditions to the closed section [0, 1], namely the condition of boundedness ($\pi(\varphi) = 0, \pi(\Theta) = 1$) and the monotonicity condition (if the subset E, F of Θ is in the relation of $E \subseteq F$, $\pi(E) \leq \pi(F)$ is satisfied). The term "non-additive probability" is derived from the fact that it does not necessarily satisfy the condition of additivity.

Even in the Ellsberg problem, if we assume the condition of boundedness, $p(\varphi) = 0$, $p(r \cup b \cup y) = 1$ and further assume the condition of monotonicity in the probability evaluation, the paradox does not necessarily occur. From the condition of monotonicity, the relationship, such as $p(r \cup b \cup y) > p(b \cup y) > p(r) > p(\varphi)$, must be satisfied; however, if $p(r) = 1/3$, $p(b \cup y) = 2/3$, $p(b) < 1/3$, $p(r \cup y) < 2/3$, the deviation from the condition of non-additive probability does not occur. In this case, there is no inconsistency in Ellsberg's problems 1 and 2. However, at this time, this non-additive probability satisfies the condition of the subadditivity of $p(y) + p(b) < p(y \cup b)$, $p(y) + p(r \cup b) < p(y \cup r \cup b)$.

4.1.2 Nonlinear Utility Theory

In expected utility theory, the expected utility maximization criterion can be grasped from the viewpoint of the Lebesgue integral with respect to the probability measure. However, for the expected utility regarding the non-additive probability defined above, there are several ways to achieve integral representation apart from the Lebesgue integral. In the field of the fuzzy measure theory in engineering, several integral representations from the viewpoint of the fuzzy integral integration have been proposed (Sugano & Murofushi, 1993), including the expected utility by the Choquet integral, which is being investigated by many researchers specializing in nonlinear utility and fuzzy theories (Choquet, 1954). Since Schmeidler (1989) first axiomatized expected utility theory by integration, it has become a representative nonlinear utility theory—one that can also explain the Allais paradox and Ellsberg's paradox.

The concept of expected utility by the Choquet integral can be explained as follows (Camerer, 1995). First, suppose that the natural condition $s_i \in \theta$ is ranked depending on the utility $u\ (f\ (s_i))$ for the result $f\ (s_i)$ by the option f, such as $u\ (f\ (s_1)) > u\ (f\ (s_2)) > \cdots > u\ (f\ (s_n))$. Hence, the expected utility $\mathrm{EU_c}$ by the Choquet integral on a finite set on a non-additive probability π is given by

$$\mathrm{EU_c} = u(f(s_1))\pi(s_1) + \sum_{i=2}^{n} u(f(s_i))\left[\pi\left(\bigcup_{j=1}^{i} s_i\right) - \pi\left(\bigcup_{j=1}^{i-1} s_i\right)\right]$$

If p is an additive measure and the natural conditions s_j are mutually contradictory, the expected utility stated above is consistent with the one by the subjective expected utility theory (Camerer, 1995). The expected utility theory by this integration is very similar to the rank-dependent utility theory. In addition, for the description in the form when the objective probability is defined, both models are matched when it is distorted by a non-additive probability. This is called a rank-dependent type because it performs ranking based on the goodness of the result and also performs integration based on it. When this is integrated and displayed, the result is shown below (Tamura et al., 1997).

$$\mathrm{EU_c} = \int_{0}^{+\infty} (1 - \pi(\{s \in \Theta : u(f(s)) \leq \tau\}))\mathrm{d}\tau - \int_{-\infty}^{0} \pi(\{s \in \Theta : u(f(s)) \leq \tau\})\mathrm{d}\tau$$

Are non-additive probabilities applicable as a measure of people's subjective uncertainty judgment? In conclusion, even non-adductive probability assuming only monotonicity without the condition of additivity is too psychologically strict. The condition of monotonicity may not be mathematically as strict as the probability measure, which assumes additivity and may generally hold in people's judgment as well. However, psychologically, we sometimes observe cases wherein the condition of monotonicity does not hold.

Tversky and Kahneman (1983) found that, when subjects are given a depiction of a 31-year-old single woman named Linda, who is active and brilliant, many subjects may highly infer the probability of an event that "she is currently a bank teller and is enthusiastic about the women's liberation movement" rather than "she is currently a bank teller." The result means that the probability $p(s_1 \cap s_2)$ for the product set of the event (s_1) as being a bank teller, and the event (s_2) as being keen on the women's liberation movement, is judged higher than $p(s_1)$ (i.e., $p(s_1 \cap s_2) > p(s_1)$). However, $s_1 \cap s_2 \subseteq s_1$ and r, $s_1 \cap s_2 \subseteq s_2$ are obvious; assuming the monotonicity of the probability to be judged, $p(s_1 \cap s_2) \leq \min(p(s_1), p(s_2))$ is thus given. Therefore, many subjects in their research have made judgments deviating from monotonic conditions.

Such a judgment is called "conjunction fallacy" as a bias in judgment on a conjunctive event and has been studied vigorously by psychologists. Past studies have found that this conjunction fallacy easily occurs even in judgment, and this allows ambiguity by fuzzy evaluation, such that the probability is expressed in sections like upper and lower limits (Takemura, 2014). In the case wherein the probability judgment does not satisfy such monotonicity, the expression by non-adductive probability assuming monotonicity has a limit as a description theory. Meanwhile, the Choquet integral on the measure with non-monotonic properties has been proposed (Murofushi, Sugeno, & Machida, 1994; Waegenaere & Wakker, 2001), and empirical studies on judgment and decision making based on this theory may be expected in the future.

The expression of expected utility by the Choquet integral is also treated in the fuzzy and expected utility theory systems and is also used in the psychological description theory system of decision making called the "prospect theory" (Kahenaman & Tversky, 1979; Tversky & Kahneman, 1992).

4.2 The Choquet Integral Model as a Nonlinear Utility Theory Under Uncertainty

4.2.1 Subjective Expected Utility Model of Anscombe and Aumann

The set F of options in decision making under uncertainty can be expressed as

$$F = \{f | f : \Theta \to X\}.$$

That is, the decision making under uncertainty has a structure wherein if an option f is selected and a condition $s \in \Theta$ occurs, then a certain result $x \in X$ can be known. If the option f and condition s are known, $x = f(s)$ is given, and we can understand that the option has the result x with the condition s. There are various cases in uncertainty: You know the mapping from Θ to X, you do not know the mapping from Θ to X, and you do not know what the Θ element is, or there are even conditions of ignorance wherein elements of X are unknown.

Now let us write the relationship $\succcurlyeq$ in the binary relationship of F. Here, $\succcurlyeq$ is a subset of $F \times F$. In this way, the set of options can be thought of as a mapping from a set of conditions to a set of results. However, when the results are considered to have uncertainty determined by objective probability, by considering the convex set P_X of the finite additive probability measure on 2^X, and considering the set of options as F_p, for $f \in F_p$, $f(\Theta) = \{f(s) \in P_x : s \in \Theta\}$ is thus given. Here, assuming that P_X is a convex set, for all $f, g \in F_p$ and $0 \leq \lambda \leq 1$, for any $s \in \Theta$, $(\lambda f + (1-\lambda)g)(s) = \lambda f(s) + (1-\lambda)g(s)$ is given, so $(\lambda f + (1-\lambda)g)(s)b$ itself is also an element of P_X. Given that $f\ (s)$ and $g(s)$ are probability measures, the fact that P_X is a convex set means that the convex combination also becomes a probability measure. Moreover, as $\lambda f + (1-\lambda)g \in F_p$ is given, we know that F_p itself is also a convex set (Tamura et al., 1997).

Before explaining the Choquet integral model, we first explain the Anscombe and Aumann theorem (Anscombe & Aumann, 1963), which is deeply related to the model. The Choquet integral model is formed based on this theorem. As a premise of this Anscombe and Aumann theorem, considering that the result has uncertainty determined by the objective probability, we think of a convex set P_X of the finite additive probability measure on 2^X, let a set of options be F_p, and let the preference relationship $\succcurlyeq$ be a subset of $F_p \times F_p$. In this theorem, the following axioms are placed (Gilboa, 2009; Tamura et al., 1997):

A1 Weak order property: Preference relationships $\succcurlyeq$ are complete and transitive.

A2 Continuity: For any $f, g, h \in F_p$, if $f \succ g \succ h$,

$$\alpha f + (1-\alpha)h \succ g \succ \beta f + (1-\beta)h$$

There exists $\alpha, \beta \in (0, 1)$, which results in the above.

A3 Independence: For any $f, g, h \in F_p$ and $\alpha \in (0, 1)$,

$$f \succ g \Leftrightarrow \alpha f + (1-\alpha)h \succcurlyeq \alpha g + (1-\alpha)h$$

is obtained.

A4 Monotonicity: On any $f, g \in F_p$, for any $s \in \Theta$, if $f(s) \succcurlyeq g(s)$ $f \succcurlyeq g$ is given.

A5 Nonobviousness: There exists $f, g \in F_p$ like $f \succcurlyeq g$.

4.2.1.1 Anscombe and Aumann theorem (Anscombe & Aumann, 1963)

When the preference relationship $\succcurlyeq$ satisfies A1 to A5, for the probability measure μ on Θ and any $f, g \in F_pX$, there exist a linear functional U on P_X and finite additive probability measure P on 2^s, which satisfy

$$f \succcurlyeq g \Leftrightarrow \int_{\Theta} U(f(s)\mathrm{d}P(s)) \succcurlyeq \int_{\Theta} U(g(s)\mathrm{d}P(s))$$

and U is unique with respect to positive linear transformations. The linear function U is the expected utility function shown in the von Neumann–Morgenstern axiomatic system, and P is an additive subjective probability measure. Therefore, satisfying axioms A1 to A5 means that subjective expected utility theory can be expressed. In the Anscombe and Aumann axiom, it is the Schmeidler's theorem (Schmeidler, 1989) that leads to expected utility theory by the Choquet integral on non-additive probability (capacity). This is achieved by relaxing the axiom of independence as explained later.

4.2.2 *Schmeidler's Theorem and the Choquet Integral Model*

Schmeidler (1989) constructed the axiomatic system by changing the conditions of the independence of the Anscombe and Aumann axiom to the conditions of the following comonotonic independence (Gilboa, 2009).

Being comonotonic means that there is no s or t wherein $f(s) > f(t)$ and $g(s) < g(t)$ are given. Meanwhile, comonotonic independence is explained as follows.

AA 3 Comonotonic independence: In a given set, for comonotonic $f, g, h \in F_p$ and $\alpha \in (0, 1)$,

$$f \succ g \Leftrightarrow \alpha f + (1-\alpha)h \succcurlyeq \alpha g + (1-\alpha)h$$

is obtained. In this axiom, independence holds only when the options are mutually comonotonic.

Schmeidler deduced the following theorem on the Choquet integral model (Gilboa, 2009; Schmeidler, 1989; Tamura et al., 1997).

When the preference relationship $\succcurlyeq$ satisfies A1, A2, A3, A4, and A5, for the probability measure μ on Θ and any $f, g \in F_p$, there exist linear functional U on P_X and a non-additive probability measure π on 2^s, which satisfies

$$f \succcurlyeq g \Leftrightarrow (C)\int_{\Theta} U(f(s)\mathrm{d}\pi(s)) \succcurlyeq (C)\int_{\Theta} U(g(s)\mathrm{d}\pi(s)).$$

However, an integral $(C)\int$ is the Choquet integral, and U is unique with respect to positive linear transformations. This means that π is a non-additive probability and U is a utility function. We can explain the preference relationship under uncertainty by the magnitude of the Choquet integrated utility. Moreover, this utility is an interval scale; hence, even if a positive linear transformation is performed, it does not change in terms of the description of preference relationships. To summarize, the axioms of the Choquet integral expected utility model are given below.

A1 Weak order property: Preference relationships $\succsim$ are complete and transitive.

A2 Continuity: For any $f, g, h \in F_p$, if $f \succ g \succ h$,

$$\alpha f + (1 - \alpha)h \succ g \succ \beta f + (1 - \beta)h$$

and there exists $\alpha, \beta \in (0, 1)$ which results in the above.

A3 Comonotonic independence: In any set, for comonotonic $f, g, h \in F_p$ and $\alpha \in (0, 1)$,

$$f \succ g \Leftrightarrow \alpha f + (1 - \alpha)h \succsim \alpha g + (1 - \alpha)h \text{ is given.}$$

A4 Monotonicity: On any $f, g \in F_p$, for any $s \in \Theta$, if $f(s) \succsim g(s)$, $f \succsim g$ is given.

A5 Nonobviousness: There exists $f, g \in F_p$ satisfying $f \succsim g$.

4.3 Qualitative Expression of Prospect Theory

Prospect theory, initially advocated by Kahneman and Tversky (Kahneman & Tversky, 1979; Tversky & Kahneman, 1992), is a comprehensive theory based on the previous findings of behavior decision and nonlinear utility theories (or generalized expected utility theory). Prospect theory was initially proposed as a descriptive theory dealing with decision making under risks (Kahneman & Tversky, 1979), but was later developed into a theory that can explain decision making under uncertainty (Tversky & Kahneman, 1992).

"Prospect" in prospect theory is a combination of various results when adopting an option and its corresponding probability, and is actually the same as "gambling" in decision making under risk. In decision making under risk, we select the desired prospect from among several prospects. That is, considering a set of occurring results $X = \{x_1,\ldots, x_j,\ldots, x_m\}$, we can replace it with the problem about which we should choose from among the probability distribution on X, $p_1 = [p_{11}, p_{12},\ldots, p_{1m}]$, $p_2 = [p_{21}, p_{22},\ldots, p_{2m}],\ldots, p_l = [p_{l1}, p_{l2},\ldots, p_{lm}]$. At this time, one prospect is expressed by $(x_1, p_{11};\ldots, x_j, p_{1j};\ \ldots, x_m, p_{mj})$. In prospect theory, we assume that this prospect is evaluated in a different way from expected utility theory.

Furthermore, in prospect theory, the decision making process is divided into an editing phase, in which the problem is perceived and the framework of decision making is determined, and an evaluation phase, in which the options are evaluated according to the problem perception (Kahneman & Tversky, 1979). The former stage is subject to the circumstances and changes according to differences in some linguistic expressions, etc. In the latter stage, once the problem is identified, evaluations that are dependent on the circumstances and decision making are made.

4.3.1 Editing Phase

The editing phase is a stage wherein options are cognitively reconstructed, and even in the extremely psychological same decision making problem, the problem is recognized differently. Moreover, the manner of framing the problem differs depending on slight differences in linguistic expression, among other factors. In the editing process, the following mental operations are done: (1) coding, (2) combination, (3) segregation, (4) cancellation, (5) simplification, and (6) detection of dominance.

(i) Coding: This is the process wherein mental operation is performed to divide the results into either gain or loss. For example, if a person who has always been working part-time for 10 dollars an hour suddenly gets 11 dollars an hour, he will recognize it as "gain," but if it is reduced to 9 dollars, he will recognize it as "loss." In this case, the regular hourly wage functions as a reference point.
(ii) Combination: This refers to the operation by which the same gains are combined and simplified. For example, the prospects (200, 0.25; 200, 0.25) with the 0.25 probability of obtaining $200 and the 0.25 probability of obtaining $200 are edited as the prospect (200, 0.50) with the 0.50 probability of obtaining $200.
(iii) Separation: This is the operation by which a reliable gain part and a risky gain part are separated. For example, the prospects (300, 0.80; 200, 0.20) with the 0.80 probability of obtaining $300 and 0.20 probability of obtaining $200 are separated into a prospect (200, 1.00) wherein one can surely obtain $200 and another prospect (100, 0.80) with the 0.80 probability of obtaining $100.
(iv) Cancellation: When comparing two prospects, they are understood with common elements ignored. For example, the prospects (200, 0.20; 100, 0.50; −50, 0.30) and (200, 0.20; 150,0.50; −100,0.30) are understood as the prospects (100,0.50; −50,0.30) and (150, 0.50; −100, 0.30) by reduction, respectively.
(v) Simplification: This is the operation that rounds out the result and its probability to simplify it. For example, we understand the prospect (101, 0.49) as (100, 0.50) by simplifying it.
(vi) Detection of dominance: This refers to the mental operation that detects superior options. For example, for the prospects (500, 0.20; 101, 0.49) and (500, 0.15; 99, 0.51), if the second elements of both prospects are simplified as (100, 0.50), they are compared between the prospects (500, 0.20) and (500, 0.15), and the former is deemed superior to the latter. The detection of dominance is a mental operation that performs simplification, among others, to detect dominance.

4.3.2 Evaluation Phase

Each prospect is reconstructed in the editing phase, and the prospect with the highest evaluation value is selected in the evaluation phase. In the evaluation phase, evaluations are made with the value function they call, a type of utility function, and the weighting function to the probability. The important thing is that, in the editing phase, the reference point, which is the origin of the value function, is determined. The manner of evaluation at this evaluation phase is basically the same as the rank-dependent type utility theory in the nonlinear utility theory.

As shown in Fig. 4.1, we can see that the value function becomes risk aversive in the gain area because it is a concave function there, and it becomes risk-seeking in the loss area because it is a convex function there. Furthermore, the slope of the value function is generally larger in the loss area than in the gain area. This means that the loss has a greater impact than the gain.

A special aspect of prospect theory is that the point corresponding to the origin of utility theory is the reference point, and it is assumed that the reference point moves easily in accordance with how the decision making problems are edited. In prospect theory, the evaluation of the result is made from the deviation from the reference point, which is the psychological origin, and the decision maker evaluates the result as either a gain or loss. In addition, prospect theory assumes that decision makers become risk aversive when evaluating gains and become risk-seeking when evaluating loss. In fact, even with the same decision making problem, because of the movement of the reference point, they become risk aversive and risk-seeking when grasping options in the gain and loss areas, respectively.

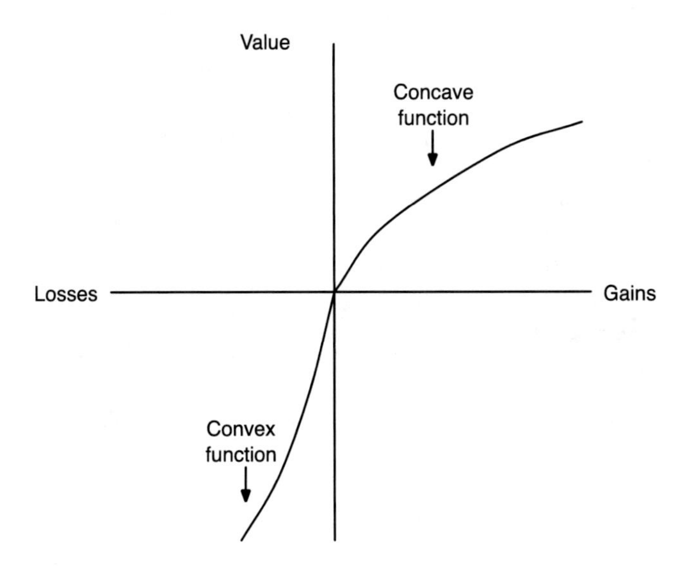

Fig. 4.1 Value function in prospect theory. *Source* Kahneman and Tversky (1979). Reproduced in part by author

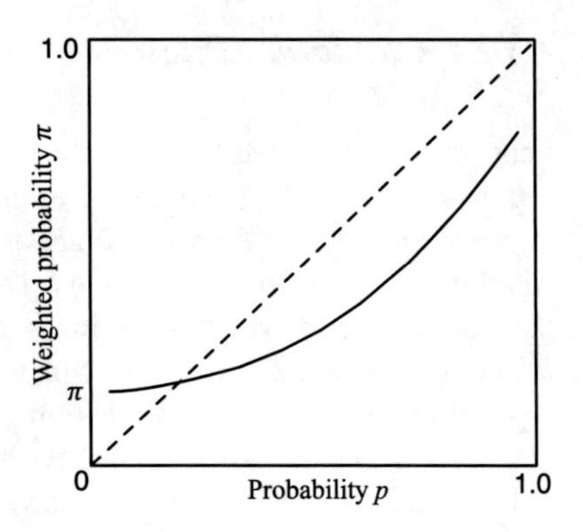

Fig. 4.2 Probability weighting function in the prospect theory. *Source* Kahneman and Tversky (1979). Reproduced in part by author

Moreover, in prospect theory, the non-additive probability weighting function is $\pi(0) = 0, \pi(1) = 1$, which has a shape shown in Fig. 4.2. Assuming that the probability weighting function is π and the objective probability is p, this probability weighting function must meet the following conditions: (1) It must be a non-unital sum of $\pi(p) + \pi(1 - p) \leq 1$; (2) in situations wherein the probability is very low, the probability is overestimated, and the relation of $\pi(p) > p$ holds; (3) the non-proportional property of $\pi(pq)/\pi(p) \leq \pi(pqr)/\pi(pr)$ is assumed; and (4) discontinuity must be observed near the end point. As can be seen, these are all qualitative characteristics.

Finally, we explain the prospect evaluation method in prospect theory. According to Kahneman and Tversky (1979), when we let x, y be the results; p, q be the probability of each result, respectively; $\pi(p), \pi(q)$ be the probability weighting values for p and q, respectively; and $v(x)$, $v(y)$ be the value of each result, the evaluation value $V(x, p; y, q)$ of the prospect is given below. However,

if one of $p + q < 1$, $x \geq 0 \geq y$, $x \leq 0 \leq y$ holds and $v(0) = 0$,

$$V(x, p; y, q) = \pi(p)v(x) + \pi(q)v(y)$$

also, if $p + q = 1$, and $x > y > 0$ or $x < y < 0$,

$$V(x, p; y, q) = v(y) + \pi(p)[v(x) - v(y)]$$

holds. The prediction by using prospect theory and subsequent research will be carried out later.

4.4 Empirical Research on Prospect Theory

4.4.1 Psychological Experiment on Value Functions and Reflection Effect

In prospect theory, we assume that the value function $v(x)$ is a concave function (downward concave function) in the area of the gain, which is higher than the reference point, and the convex function (downward convex function) in the area of the loss is lower than the reference point. Moreover, if $x \geq 0$, $v''(x) < 0$ and if $x < 0$, $v''(x) > 0$. This means that decision makers become risk aversive in the area of gain and risk-seeking in the area of loss.

Kahneman and Tversky (1979) examined this assumption on the value function by distributing questionnaires to university students and teachers in Israel, the USA, and Sweden.

Problem 1: Which do you choose?

A. You get \$4000 with a probability of 80% (prospect A = (4000, 0.80)).
B. You get \$3000 for sure (prospect B = (3000, 1.00)).

Problem 2: Which do you choose?

C. You lose \$4000 with a probability of 80% (Prospect C = (−4000, 0.80)).
D. You lose \$3000 for sure (prospect D = (−3000, 1.00)).

In Problem 1, 20% of 95 people chose A and 80% chose B. In Problem 2, 92% of 95 people chose C, and 8% chose D. This majority choice pattern is consistent with the prediction via prospect theory, which states that such people tend to be risk aversive in the area of gain and risk-seeking in the area of loss. Such a phenomenon wherein the risk attitude is reversed in the area of gain and loss is called a "reflection effect."

In addition, from the answer pattern of the following problems, they also report the phenomenon wherein the preference is reversed by the reflection effect even if the prospect has the same result.

Problem 3: You received \$1000 first. Choose one of the following options.

A. You get \$1000 with a probability of 50% (prospect A = (1000, 0.50)).
B. You get \$500 for sure (prospect B = (500, 1.00)).

Problem 4: You received \$2000 first. Choose one of the following options.

C. You lose \$1000 with a probability of 50% (prospect C = (−1000, 0.50)).
D. You lose \$500 for sure (prospect D = (−500, 1.00)).

When 70 subjects answered Problem 3, 16% chose A and 84% chose B. When 68 subjects answered Problem 4, 69% chose C and 31% chose D. The majority choice pattern, again, is consistent with the prediction via prospect theory, which states that such people tend to be risk aversive in the area of gain and risk-seeking in the area of loss.

Here, Problems 3 and 4 are identical when focusing only on the final results. That is, the final results of A (=(2000, 0.50; 1000, 0.50)) and C, B (=(1500, 1.00)), and D are the same. This indicates that the subjects did not judge the first $1000 and $2000 they received in an integrated manner. The results of this experiment indicate that people do not make decisions by considering the final amount of assets but do so based on the change of their reference point. In other words, depending on the situation, they can either become risk-seeking or risk aversive.

4.4.2 *Explanation by Value Functions of Prospect Theory of Economic Phenomena*

One example showing the value function of prospect theory is presented through Toshino's (2004) explanation of the investment behaviors of stocks. In his sample case, when the share price purchased at 10,000 dollars per share rose to 20,000 dollars and then dropped to 15,000 dollars, despite the fact that they have earned 5000 dollars per share compared with the purchase price, the buyers will keep holding the share without selling it. In this case, if the reference point is 20,000 dollars, they would judge whether it is a loss of 5000 dollars, and are likely to keep holding the share without selling it. Holding stocks is a form of risk-seeking behavior compared with selling stocks and determining profits and losses, so this is consistent with prospect theory. Meanwhile, also according to prospect theory, if the reference point is 10,000 dollars, they would then judge it as a gain of 5000 dollars, and they become risk aversive and tend to sell the share.

Various phenomena that demonstrate the reflection effect based on such a value function have been observed in actual stock markets. Odean (1998) analyzed stock trading data and reported that the median holding periods of stocks when profiting and losing are 104 and 124 days, respectively. This result can be interpreted as evidence that investors tend to become risk aversive when profiting (preferring to sell stocks quickly) and become risk-seeking when losing (preferring to hold stocks longer) (Camerer, 2000). Such a pattern of investment behavior is known as a disposition effect in the field of finance (Sheflin & Statman, 1985; Toshino, 2004). This disposition effect is found not only in the field of finance but also in the housing market. That is, when the owner of a house loses with the fall of the house price, he does not sell the house but holds it longer; this phenomenon can be explained through the disposition effect (Camerer, 2000).

4.4.3 Empirical Research on Value Functions and Loss Aversion

In the value function of prospect theory, the slope of the loss region is steeper than the gain region, that is, $v'(x) < v'(-x)$ when $x > 0$. This indicates that loss has more impact than gain, and this property is called "loss aversion."

From the property of loss aversion assumed in prospect theory, gambling with zero expected value is avoided. For example, in the case of whether you gamble on getting 1 million dollars with 50% probability, losing 1 million dollars with 50% probability (expected value is 0 dollar)), or not gambling at all (expected value is 0 dollar), it is concluded that you do not gamble from the property of loss aversion. In addition, when $x > y > 0$, the prospect $(y, 0.50; -y, 0.50)$ is preferred to the prospect $(x, 0.50; -x, 0.50)$ (Kahnaman and Tversky, 1979). That is, $v(y) + v(-y) > v(x) + v(-x)$ and $v(-y) - v(-x) > v(x) - v(y)$ hold. Furthermore, if $y = 0$, then $v(x) < -v(-x)$ can be obtained.

As a phenomenon related to this notion, we can point out the problem of the equity premium in the field of finance. As the price fluctuation of the stock market is larger than that of the bond market, assuming that the same earnings can be expected, we can consider that people generally prefer bond investments. This assumption is consistent with the property of the loss aversion described above. Benartzi and Thaler (1995) hypothesized that from the perspective of the loss aversion property in prospect theory, such investments as stocks that may lead to either great benefits or heavy losses would be evaluated with a low value; they also explained why a very large excess earning rate (premium) on the stock market occurs.

As a phenomenon derived from the loss aversion property, the endowment effect can be pointed out (Kahneman, Knetch, & Thaler, 1990, 1991). In this phenomenon, when a good is given and an individual decides to hold it, the selling price of that good becomes higher than the buying price in the case where the good is not given. Simply put, it is a phenomenon wherein it is difficult to part with the goods that one initially owned. Sometimes, this phenomenon is interpreted as representing the status quo bias. Kahneman and his colleagues conducted a series of experiments to confirm the endowment effect. In one such experiment, they first randomly divided 77 students at Simon Fraser University into three groups of "selling" conditions, "buying" conditions, and "choosing" conditions (Kahneman et al., 1990). To the subjects with the conditions to sell, they gave coffee mugs and investigated the price at which the subjects are willing to part with the mug. Meanwhile, for the subjects with the conditions to buy, they investigated the price at which the subjects are willing to pay. For both cases, in the conditions to choose, the authors presented various prices and had the subjects choose between getting the mug and gaining some cash. As a result, the medians of the pricing were \$7.12, \$2.87, and \$3.12 for the selling, buying, and choosing conditions, respectively. In the selling conditions, the reference point is in the condition of possessing the mug, and in the buying and choosing conditions, the reference point is in the condition of not having the mug, which we can think caused such differences of pricing.

4.4.4 *Empirical Research on the Probability Weighting Functions*

According to prospect theory, in a situation wherein the probability is very low, the probability is overestimated, and the relation $\pi(p) > p$ holds. Kahneman and Tversky (1979) examined this assumption on the probability weighting function by asking the following question to university students and teachers who were their participants.

Problem 1: Which do you choose?

A. You get \$5000 with a probability of 0.1% (prospect A = (5000, 0.001)).
B. You get \$5 for sure (prospect B = (5, 1.00)).

Problem 2: Which do you choose?

C. You lose \$5,000 with a probability of 0.1% (prospect C = (−5000, 0.001)).
D. You lose \$5 for sure (prospect D = (−5, 1.00)).

Of 72 subjects, in Problem 1, 72% chose A and 28% chose B; in Problem 2, 17% chose C and 83% chose D. These results indicate that, in Problem 1, the participants prefer a gamble of gain with a very low probability to its expected value amount, and in Problem 2, participants prefer the expected value amount to a gamble of loss with a very low probability.

The relationship $\pi(0.001)v(5000) > v(5)$ is shown from the answer pattern of a large number of participants in Problem 1. Assuming that the value function v in prospect theory is a concave function in the gain region, we can see that the following relationship holds:

$$\pi(0.001) > \frac{v(5)}{v(5000)} > 0.001$$

Similarly, from the result of Problem 2, the relationship $\pi(0.001)v(-5000) < v(-5)$ is shown. Assuming that the value function v in prospect theory is a convex function in the loss region, we can see that the following relationship holds:

$$\pi(0.001) > \frac{v(-5)}{v(-5000)} > 0.001$$

From this result, we can see that when the probability is very low, it tends to be overestimated, and the relationship $\pi(p) > p$ holds.

Next, in prospect theory, it is assumed that for a probability $0 \le p, q, r \le 1$, the non-proportionality of $\pi(pqr)/\pi(pr) > \pi(pq)/\pi(p)$ holds. Kahneman and Tversky (1979) examined this property by asking their participants the following questions.

Problem 3: Which do you choose?

A. You get \$6000 with a probability of 45% (prospect A = (6000, 0.45)).

B. You get $3000 with a probability of 90% (prospect B = (3000, 0.90)).

Problem 4: Which do you choose?

C. You get $6000 with a probability of 0.1% (prospect C = (6000, 0.001)).
D. You get $3000 with a probability of 0.2% (prospect D = (3000, 0.002)).

In Problem 3, 66 participants answered the problem, of whom 14% chose A and 86% chose B. Similarly, 66 participants answered Problem 4, and 73% chose C and 27% chose D. From the results of Problem 3 and Problem 4, when we assume that the value function v is a concave function in the gain region, we obtain

$$\frac{\pi(0.001)}{\pi(0.002)} > \frac{v(3000)}{v(6000)} > \frac{\pi(0.45)}{\pi(0.90)}$$

Here, if $p = 9/10$, $q = 1/2$, and $r = 1/450$ are assumed, we can see that $\pi(pqr)/\pi(pr) > \pi(pq)/\pi(p)$ holds.

There are several phenomena that can be explained using the properties of the probability weighting function of prospect theory (Camerer, 2000; Tada, 2003). First of all, we can point out the great hit bias of horse racing. As reported by Thaler and Ziemba (1988), the expected dividend rate of a great hit with a very low probability of winning is considerably lower than that of betting on a favorite horse, but at a race place people tend to prefer betting on a great hit. This can be explained by referring to the property of the probability weighting function, which shows the overestimated probabilities in situations wherein the probabilities are very low.

Similarly, it can be explained from the property of probability weighting function that many people participate in public lotteries and lotto. Furthermore, the property of this probability weighting function can explain why many people buy insurance. For example, the telephone line repair insurance is 45 cents a month, but the repair cost is $60 and the expected cost of repair is only 26 cents a month (Chiccheti & Dubin, 1994). In this way, based on prospect theory, we can explain the phenomenon of insurance purchase by understanding how the weights of events with a low probability become larger.

4.5 Cumulative Prospect Theory

4.5.1 Assumption of Cumulative Prospect Theory

In the original paper published in 1979, prospect theory was a model expressing decision making under risk (Kahneman & Tversky, 1979), but in the 1992 paper, it was renamed into cumulative prospect theory and was expanded into a model expressing decision making, including ambiguity and risk under uncertainty (Tversky & Kahneman, 1992). Cumulative prospect theory can be interpreted as a type of rank-dependent nonlinear expected utility theory (e.g., Quiggin, 1993; Starmer, 2000; Tamura et al., 1997).

First, we define the elements of the decision making problems. Let X be a set of results and Θ be a set of natural conditions. In addition, let the prospect (option) under uncertainty be $f : \Theta \rightarrow X$. In other words, if there occurs the result of $x \in X$ under some natural conditions $s \in \Theta$, we think that there is a function $f\ (s)= x$. However, for the sake of simplicity, we consider the result $x \in X$ to be a monetary value. For example, f is a kind of lottery, in which you roll a die and get 10,000 dollars (x_1) when an "odd number spot" (s_1) comes up, and you get 20,000 dollars (x_2) when an "even number spot" (s_2) comes up.

As a preparation for considering cumulative prospect theory, we rank the results in order of the increasing desirability (e.g., depending on the result, 10,000 dollars, 20,000 dollars, 40,000 dollars, and so on) and order them in increasing order. The way to obtain the comprehensive evaluation value by the ranking of the desired result is basically the same as the one explained earlier when obtaining the rank-dependent type nonlinear expected utility by the Choquet integral (Choquet, 1954). Actually, cumulative prospect theory also uses the Choquet integral.

In addition, suppose that $\{s_i\}$ is a subset of Θ, and when s_i occurs, the result becomes x_i; as such, the prospect f can be expressed as the sequence of pairs of (x_i, s_i). In the example of rolling a die above, it can be expressed as prospect $f =$ (10,000 dollars, odd number spot; 20,000 dollars, even number spot). Again, by the ascending order of the desirability of the results, we order it in the natural condition corresponding to the results. As cumulative prospect theory assumes that the value functions are different in the gain and loss regions, we distinctly treat f^+ and f^- as positive and negative results, respectively. That is, $f^+(s) = f(s)$ if $f(s) > 0, f^+(s) = 0$ if $f(s) \leq 0$, $f^-(s) = f(s)$ if $f(s) < 0$, $f^-(s) = 0$ if $f(s) \geq 0$. In the example of the die spot above, $f^+(s_1) =$ 10,000 dollars, $f^+(s_2) =$ 20,000 dollars, $f^-(s_1) = 0$ dollar, and $f^-(s_2) = 0$ dollar. In the same way as the expected utility theory, we consider such a function that gives $V(f) \geq V(g)$ if the prospect f is strongly preferred to the prospect g or is indifferent. The overall utility is determined by assuming that

$$V(f) = V\left(f^+\right) + V\left(f^-\right), V(g) = V\left(g^+\right) + V\left(g^-\right)$$

and by referring to the functions of the gain region prospect and the loss region prospect.

Box 1: Daniel Ellsberg

Born in 1931. He earned a doctoral degree from Harvard University. His studies can be considered the pioneer of the current nonlinear utility theory and behavioral decision making theory. Many of his numerous papers are still studied today. In fact, Ellsberg is well known also as a historic anti-war activist who made an accusation against the problems of the Vietnam War. His Web site is http://www.ellsberg.net/.

https://www.flickr.com/photos/smallcurio/29427033337/in/photolist-LQneMK-7ErECR-dT2EUG-eA5BiF-rXMSkE-oSX2Q5-DLZaK-efpuv8-aNWpNH-8CwwS-7Evu1u-aYvZUB-6ZEjtN-7Evv2h-7ErEjg-7ErBXe-7cwuxM-7ErEtp-7EPZ2e-99Wx9n-7EvtMU-7Evwi9-7EPYW6-6KivZh-eKhaED-arvijt-2iWvC-dT2EXb-a4HJCF-eA8LcL-efui3b-dT2FNj-8ZPVjs-eQnff2-eA5BFt-a4HKj4-4zKjm3-9TxUbP-Fzfn9H-9JoNDY-29UkcNB-afsQd5-9JoNuA-9Q1uJp-7HRw3c-93SN9r-aVF9jn-M2znaF-aVFjLM-eBvbBz

4.5.2 Model of Decision Making in Cumulative Prospect Theory

In expected utility theory, we consider an additive set function related to the set of natural conditions, such as the subjective expected utility theory system proposed by Savage (1954). In comparison, in cumulative prospect theory, we consider a non-additive set function with the generalized probability measure. This is the same as the capacity and fuzzy measure described in nonlinear expected utility theory by the previous Choquet integral. In other words, it is the set function $W : 2^{\Theta} \rightarrow [0, 1]$

from the aggregate, which consists of the subsets of the non-empty natural condition set Θ to the closed section [0, 1]. In addition, the condition of boundedness ($W(\varphi) = 0$, $W(\Theta) = 1$) and monotonic condition (when the subset A_i of Θ is a subset of A_j, that is, if $A_i \subseteq A_j$, $W(A_i) \leq W(A_j)$) are satisfied. For example, when you roll a die, if each degree of belief that the 1, 2, or 3 spot comes up is 0.1 and the degree of belief that an odd number spot comes up is 0.4, then the additivity condition of the probability measure is not satisfied, but it can be said that the condition of monotonicity is satisfied.

In cumulative prospect theory, as a value function, the monotonic increase function $v : X \to R_e$ in a narrow sense is considered and standardized so as to satisfy $v(x_0)= v(0) =0$. As a concrete example, we assume a function, such as $v(x) = 2x^{0.8}$, but the value function is often discussed on a common basis in the same way as the explanation of the utility function. In addition, the comprehensive evaluation value $V(f)$ of the prospect is explained by the sum of $V(f^+)$ and $V(f^-)$, as shown earlier, and $V(f^+)$ and $V(f^-)$ are defined as

$$V(f) = V(f^+) + V(f^-),$$

$$V(f^+) = \sum_{i=0}^{n} \pi_i^+ v(x_i),\ V(f^-) = \sum_{i=-m}^{0} \pi_i^- v(x_i),$$

at this time, $f^+ = (x_0, A_0; x_1, A_1; \ldots; x_n, A_n)$, $f^- = (x_{-m}, A_{-m}; x_{-m+1}, A_{-m+1}; \ldots; x_0, A_0)$ is gained.

In addition, $\pi_0^+, \ldots, \pi_n^+$ are the weights of the gain region and $\pi_{-m}^-, \ldots, \pi_0^-$ are the weights of the loss region. We note that the weights are determined based on the desirability ranking of the results. In cumulative prospect theory, weights are defined as follows:

$$\pi_n^+ = W^+(A_n),\ \pi_{-m}^- = W^-(A_{-m}),$$

$$\pi_i^+ = W^+(A_i \cup \ldots \cup A_n) - W^+(A_{i+1} \cup \ldots \cup A_n), \quad 0 \leq i \leq n-1,$$

$$\pi_i^- = W^-(A_{-m} \cup \ldots \cup A_i) - W^-(A_{-m} \cup \ldots \cup A_{i-1}), \quad 1-m \leq i \leq 0$$

A little more explanation is made on the above equations. First, the decision making weight π_i^+ relates to the gain region, where the result becomes positive, and is the difference between the non-additive probability of an event that yields a result that is at least as good as x_i and that of an event that yields a result that is better than x_i. In addition, the decision making weight π_i^- relates to the negative result and is the difference between the non-additive probability of an event that yields a result that is at least as good as x_i and that of an event that yields a result that is not as desirable as x_i. If each W is additive, W is the probability measure, and π_i is simply the probability of A_i.

In order to simplify the expression, if the expressions are modified to $\pi_i = \pi_i^+$ if $i \geq 0$, and $\pi_i = \pi_i^-$ if $i < 0$,

$$V(f) = \sum_{i=-m}^{n} \pi_i v(x_i)$$

is obtained.

Next, we explain cumulative prospect theory under risk. If the prospect $f = (x_i, A_i)$ is given by the probability distribution $p(A_i) = p_i$, then a decision making problem under risk is made, and the prospect can be expressed as $f = (x_i, p_i)$. In the case of this decision making problem under this risk, the decision weight is given as follows:

$$\begin{aligned}
\pi_n^+ &= W^+(p_n), \pi_{-m}^- = W^-(p_{-m}), \\
\pi_i^+ &= W^+(p_i + \cdots + p_n) - W^+(p_{i+1} + \cdots +), \quad 0 \leq i \leq n-1, \\
\pi_i^- &= W^-(p_{-m} + \cdots + p_i) - W^-(p_{-m} + \cdots + p_{i-1}), \quad 1-m \leq i \leq 0
\end{aligned}$$

However, W^+ and W^- are monotonic increase functions in a narrow sense and are standardized as $W^+(0) = W^-(0) = 0$ and $W^+(1) = W^-(1) = 1$, respectively. Similar to cumulative prospect theory under uncertainty, if it is expressed as $\pi_i = \pi_i^+$ if $i \geq 0$, $\pi_i = \pi_i^-$ if $i < 0$,

$$V(f) = \sum_{i=-m}^{n} \pi_i v(x_i)$$

is obtained.

To show an example of prospect theory under risk, we consider the following situation (Tversky & Kahneman, 1992) . If you roll a die once and let the spot coming up be x, then $x = 1,\ldots,6$ is obtained. Consider a game wherein you gain x dollars if x is an even number and you pay x dollars if an odd number. Then, f can be considered a prospect that causes the result of $(-5, -3, -1, 2, 4, 6)$ with the objective probability 1/6 of each result. By this, expressions $f^+ = (0, 1/2; 2, 1/6; 4, 1/6; 6, 1/6)$, $f^- = (-5, 1/6; -3, 1/6; -1, 1, 1/6; 0, 1/2)$ are given. As the probability causing \$0 by f^+ is 1/2 because it is the probability of odd spots coming up, the distinct probabilities of obtaining \$2, \$4, and \$6 are 1/6 individually; the probabilities of obtaining \$−5, \$−3, and \$−1 by f^- are 1/6 individually; and the probability of obtaining \$0 is 1/2 because it is the probability of even spots coming up. Therefore, we obtain the values below.

$$\begin{aligned}
V(f) &= V(f^+) + V(f^-) \\
&= v(2)\big[W^+(\text{events with a spot of 6, 4, or 2 coming up}) \\
&\quad - W^+(\text{events with a spot of 6 or 4 coming up})\big] \\
&\quad + v(4)\big[W^+(\text{events with a spot of 6 or 4 coming up})
\end{aligned}$$

$$
\begin{aligned}
&-W^{+}(\text{events with a spot of 6 coming up})\big]\\
&+v(6)\big[W^{+}(\text{events with a spot of 6 coming up})\\
&-W^{+}(\text{events with nothing})\big]\\
&+v(-5)\big[W^{-}(\text{events with a spot of 5 coming up})\\
&-W^{-}(\text{events with nothing})\big]\\
&+v(-3)\big[W^{-}(\text{events with a spot of 5 or 3 coming up})\\
&-W^{-}(\text{events with a spot of 5 coming up})\big]\\
&+v(-1)\big[W^{-}(\text{events with a spot of 5, 3, or 1 coming up})\\
&-W^{-}(\text{events with a spot of 5 or 3 coming up})\big]\\
=\;&v(2)\left[W^{+}\left(\frac{1}{6}+\frac{1}{6}+\frac{1}{6}\right)-W^{+}\left(\frac{1}{6}+\frac{1}{6}\right)\right]\\
&+v(4)\left[W^{+}\left(\frac{1}{6}+\frac{1}{6}\right)-W^{+}\left(\frac{1}{6}\right)\right]+v(6)\left[W^{+}\left(\frac{1}{6}\right)-W^{+}(0)\right]\\
&+v(-5)\left[W^{-}\left(\frac{1}{6}\right)-W^{-}(0)\right]+v(-3)\left[W^{-}\left(\frac{1}{6}+\frac{1}{6}\right)-W^{-}\left(\frac{1}{6}\right)\right]\\
&+v(-1)\left[W^{-}\left(\frac{1}{6}+\frac{1}{6}+\frac{1}{6}\right)-W^{-}\left(\frac{1}{6}+\frac{1}{6}\right)\right]\\
=\;&v(2)\left[W^{+}\left(\frac{1}{2}\right)-W^{+}\left(\frac{1}{3}\right)\right]+v(4)\left[W^{+}\left(\frac{1}{3}\right)-W^{+}\left(\frac{1}{6}\right)\right]\\
&+v(6)\left[W^{+}\left(\frac{1}{6}\right)-W^{+}(0)\right]+v(-5)\left[W^{-}\left(\frac{1}{6}\right)-W^{-}(0)\right]\\
&+v(-3)\left[W^{-}\left(\frac{1}{3}\right)-W^{-}\left(\frac{1}{6}\right)\right]+v(-1)\left[W^{-}\left(\frac{1}{2}\right)-W^{-}\left(\frac{1}{3}\right)\right]
\end{aligned}
$$

This relation is expressed in Fig. 4.3. The $V(f^{+})$ is the area on the left-hand side of Fig. 4.3, and $V(f^{-})$ is the product of the area on the right-hand side of Fig. 4.3 multiplied by -1. When this is expressed linguistically, the comprehensive assessment in cumulative prospect theory can be derived as follows: First, the weight of the value of \$2, π , is obtained from the difference between the weight, w, of the probability of gaining \$2 or more and the weight, w, of the probability of gaining \$4 or more. Other weights π are obtainable in the same manner. The sum of products of this π and the value v engenders the comprehensive assessment. When expressing this fact by words, the comprehensive evaluation value in cumulative prospect theory is to be obtained as follows. First, the weight π on the value of \$2 is obtained by the difference between the weight w of the probability of obtaining \$4 or more and the weight w for the probability of obtaining \$2 or more. Similarly, another weight π can be obtained. The comprehensive evaluation value is obtained by the sum of the products of π and value v.

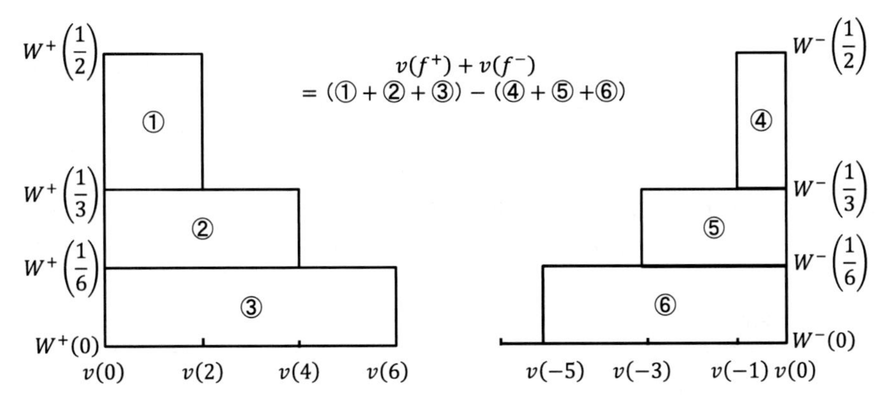

Fig. 4.3 Calculation method of $v(f)$ in cumulative prospect theory. *Note* $v(f) = v(f^+) + v(f^-)$

4.6 Experiments on Cumulative Prospect Theory

4.6.1 Parameter Estimation Experiments of Value Functions and Probability Weighting Functions of Cumulative Prospect Theory

Tversky and Kahneman (1992) presented various prospects to a total of 25 graduate students from Stanford and Berkeley by computers for choice experiments and estimated the value function of cumulative prospect theory. The prospects they presented are of the type wherein the probability of getting \$150 is 25% and the probability of getting \$50 is 75%. They then made the students compare such prospects with reliable prospects to conduct choice experiments of choosing a desirable alternative.

$$v(x) = \begin{cases} x^\alpha & (if\ x \geqq 0) \\ -\lambda(-x)^\beta & (if\ x < 0) \end{cases}$$

Based on the choice results of this experiment, they performed nonlinear regression analysis and estimated 0.88 for both α and β and 2.25 for λ. The fact that the estimated values of α and β are 1 or less indicates that the value function is concave downward in the gain region and convex downward in the loss region. In addition, the estimated value of λ indicates that the loss has an impact that is about twice as large as the gain, indicating that the property of loss aversion is strong.

They also consider the following function as the concrete decision weight function W^+, W^- of the cumulative prospect theory, and by this choice experiment, as shown in Fig. 4.2, they are able to estimate the shape of the decision weight function.

$$W^+(p) = \frac{p^\gamma}{(p^\gamma(1-p)^\gamma)^{1/\gamma}},\ W^-(p) = \frac{p^\delta}{\left(p^\delta(1-p)^\delta\right)^{1/\delta}}$$

The estimated value of γ is 0.61, and the value of δ is 0.69. The value of δ is slightly larger than the value of γ, indicating that the probability weighting function related to the positive result is slightly larger in the degree of the curve.

4.6.2 Experimental Research on Various Probability Weighting Functions in Cumulative Prospect Theory

Thus far, in the literature on cumulative prospect theory, various metric models have been proposed in the evaluation formula of the probability weighting function, and their comparisons have been made (see, e.g., Gonzalez & Wu, 1999; Murakami, Ideno, Tamari, & Takemura, 2014; Takemura, 2014; Wu & Gonzalez, 1996).

In the original prospect theory, the probability weighting function showed non-additivity, the overestimation of low probability events, non-proportionality, and discontinuity near the end point. However, in cumulative prospect theory, this is formulated below. Here, only the probability weighting function in the gain region is shown.

$$W(p) = \frac{p^{\gamma}}{(p^{\gamma} + (1 - p)^{\gamma})^{1/\gamma}}$$

In addition, p expresses a probability, $W(p)$ is a subjective weight to the probability p, and γ is a parameter. Parameter γ takes values from 0 to 1, and the shape for each parameter of this model is shown in Fig. 4.4.

In addition, we also have Prelec's model, which we modified considering another axiomatic basis (Prelec, 1998). The simple formula of Prelec's model is often used in the fields of behavioral economics, behavior decision making research, and neuroscience as well. This simple formula is given by

$$W(p) = \exp\{-(-\ln p)^{\alpha}\}$$

In addition, p is a probability, $W(p)$ is a subjective weight to, p and α is a free parameter, which takes the interval (0, 1). As shown in Fig. 4.5, the fixed point is $1/e$, or about 0.36 regardless of the value of the parameter α (Wu & Gonzalez, 1996).

Takemura et al. (1998, 2001, 2013, 2016) developed the psychometric models of probability weighting functions by using a model derived from the delayed value discounting (Mazur, 1987) and a model derived from Takemura's mental ruler theory (Takemura, 1998). Meanwhile, Takemura, Murakami, Tamari, & Ideno (2013, 2014) and Murakami et al. (2014) have compared and examined those models with prospect theory of Tversky and Kahneman (1992) and the simplified model by Prelec (1998), thus improving the metric model of Tversky and Kahneman under almost the same experimental conditions. These studies are presented below. First, the model that Takemura et al. (2013) deduced from the delayed value discounting (Mazur, 1987)

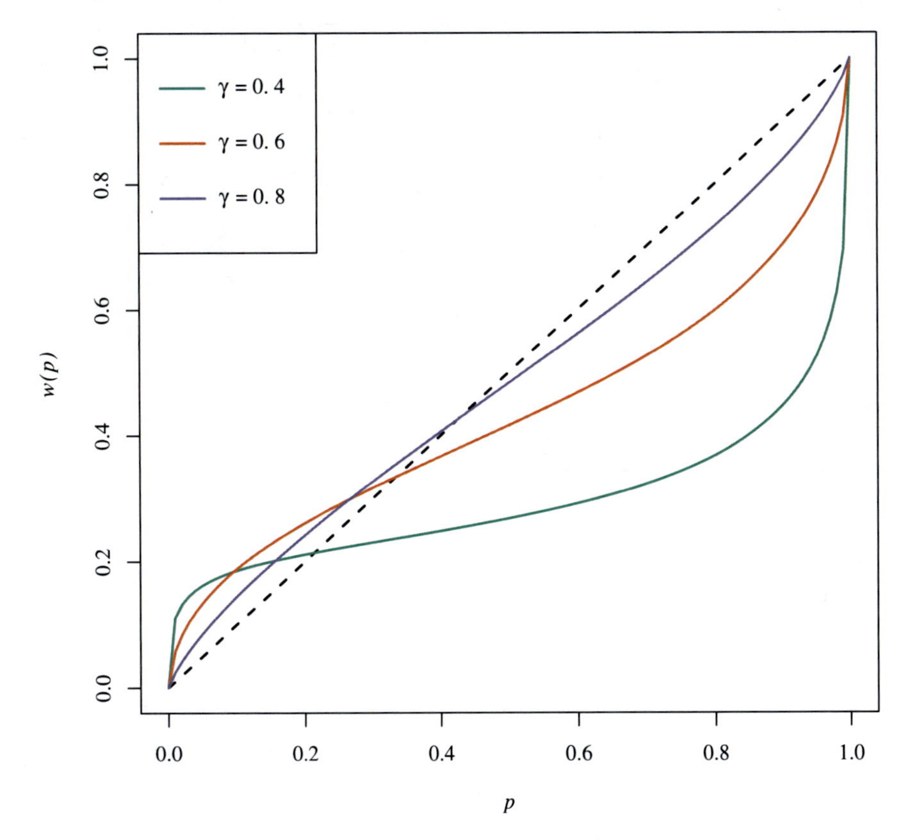

Fig. 4.4 Probability weighting function in cumulative prospect theory (Constructed based on the Tversky and Kahneman (1992) Model)

is shown in the following formula, and then an example of the model derived from the mental ruler theory (Takemura, 1998, 2001) is shown below.

4.6.2.1 Model Derived from Delayed Value Discounting

Takemura et al. (2013, 2016) deduced a probability weighting function as a model of human sensibilities to probability from a hyperbolic discount model in the research of delayed value discounting, which is often used in behavioral analysis and behavioral economics. The model is expressed as

$$V = \frac{A}{1 + kD}$$

where V expresses the value of reinforcement, A is the reinforcement amount, and D is the reinforcement delay. In addition, k is a parameter expressing the discount

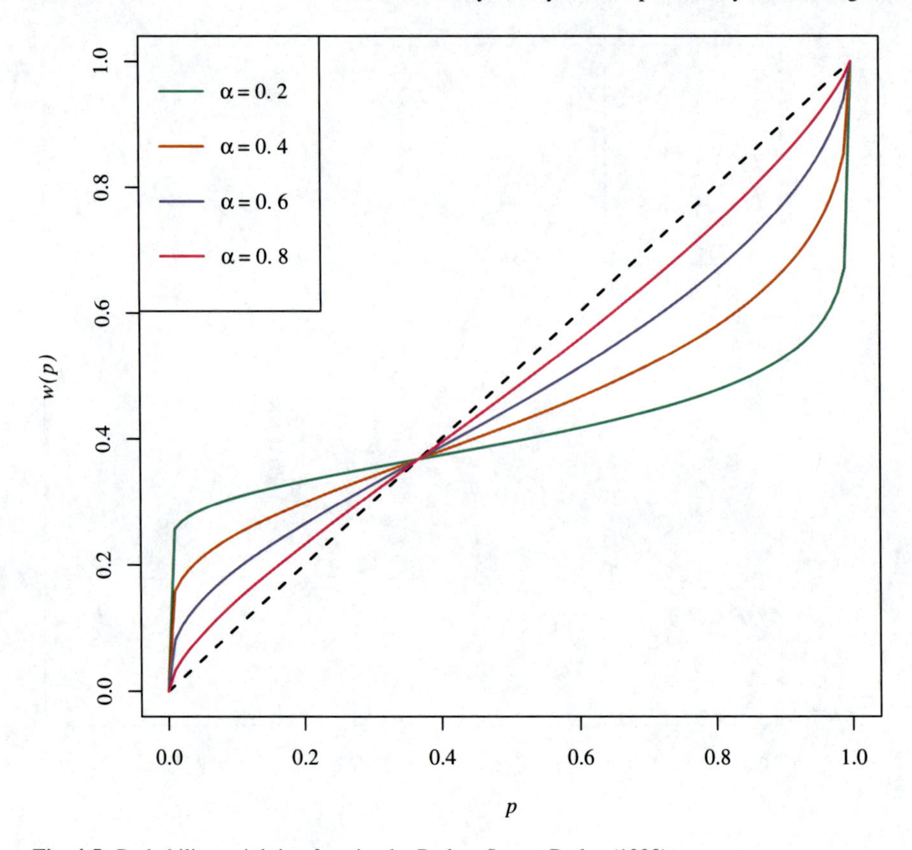

Fig. 4.5 Probability weighting function by Prelec. *Source* Prelec (1998)

rate, and as this value increases, more value is discounted. This formula is based on the idea that the value of reinforcement is proportional to the inverse of the delay.

First, replace D, which expresses the reinforcement delay of formula (5) with the probability p. Given that reinforcement is not performed until an independent trial is considered, a geometric distribution is given, and the average waiting trial number becomes $1/p$ of the inverse of the probability p. This $1/p$ expresses the average number of trials before a certain amount of reinforcement occurs at a certain time. In addition, considering that the so-called Fechner's law works, and by taking the logarithm of $1/p$, the above formula becomes

$$W(p) = \frac{1}{1 + k \log \frac{1}{p}}$$

When the common sense is rewritten, it becomes the following expression, which is the probability weighting function by the delayed value discounting model and is given by

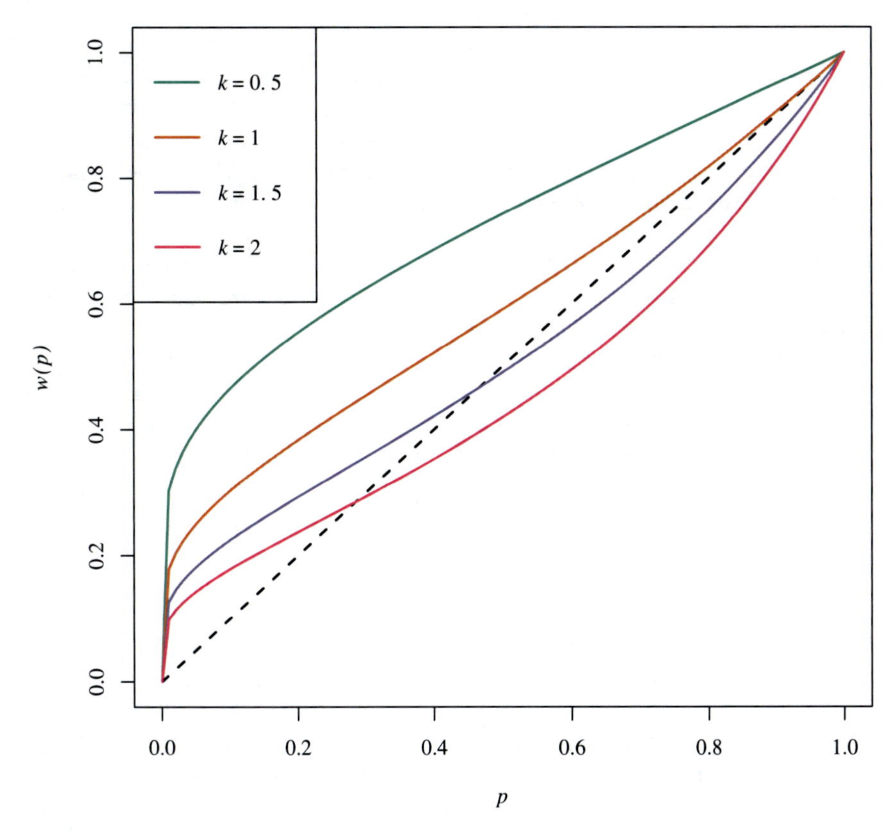

Fig. 4.6 Probability weighting function derived from delayed value discounting. *Note* Shape for parameter *w* in the model

$$W(p) = \frac{1}{1 - k \log p}$$

In the model of the probability weighting function derived from this delayed value discounting, the value of the parameter k takes a value of 0 or more. If the value of the parameter k is smaller than 1, it becomes a concave function, and when it is larger than 1, it becomes a convex function. In addition, the characteristic difference from Tversky and others' original metric model is that the subjective probability variation range is large for the objective probability. The transition of the shape for each parameter of this model is shown in Fig. 4.6.

4.6.2.2 Model in the Mental Ruler Theory Takemura (1998, 2001)

In the model of the mental ruler, both in human judgment and decision making, as it is easy for a person to make a one-dimensional mental ruler, it is assumed that he subjectively constructs the situation, makes a base set from the subjective situation, and constructs a mental ruler on it to make judgments (Takemura, 1998). In addition, there are six basic properties for a mental ruler. First, people construct an appropriate mental ruler according to the situation. Second, the positions to which the reference point and the end point of the ruler are applied may change depending on the situation. Third, the graduation of a ruler narrows particularly in the vicinity of the end point and the reference point. Fourth, high knowledge and high involvement narrow the graduation of a ruler. Fifth, even if information is given in a multidimensional manner, one-dimensional judgment is made by the mental ruler. Sixth, making a comparison among different mental rulers is difficult.

The metric model of mental ruler theory with such assumptions and basic functions is the linear sum of a concave function (downward concave function) and a convex function (downward convex function). The logarithmic function in Fechner's psychophysics and a model wherein Stevens' power exponent is less than 1 are both concave functions, and when the power exponent of Stevens' model exceeds 1, it is a convex function. One expression type of this model is given below.

$$W(p) = wp^{\beta} + (1 + w)\left(1 - (1 - p)^{\beta}\right)$$

Here, if the value of w of the relative weight is closer to 1, it means that the position of the fixed point becomes higher; if the value of β is closer to 1, then it becomes more linear. By using two parameters in this model, we can see the degree of overestimation or underestimation for the objective probability and decide the fixed point. Therefore, the mental ruler model can express the human sensibilities to the probability judgment in a more flexible manner than the conventional models. The transition of shape for each parameter of this model is shown in Fig. 4.7a, b.

Next, the following expression is a model in which the delayed value model is described in the category of mental ruler theory.

$$w(p) = \alpha\left(\frac{1}{1 - k\left(\frac{1}{\log(1-p)}\right)}\right) + (1 - \alpha)\left(1 - \frac{1}{1 - k\left(\frac{1}{\log p}\right)}\right)$$

Here, α is a relative weight, and k ($k > 0$) is a parameter related to distortion. Figure 4.8a, b shows the shape of the probability weighting function when k and α are changed. Many models of such probability weighting functions have been submitted, but the problem of empirically determining what kind of model is better among various options remains.

Takemura et al. (2013) and Murakami et al. (2014) conducted psychological experiments to estimate a certainty equivalent. A certainty equivalent is an amount that is equivalent to options (lottery) specifying multiple results and the probabilities that those results will occur. Using the method developed by Gonzalez and Wu (1999) and the certainty equivalent estimated by psychological experiments, Takemura et al. (2013) and Murakami et al. (2014) estimated the free parameters of the probability

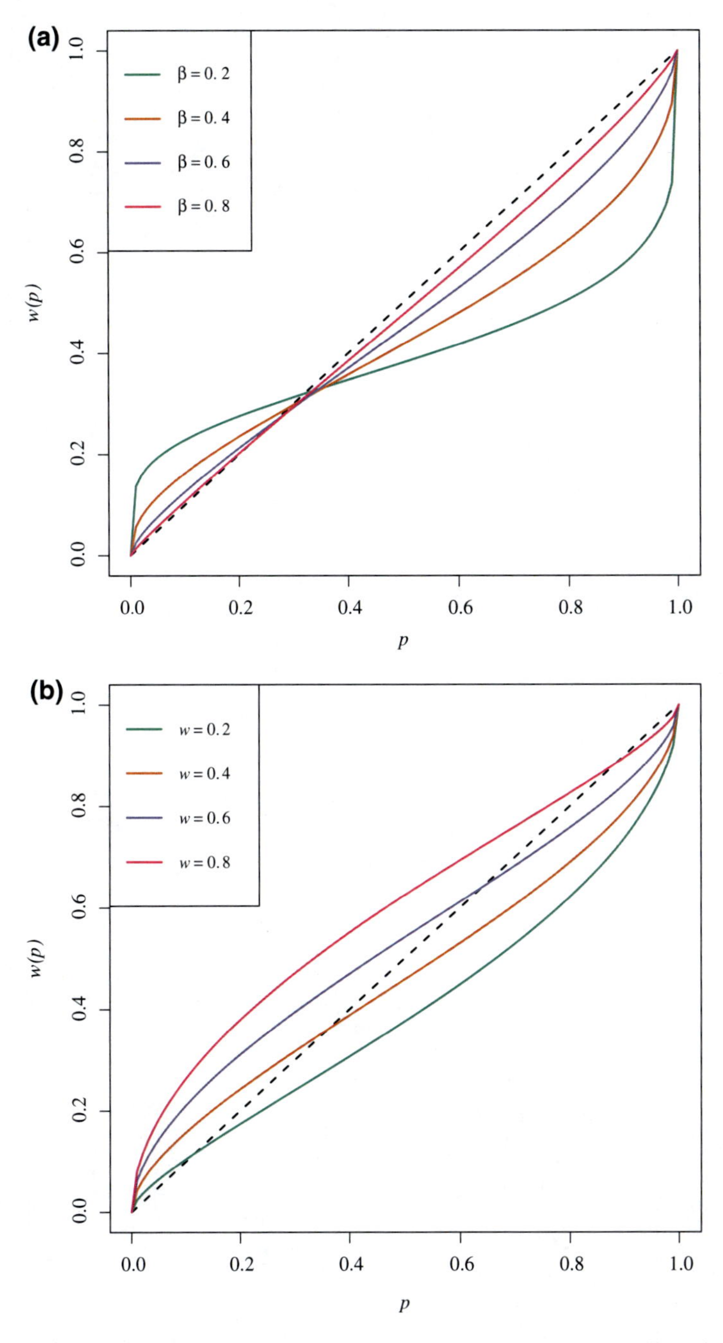

Fig. 4.7 **a** Probability weighting function derived from mental ruler theory. *Note* Shape for parameter β in the model ($w = 0.34$). **b** Probability weighting function derived from mental ruler theory. *Note* Shape for parameter w in the model ($\beta = 0.5$)

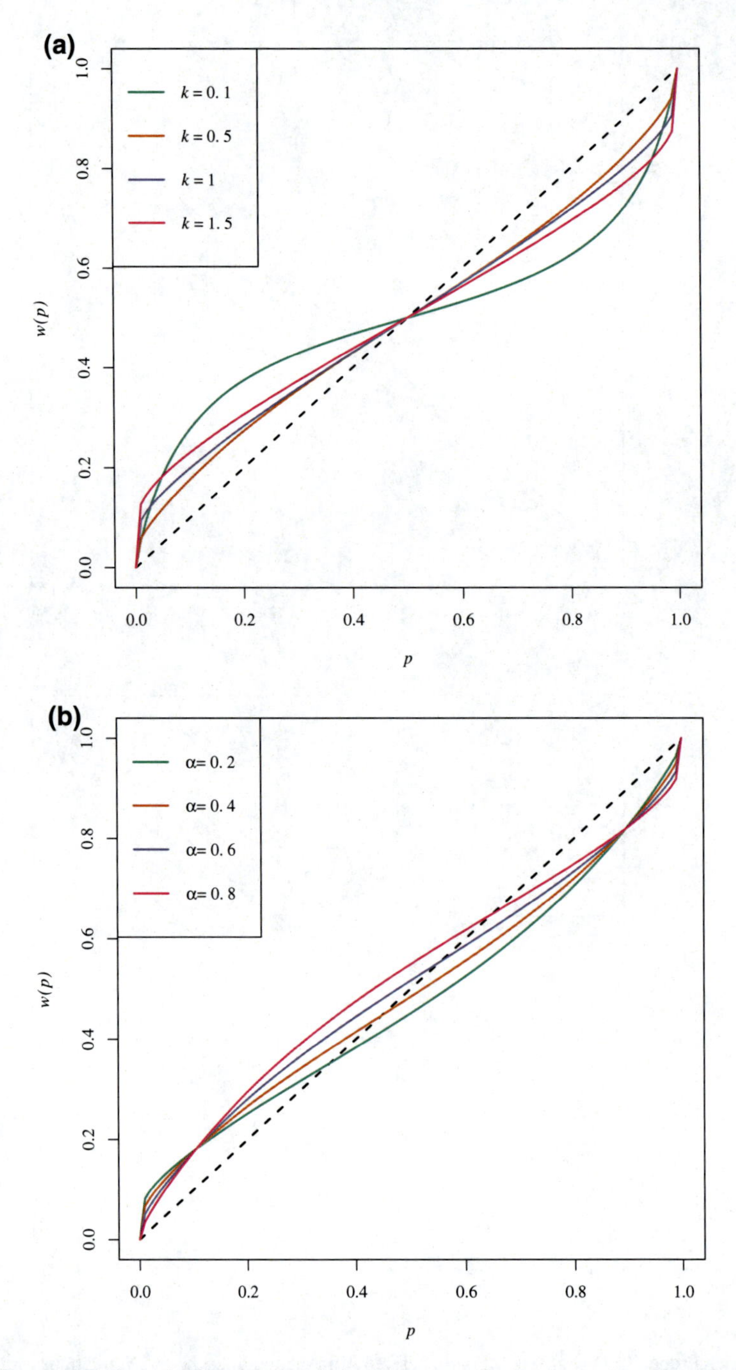

Fig. 4.8 **a** Probability weighting function derived from mental ruler theory. *Note* Shape for parameter k in the model ($\alpha = 0.5$). **b** Probability weighting function derived from mental ruler theory. *Note* Shape for parameter α in the model ($k = 0.5$)

weighting function. They recruited 100 university students (40 males, average age of 21.9 years old), who participated in the experiment. As 18 of 100 participants had low reliability of data, they estimated a certainty equivalent of 82 people.

The outline of the experimental task flow (Murakami et al. 2014) is shown in Fig. 4.8. As can be seen, the task was to present a lottery and six different money amounts in ascending order that students can securely gain in pairs. They were asked to choose the option they prefer: to draw the lottery or to securely gain an amount of money (they can choose for each amount of six amounts of money they can securely gain).

The content of the lottery used as an experimental stimulus had 15 kinds of results and 11 kinds of probabilities, whose combinations reached a total of 165 options. In order to secure reliability to some extent, nine kinds of gambles were randomly selected from 165 kinds of gambles and these were repeatedly presented. In total, 174 sets of gamble combinations were presented and the money amounts were deemed obtainable. In addition, regarding nine kinds of gambles presented repeatedly, they operated them so as not to continuously present the same stimulus. Two patterns of stimuli were prepared, one with securely obtainable money amounts in ascending order and the other with those in descending order.

The parameters of each model were estimated from the median of the certainty equivalent obtained by the experiment. Given that each model has a different number of parameters, in order to compare models, it would be preferable to use a fit index, in which the effect of the number of parameters is discounted. Therefore, they used AIC as an index representing the good fitness of the model for the data. The AIC values of each model were different for each experiment participant. However, when compared as a whole on an average basis, the delayed value discounting model using the mental ruler theory had the best fit, followed by Prelec's model.

4.6.2.3 Individual Analysis of Probability Weighting Functions

Takemura and Murakami (2016) performed an experimental study and individual analysis study in order to illustrate the fitness of our models as well as previous models. Concerning the empirical research on probability functions, important research has been conducted by Stott (2006), who reviewed eight different forms of the probability weighting function (linear model, power model, log-odds model, Tversky-Kahneman model, Wu-Gonzalez model, two versions of Prelec's model, and a non-parametric model) and reported parameters estimated from multiple empirical papers over a period of 10 years. In addition, he reported an extensive empirical study for 96 participants by utilizing 90 different gamble stimuli. His study compared fits on a total of 256 combinations of cumulative prospect theory functional forms, including eight probability functions, eight value function forms, and four choice functions and concluded that the best model has a risky weighting function of the simple version of Prelec's (1998) model, a power value function, and a logit choice function. We also examined the Prelec (1998) model using the power function for a value function by comparing the proposed model and some previous models. Although the number

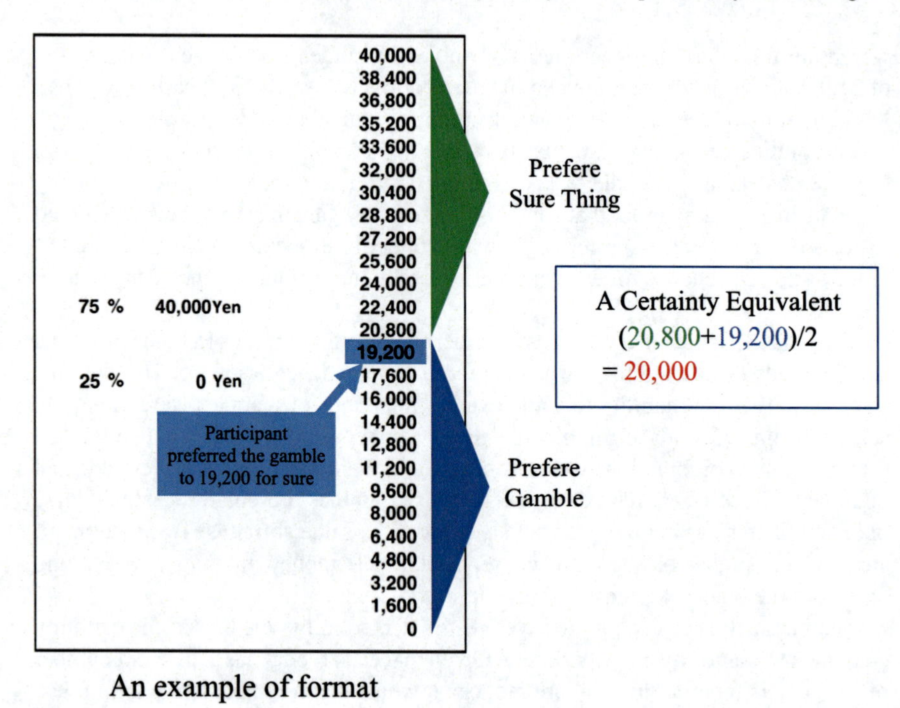

Fig. 4.9 Probability weighting function estimation experiment flow. *Source* Takemura and Murakami (2016)

of participants was limited (50 participants), Prelec's (1998) general version of the probability weighting model fitted our data better than the other models did. This finding shows that the best model is the probability weighting function based on Loewenstein and Prelec's (1992) generalized hyperbolic time discounting model.

We proposed a probability weighting function derived from a hyperbolic time discounting model by assuming a geometric distribution. Moreover, our probability weighting function was derived from Loewenstein and Prelec's (1992) generalized hyperbolic time discounting model. The present study derived this hyperbolic logarithmic model from the generalized hyperbolic time discounting model assuming Fechner's (1860) psychophysical law of time and a geometric distribution of trials. Because the geometric distribution is skewed, a logarithmic psychophysical function ($-\log p$) was considered to be an approximation to the median of trials. Under this interpretation, the probability of the weighting function was derived from the generalized hyperbolic model using the median of geometrically distributed trials.

There are two primary contributions of this study. First, Takemura and Murakami (2016) derived the probability weighting function on the basis of the generalized hyperbolic time discounting function. Second, they demonstrated the empirical study comparisons that fitted for six different probability weighting functions for 50 participants, each corresponding to 165 unique gambles. The outline of the experimental

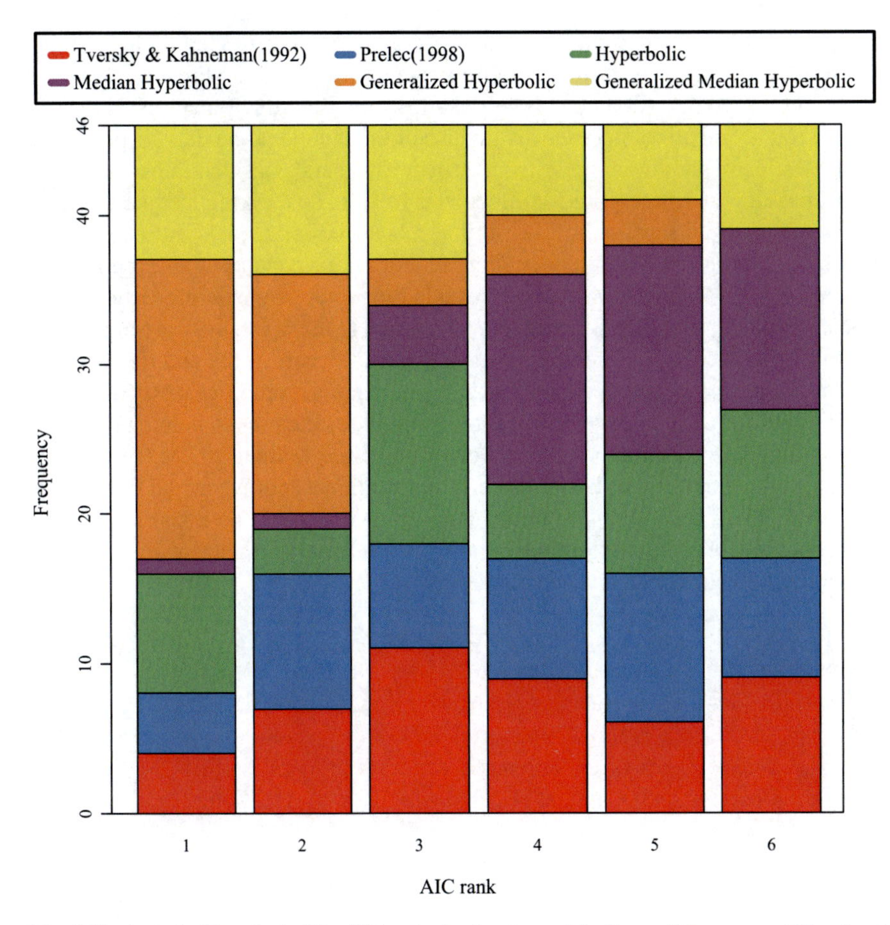

Fig. 4.10 A stacked bar chart of the AIC ranks for the six models. *Source* Takemura and Murakami (2016)

task flow (Takemura & Murakami, 2016) is shown in Fig. 4.9. As can be seen, the task was similar to the above-mentioned experimental study in Murakami et al.'s (2014) research.

The results of the psychological experiment indicated that the expected value model of generalized hyperbolic discounting was a better fit of AIC measure than previous probability weighting decision making models as shown in Fig. 4.10. Further theoretical and empirical studies will be required to examine the shape of the probability weighting function.

Takemura and Murakami (2016) provided theoretical and empirical support of the individual level analysis that assists a psychological interpretation of the probability weighting function from a time discounting perspective. However, the empirical method used in their research was a psychometric nonlinear regression analysis. Although there are several empirical psychometric tests available to examine probability weighting functions, there is no concrete method to examine the axiomatic properties of the probability weighting functions. Prelec (1998) had already proposed the axiomatic properties for some weighting functions. However, no concrete axiomatic properties distinguished the individual models he proposed, and no testing method was suggested. Per their axiomatic considerations, Takemura and Murakami (2018) proposed axiomatic properties and a testing method to examine the generalized hyperbolic logarithmic model, power model, and exponential power model of the probability weighting functions and conducted an experimental research using 14 participants, which revealed that the axiomatic properties of the probability weighting functions did not correspond to the psychometric fitting result of probability weighting functions. A similar result occurs in the additive conjoint systems in judgment and decision making. For example, empirical tests of double cancellation for the conjunctive measurement had rejected the double cancellation axiom (Gigerenzer & Strube, 1983; Levelt, Riemersma, & Bunt, 1972). On the other hand, psychometric studies have indicated that the linear additive model fitted better (Dawes, 1979). There are some contradictions between psychometric and axiomatic studies. This case is the same as that in previous research. Further research is needed to identify the reason behind such discrepancies.

Box 2: Daniel Kahneman

Born in 1934. He is currently working as a professor at Princeton University. He was awarded the Nobel Memorial Prize in Economic Sciences in 2002 for his achievements in the application of psychological studies—including decision making and judgment under uncertainty—to economics.

Photograph by Buster Benson:

https://www.flickr.com/photos/erikbenson/9585678357/in/photolist-61c8x8-61c8vr-5Wmq5x-5Wmqdi-7JDXjT-6E6mgM-fB47dv-61gkij-aHxVUD-2dKjP1m/

Box 3: Amos Tversky

Born in 1937; deceased in 1996. He spent many years studying decision making under risk and uncertainty with Daniel Kahneman, the winner of the 2002 Nobel Prize in economics. There must be many researchers who think that he would have received the Nobel Prize jointly with Kahneman if he had lived until 2002.

Photograph: Ed Souza/Stanford News Service

4.6.2.4 Summary

- The Allais paradox and Ellsberg's paradox are presented as phenomena that are contrary to the independence axioms of expected utility theory. Various nonlinear utility theories explaining these paradoxes are proposed.
- In nonlinear utility theories, a theory is used to explain these paradoxes by the Choquet integral on non-additive probability. A representation theorem in the expected utility model by this Choquet integral is also presented.
- In the expected utility model of Schmeidler's Choquet integral type, when weak order property, continuity, independence, comonotonic independence, and nonobviousness hold in decision making under uncertainty, only then can an integral type nonlinear expected utility model hold. Moreover, the model has uniqueness to a positive linear transformation. In the expected utility model of the Choquet integral type, utility is an interval scale.
- Prospect theory is considered as a type of nonlinear utility theory and is a code-dependent model assuming different value function shapes in the gain region and the loss region. It also assumes utility evaluation by the Choquet integral on the shift of the reference point and the non-additive probability.
- Prospect theory assumes that the value functions concave downward for gains and convex downward for losses. This allows us to explain the properties of various types of decision making scenarios, such as endowment effect, reflection effect, and so on.
- With respect to the value function and probability weighting function of prospect theory, various metric models have been proposed and many empirical studies have been done, and these have been applied to neuroscience, marketing, and economics, etc.

4.6.2.5 Reading Guide for More Advanced Learning

– Fishburn (1988). Nonlinear preference and utility theory (John Hopkins series in the mathematical sciences). Baltimore MD: John Hopkins University Press.

This book provides the axiomatic theory base about expected utility theory and nonlinear expected utility theory in detail. In addition, the problem of the counter-example of expected utility theory is explained in detail in relation to the axioms. This book is somewhat difficult to understand but is very important as a specialized book in this area.

– Gilboa (2009). Theory of decision under uncertainty (Econometric Society Monographs). Cambridge, UK: Cambridge University Press.

This book provides in-depth explanations of the epistemological base and the axiomatic theory base about the nonlinear expected utility theory and prospect theory. This book is somewhat difficult to understand, but it is considered very important as a specialized book in the area of the theoretical research on decision making under risk and uncertainty.

– Kazuhisa Takemura (2014). Behavioral Decision Theory: Psychological and mathematical representations of human choice behavior, Tokyo, JP: Springer.

The nonlinear expected utility theory and prospect theory are explained clearly while comparing the findings of different behavioral experiments.

– Dhami, S. (2017). The foundations of behavioral economic analysis. Oxford, UK: Oxford University Press.

This book also provides an explanation of nonlinear expected utility theory and the empirical research on some paradoxes of utility theory while comparing the findings of past behavioral experiments.

References

Allais, M. (1953). Le comportement de l'homme rationnel devant le risque: Critique des postulates et axiomes de l'ecole Americaine. *Econometrica, 21,* 503–546.

Anscombe, F. J., & Aumann, R. J. (1963). A definition of subjective probability. *The Annals of Mathematical Statistics, 34,* 199–205.

Benartzi, S., & Thaler, R. H. (1995). Myopic loss aversion and the equity premium puzzle. *Quarterly Journal of Economics, 110,* 73–92.

Camerer, C. (1995). Individual decision making. In J. H. Hagel & A. E. Roth (Eds.), *Handbook of experimental economics* (pp. 587–703). Princeton, NJ: Princeton University Press.

Camerer, C. F. (2000). Prospect theory in the wild: Evidence from the field. In D. Kahneman & A. Tversky (Eds.), *Choices, values, and frames* (pp. 288–300). Cambridge, UK: Cambridge University Press.

Camerer, C. F., Loewenstein, G., & Rabin, M. (Eds.). (2004). *Advances in behavioral economics.* Princeton, NJ: Princeton University Press.

Chiccheti, C., & Dubin, J. (1994). A microeconometric analysis of risk aversion and the self-insure. *Journal of Political Economy, 102,* 169–186.
Choquet, G. (1954). Theory of capacities. *Annales de l'institut Fourier, 5,* 131–195.
Dawes, R. M. (1979). The robust beauty of improper linear models in decision making. *American Psychologist, 34*(7), 571–582.
De Waegenaere, A., & Wakker, P. P. (2001). Nonmonotonic Choquet integrals. *Journal of Mathematical Economics, 36,* 45–60.
Edwards, W. (Ed.). (1992). *Utility theories: Measurement and applications*. Boston, MA: Kluwer Academic Publishers.
Einhorn, H., & Hogarth, R. (1985). Ambiguity and uncertainty in probabilistic inference. *Psychological Review, 92,* 433–461.
Einhorn, H., & Hogarth, R. (1986). Decision-making under ambiguity. *Journal of Business, 59,* 225–250.
Ellsberg, D. (1961). Risk, ambiguity, and the Savage axiom. *Quarterly Journal of Economics, 75,* 643–669.
Fechner, G. (1860). *Elemente der psychophysik*. Leibzing, DE: Breitkopf & Hartel.
Fishburn, P. C. (1988). *Nonlinear preference and utility theory*. Sussex, UK: Wheatsheaf Books.
Gigerenzer, G., & Strube, G. (1983). Are there limits to binaural additivity of loudness? *Journal of Experimental Psychology: Human Perception and Performance, 9*(1), 126–136.
Gilboa, I. (2009). *Theory of Decision under Uncertainty*. Cambridge, UK: Cambridge University Press.
Gonzalez, R., & Wu, G. (1999). On the shape of the probability weighting function. *Cognitive Psychology, 38*(1), 129–166.
Kahneman, D., Knetsch, J. L., & Thaler, R. H. (1990). Experiment tests of the endowment effect and coase theorem. *Journal of Political Economy, 98,* 1325–1348.
Kahneman, D., Knetsch, J. L., & Thaler, R. H. (1991). The endowment effect, loss aversion, and status quo bias. *Journal of Economic Perspectives, 5,* 193–206.
Kahneman, D., & Tversky, A. (1979). Prospect theory: an analysis of decision under risk. *Econometrica, 47,* 263–292.
Levelt, W. J., Riemersma, J. B., & Bunt, A. A. (1972). Binaural additivity of loudness. *British Journal of Mathematical and Statistical Psychology, 25,* 51–68.
Loewenstein, G., & Prelec, D. (1992). Anomalies in intertemporal choice: evidence and an interpretation. *The Quarterly Journal of Economics, 107*(2), 573–597.
Mazur, J. E. (1987). An adjusting procedure for studying delayed reinforcement. In M. L. Commons, J. E. Mazur, J. A. Nevin, & H. Rachlin (Eds.), *The effect of delay and of intervening events on reinforcement value* (pp. 55–73). Hillsdale, NJ: Erlbaum.
Murakami, H., Ideno, T., Tamari, Y., & Takemura, K. (2014). Kakuritsu kajū kansū ni taisuru moderu no teian to sono hikaku kentō [Proposal of probability weighting function model and psychological experiment for the comparisons]. Poster session presented at The Proceedings of the 78th Annual Convention of the Japanese Psychological Association, Kyoto, JP.
Murofushi, T., Sugeno, M., & Machida, M. (1994). Nonmonotonic fuzzy measures and choquet integral. *Fuzzy Sets and System, 64,* 73–86.
Nakamura, K. (1992). On the nature of intransitivity in human preferential judgments. In V. Novak, J. Ramik, M. Mares, M. Cherny, & J. Nekola (Eds.), *Fuzzy approach to reasoning and decision making* (pp. 147–162). Dordrecht, NL: Kluwer.
Odean, T. (1998). Are investors reluctant to realize their losses? *Journal of Finance, 53,* 1775–1798.
Payne, J. W., Bettman, J. R., & Johnson, E. J. (1993). *The adaptive decision maker*. New York, NY: Cambridge University Press.
Prelec, D. (1998). The probability weighting function. *Econometrica, 66*(3), 497–527.
Quiggin, J. (1993). *Generalized expected utility theory: The rank dependent model*. Boston, MA: Kluwer Academic Publishers.
Savage, L. J. (1954). *The foundations of statistics*. New York, NY: Wiley.

Schmeidler, D. (1989). Subjective probability and expected utility without additivity. *Econometrica, 57,* 571–587.
Seo, F. (1994). *Shikō no Gijutsu: Aimai kankyōka no keiei ishikettei [Thinking techniques: Management decision making under ambiguity environment].* Tokyo, JP: Yuhikaku Publishing.
Shefrin, H. M., & Statman, M. (1985). The disposition to sell winners too early and ride losers too long. *Journal of Finance, 40,* 777–790.
Starmer, C. (2000). Developments in non-expected utility theory: The hunt for descriptive theory of choice under risk. *Journal of Economic Literature, 38,* 332–382.
Stott, H. P. (2006). Cumulative prospect theory's functional menagerie. *Journal of Risk and Uncertainty, 32*(2), 101–130.
Sugeno, M., & Murofushi, T. (1993). *Kōza fajī 3: Fajī sokudo [Fuzzy theory 3: Fuzzy measure].* Tokyo, JP: Nikkan Kogyo Shimbunsha.
Tada, Y. (2003). *Kōdō Keizaigaku Nyūmon [Introduction to behavioral economics].* Tokyo, JP: Nikkei Inc.
Takemura, K. (1998). Jōkyō izonteki ishikettei no teiseiteki moderu: Shinteki monosashi riron niyoru setsumei [Qualitative model of contingent decision-making: An explanation of using the mental ruler theory]. *Ninchi Kagaku [Cognitive Studies], 5*(4), 17–34.
Takemura, K. (2000). Vagueness in human judgment and decision making. In Z. Q. Liu & S. Miyamoto (Eds.), *Soft computing for human centered machines* (pp. 249–281). Tokyo, JP: Springer.
Takemura, K. (2001). Contingent decision making in the social world. In C. M. Allwood & M. Selart (Eds.), *Decision making: Social and creative dimensions* (pp. 153–173). Dordrecht, NL: Kluwer Academic Publishers.
Takemura, K. (2011). Tazokusei ishikettei no shinri moderu to "yoi ishikettei" [Psychological model of multi-attribute decision making and good decision]. *Opereshonzu risachi [Journal of the Operations Research Society of Japan], 56,* 583–590.
Takemura, K. (2014). *Behavioral decision theory: Psychological and mathematical representations of human choice behavior.* Tokyo, JP: Springer.
Takemura, K., & Murakami, H. (2016). Probability weighting functions derived from hyperbolic time discounting: psychophysical models and their individual level testing., *Frontiers in Psychology, 7,* 778. https://doi.org/10.3389/fpsyg.2016.00778.
Takemura, K., & Murakami, H. (2018). A testing method of probability weighting functions from an axiomatic perspective. *Frontiers in Applied Mathematics and Statistics, 4,* 48. https://doi.org/10.3389/fams.2018.00048.
Takemura, K., Murakami, H., Tamari, Y., & Ideno, T. (2013). *Probability weighting function and value function based on unified psychological model.* Paper presented at 2013 Asia-Pacific Meeting of the Economic Science Association, Tokyo, JP.
Takemura, K., Murakami, H., Tamari, Y., & Ideno, T. (2014). *Probability weighting function models and psychological experiment for the comparisons.* Paper presented at Soft Science Workshop of Japan Society for Fuzzy Theory and Intelligent Infomatics, Fukuoka, JP.
Tamura, H., Nakamura, Y., & Fujita, S. (1997). *Kōyō bunseki no sūri to ōyō [Mathematical principles and application of utility analysis].* Tokyo, JP: Corona Publishing.
Thaler, R. H., & Ziemba, W. T. (1988). Anomalies: Parimutuel betting markets: Racetracks and lotteries. *Journal of Economic Perspectives, 2*(2), 161–174.
Toshino, M. (2004). *Shōken shijō to kōdō fainansu [Security market and behavioral finance].* Tokyo, JP: Toyo Keizai.
Tversky, A., & Kahneman, D. (1983). Extensional versus intuitive reasoning: The conjunction fallacy in probability judgment. *Psychological Review, 90*(4), 293–315.
Tversky, A., & Kahneman, D. (1992). Advances in prospect theory: Cumulative representation of uncertainty. *Journal of Risk and Uncertainty, 5,* 297–323.
Wu, G., & Gonzalez, R. (1996). Curvature of the probability weighting function. *Management Science, 42*(12), 1676–1690.

Chapter 5
Mental Accounting and Framing: Framework of Decisions in Consumer Price Judgment

Keywords Mental accounting · Framing · Reference price · Mental ruler · Psychophysical law

In decision making in economic behavior, information about price and quality is compiled and evaluated. This chapter focuses on the psychology behind price judgment. How do people organize and understand information such as prices psychologically? The concept of mental accounting is relevant to answer this question. This chapter addresses phenomena related to mental accounting. The problem of mental accounting is also closely related to the concepts of psychological purse and mental ruler. Furthermore, the concepts of mental accounting, psychological purse, and mental ruler can be interpreted as a manifestation of a phenomenon called "framing" in decision making. In this chapter, the problems of mental accounting and framing are interpreted especially by using the mental ruler model.

5.1 Consumers' Price Judgment

5.1.1 Revealed Preference and Price

As described in Chap. 2, according to the concept of revealed preference, price information is useful in knowing consumer preferences. Moreover, price information leads to inferring the utility function. We assume that an indifference curve of goods x and y exists. When the price per unit of a certain item x is p_1 and the price per unit of y is p_2, the budget m is defined as

$$m = p_1 x + p_2 y.$$

K. Takemura, *Foundations of Economic Psychology*,
https://doi.org/10.1007/978-981-13-9049-4_5

We suppose that a consumption vector (x_1, y_1) is selected under the price (p_1, p_2). We also suppose that (x_2, y_2) is another consumption vector that satisfies $p_1x_1 + p_2y_1 \geq p_1x_2 + p_2y_2$. According to the weak axiom of revealed preference, when a consumption vector (x_2, y_2) can be purchased, and if (x_1, y_1) and (x_2, y_2) are purchased, then (x_1, y_1) is not available for purchase. That is, if you have bought a combination of goods at a price higher than that of a combination of other goods, then you prefer that combination, and if you have bought a cheaper combination, then you do not have enough money to buy the more expensive goods. Also, according to the strong axiom in revealed preference, if a consumption vector (x_1, y_1) is directly or indirectly revealingly preferred to (x_2, y_2), and if these consumption vectors are not identical, then (x_2, y_2) is not directly or indirectly revealingly preferred to (x_1, y_1). When the strong axiom of this revealed preference is satisfied, the transitivity of preference holds, and the consumers are able to maximize utility. Based on the axiom of revealed preference, preference can be inferred from actual purchasing patterns of consumers. Then, based on the consumers' selection patterns under a certain price, we are able to estimate utility functions by examining the preference relationship of consumers by checking whether the strong axiom of revealed preference is violated or not. However, what is important is that such estimation is possible only when consumers' decision making is rational such as when satisfying a weak order. Therefore, we have to investigate how consumers grasp and judge it and decide when they are actually giving price information on a certain good.

5.1.2 Price Judgment

In human economic behavior, price is an important judgment factor. Even brands that are judged to be of high quality are not purchased if they are judged to be expensive, and luxury brands can be purchased if judged to be cheap. For example, while you think that a bottled drink worth 100 yen is expensive in a 100-yen store, you may judge a 500-yen bottled drink as cheap in a department store.

Aoki (2004) studied consumers' information processing of price as shown in Fig. 5.1. According to this figure, the price information on a brand is interpreted through the perception of consumers. First, the price information functions as a clue of "how good [a] brand's quality is." Second, it functions as a clue on "how cheap [the] brand is." Considering these clues, we obtain "reaction to price," which is the comprehensive evaluation of the price by consumers. Based on this reaction, the decision making on purchase behavior is conducted. This idea is different from the tendency assumed for the principle of revealed preference. For example, if some consumers think that the higher the price, the better is the quality, then it becomes impossible to consider whether the revealed preference is established. As a practical example, McDonald's Japan halved the price of its hamburgers on weekdays to 65 yen from 130 yen in February 2000. At this time, the clue of savings as "the price halved," the second function, was present, but we can also think that the low estimation that "McDonald's hamburgers are approximately 65 yen in quality" was due to the first

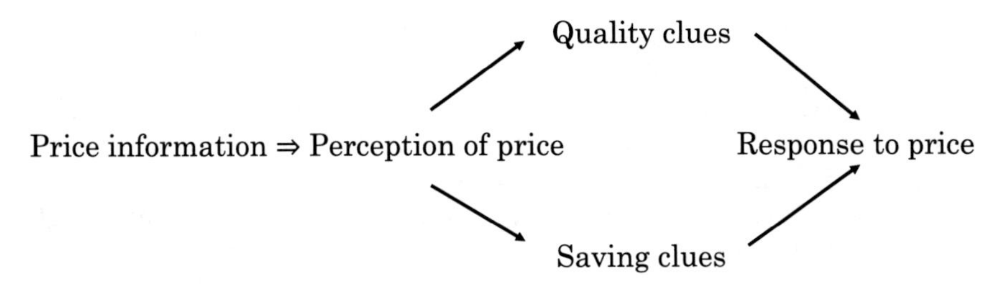

Fig. 5.1 Consumer's information processing process with price. *Source* Aoki (2004)

function. With this price reduction, the sales of hamburgers on weekdays increased five times, but dissatisfaction with other types of hamburgers (from 190 yen to 280 yen at the time) grew further, and the price satisfaction level, which had previously reached 30% or higher, decreased to 10% or lower (Aoki, 2004). An interpretation of this result is that the overall quality evaluation of McDonald's Japan's hamburgers has dropped due to the price cut. Such examples show the problems of price strategies in marketing, which one can frequently find in the real world.

5.2 Consumers' Actual Price Judgment

5.2.1 Consumers' Price Judgment

According to the survey results published by the Office of Public Relations, Minister's Secretariat, Cabinet Office in 2005, among the attributes emphasized most in store selection, price (61.6%) was the most prominent. In particular, the 20–40 age-group had a tendency to consider price as important. Thus, price is a key factor in consumers' decision making.

Dickson and Sawyer (1990) investigated the kind of price perception that consumers have in supermarkets. In their study, 30s after a consumer placed an item in her cart, the consumer was asked, "Can you state the price of the item you have chosen without checking the cart and with your eyes closed?" The results showed that 55.6% of consumers could estimate the price within 5% of the actual price. According to a survey on the price perception of drinks in Japanese supermarkets, which was conducted with a similar method, 50.6% of consumers were able to estimate the price without any errors (Takemura, 1996). Considering that in this survey, consumers who were able to recall both brand names and company names were 37.3%, and that 51.7% could not identify any ingredients (Takemura, 1996), although some deviation exists in the actual price information, we can see that the perception of price is relatively accurate compared with other attributes. However, both surveys show that only a few consumers remember the exact price. Dickson and Sawyer (1990) reported that the estimated value of the price decreased and that most consumers grasped the relative relationship, such as higher, lower, or the same, when comparing a product with other brands in the same category.

Wakayama (2014) conducted a survey on university students to compare the average payment intention amount at shops that assumed the lowest price guarantee and at shops that did not; the results showed that students' payment intention amount was higher at shops that assumed the lowest price guarantee and that the students thought they would pay a higher price. Wakayama (2014) pointed out that even if the lowest price guarantee was available, the students did not negotiate prices by bringing the flyer of another store unless the actual payment was reduced by approximately 30%. The reason for this behavior, as concluded from the survey results, is that the students considered the potential embarrassment of negotiating a price discount by bringing the flyer of another store. The results of this survey indicate that university students are self-conscious in asking for a discount even at stores that guarantee the lowest price in the same way as general stores do.

5.2.2 Price Judgment and Psychophysics

In considering consumer price judgment, research findings in psychophysics, an old subfield of psychology, can be the basis for learning. Fechner (1860), the founder of psychophysics, proposed a method that sometimes translates as a psychophysical measurement approach, developed a constant measurement method and a scale construction method to specify the functional relationship with psychological quantity produced through stimulus intensity and judgment, and derived the theory of sensory quantity expressed by logarithmic functions.

Based on the so-called Weber law that $\Delta I/I$, the ratio of stimulus intensity I to its difference threshold ΔI, is constant, a condition called Weber's law, a theory called Fechner's law was proposed, where the magnitude S of the judged sensation is proportional to the logarithm of the stimulus intensity I ($S = k \log I$, where k is a positive constant) (Indo, 1977; Wada, Oyama, & Imai, 1969).

Weber's law states that the just recognizable increment of the stimulus of the standard stimulus and the experiment stimulus, that is, the magnitude of the difference threshold, is proportional to the initial strength of the stimulus. This law further states that the difference is proportional to the magnitude of the standard stimulus when detecting the difference between the two stimulus amounts. For example, according to this law, when judging the weight of a weight, the difference that has been discriminable becomes undetectable if the weight of the standard stimulus is large and the difference threshold can be expanded. This law concerning difference thresholds is known to hold in various sensory areas such as auditory, visual, and tactile (Wada et al., 1969), and it holds not only in such basic senses but also in the sense of low price of discounted goods (Kojima, 1986). For example, according to Weber's law, when comparing a discount of 30 yen from goods with a fixed price of 100 yen and a discount of 30 yen from goods with a fixed price of 10,000 yen, one cannot have a similar sense of low price, but can have the same sense of low

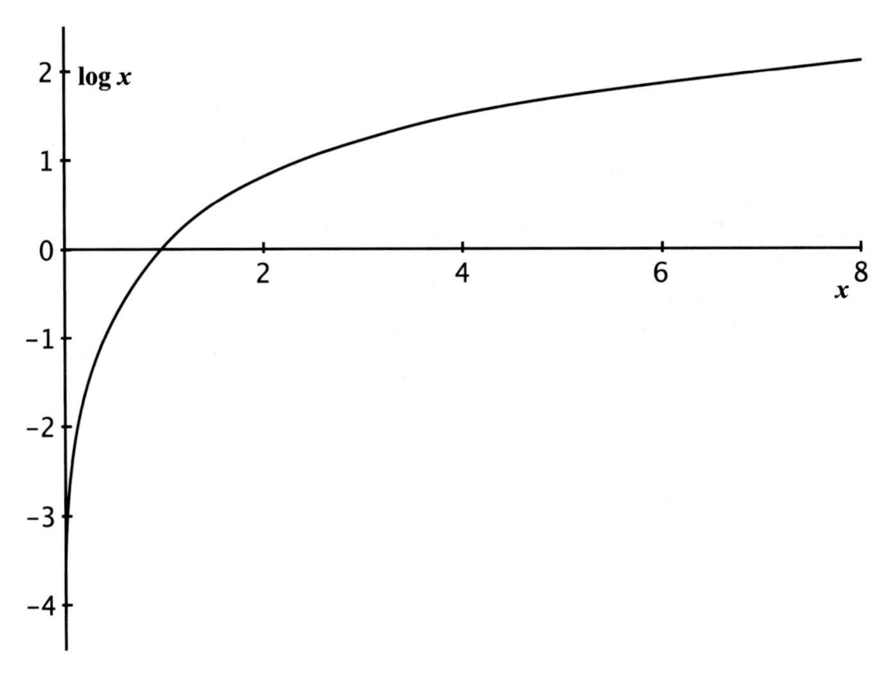

Fig. 5.2 Fechner's logarithmic function

price when comparing a discount of 30 yen from 100 yen and a discount of 3000 yen from 10,000 yen. In Fechner's law, ΔI is considered by differential and $\Delta I = \mathrm{d}I$ is assumed, which is considered to be proportional to the minimum unit of a sense $\Delta S = \mathrm{d}S$, where $\mathrm{d}S = k\mathrm{d}I/I$ (k is a constant). Both sides of this equation are integrated to obtain $S = k \log I + C$ (where C is a constant). Based on the assumption that the stimulus intensity at $S = 0$ is I_0, $C = -k \log I_0$ can be considered, $S = k \log I - k \log I_0 = k \log I/I_0$. Here, if I/I_0 is considered to be the stimulus intensity standardized by the stimulation threshold value I_0, then Fechner's law is obtained. The logarithmic function deduced by Fechner is shown in Fig. 5.2. When the sense of low price of the discount rate is predicted on the basis of Fechner's law, as the discount rate increases to 10, 20, and 30%, the increment of the sense of low price decreases.

$$\text{Sense amount: } S = k \log I$$

The law on the relationship between physical quantity and psychological quantity, such as Fechner's law, is called psychophysical law. Various studies have been conducted on this subject. A different view exists as to whether the psychophysical function of the logarithmic function by Fechner is valid or not, and deduction from Weber's law has been criticized as a leap. The theory by Stevens (1975) states that the power function is more reasonable than the logarithmic function ($S = \alpha I^{\beta}$, where α and β are constants) (Wada et al., 1969). Many theories using value and utility functions similar to those of Fechner and Stevens' psychophysical functions also appear in the theories of value and utility regarding price. For example, value functions are estimated by power functions in the nonlinear utility theory, such as the prospect theory (Tversky & Kahneman, 1992), which describes the evaluation of monetary gain.

In addition, Takemura (1998, 2001) proposed the mental ruler theory that the evaluation function in consumer price judgment is characterized by a concave downward near the lower limit and convex downward near the upper limit of the judgeable stimulus, and performed formulation including Fechner's and Stevens' laws as special examples. This model was also used to estimate the probability weighting function and is a psychological evaluation function that does not particularly distinguish the probability and the result.

Basically, the evaluation function of the mental ruler is the following function:

Comprehensive evaluation = weighting to the left end point x evaluation function based on the left end point (downward concave function) + weighting to the right end point x evaluation function based on the right end point (downward convex function).

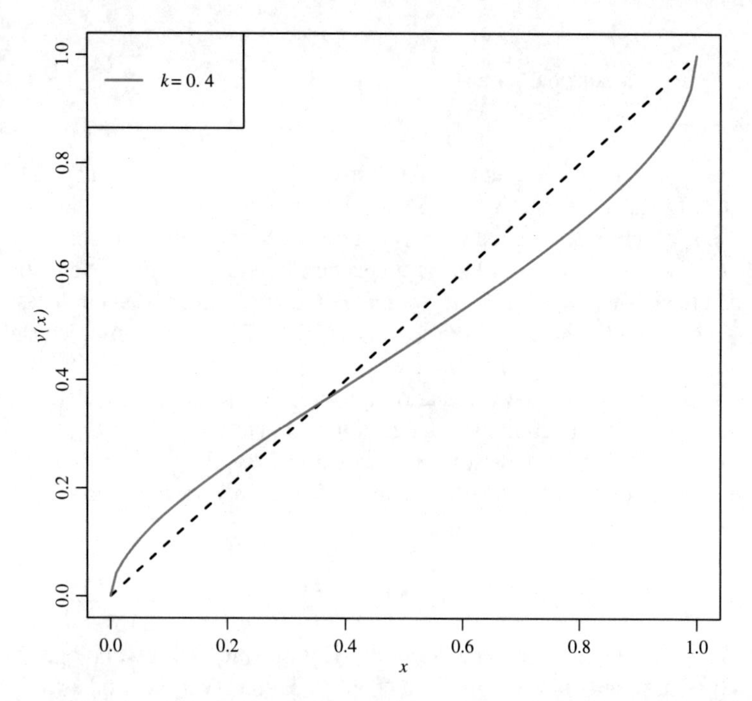

Fig. 5.3 **a** Inverted *S*-shaped evaluation function of mental ruler model. **b** Evaluation function of mental ruler model in which the reference point is located on the left end of the ruler. **c** Evaluation function of mental ruler model in which the reference point is located on the right end of the ruler

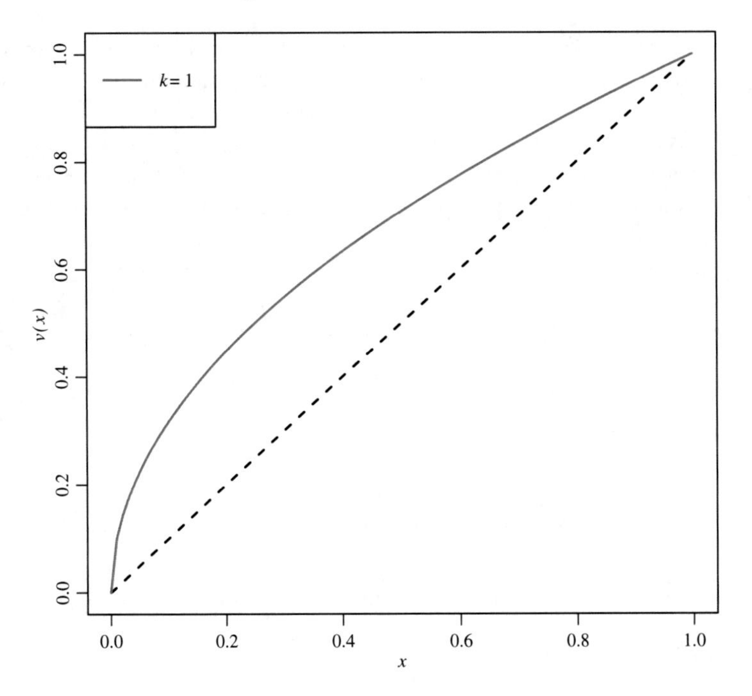

Fig. 5.3 (continued)

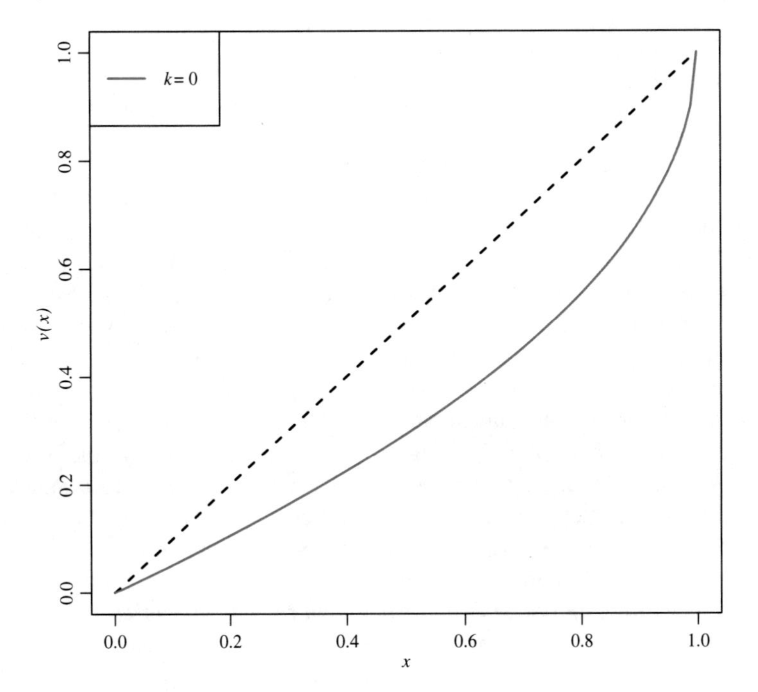

Fig. 5.3 (continued)

Therefore, the evaluation function of the mental ruler is the linear sum of the concave and convex functions, and it is generally considered to have an inverted S shape as a result. The property of the evaluation function of this mental ruler is completely opposite to the property of diminishing marginal utility in the utility theory and the prospect theory in the vicinity of the maximum value of the evaluation area. The utility and prospect theories always assume a downward concave function, but the mental ruler model predicts a downward convex function near the upper limit. For example, in the negotiation of price discount and others, if one is acting with a goal, the sensitivity would increase when the targeted price is near (see Fig. 5.3).

According to this formulation, let the intensity of sensation be S, weight ($0 \leq k \leq 1$) be k, the stimulus be $x \in X$, and the psychophysical function u of the concave function $X \rightarrow [0, 1]$. Then, the psychophysical function v of the convex function $X \rightarrow [0, 1]$ is set.

$$S = k\,u\left(\frac{x}{\text{possible maximum value of } x - \text{possible minimum value of } x}\right) + (1-k)v\left(\frac{x}{\text{possible maximum value of } x - \text{possible minimum value of } x}\right)$$

are indicated. For example,

$$S = k\left(\frac{x}{\text{possible maximum value of } x - \text{possible minimum value of } x}\right)^{\alpha} + (1-k)\left(\frac{x}{\text{possible maximum value of } x - \text{possible minimum value of } x}\right)^{\beta}$$

However, let the intensity of sensation be S, a weight ($0 \leq k \leq 1$) be k, the stimulus be $x \in X$, and a sensitivity parameter with $0 \leq \alpha \leq 1$ and $0 \leq \beta \leq 1$.

Regarding this model, the researchers conducted a survey that the desirability of discount on goods is numerically evaluated by the magnitude estimation method and obtained the results shown in Fig. 5.4 (Takemura, 2001). In this survey, by using mobile phones with a standard retail price of 8800 yen, the researchers set discount display groups that were classified into (1) a group that displays the discount rate by percentage (e.g., a 10% discount from the standard retail price is shown); (2) a group that displays the absolute amount of the discount (e.g., a discount of 880 yen from the standard retail price is shown); (3) a group that displays the discounted rate by percentage (e.g., a price with 90% of the standard retail price is shown); and (4) a group that displays the absolute price after discounting (e.g., 7920 yen, the price after discount, is shown). In each indication, as shown in Fig. 5.4, an inverse S-shaped evaluation function has been found. In other words, even a slight discount from the standard retail price is evaluated with a considerable effect, but this effect is not highly sensitive in the region of the medium discount rate, and it becomes considerably sensitive in the region of the high discount rate.

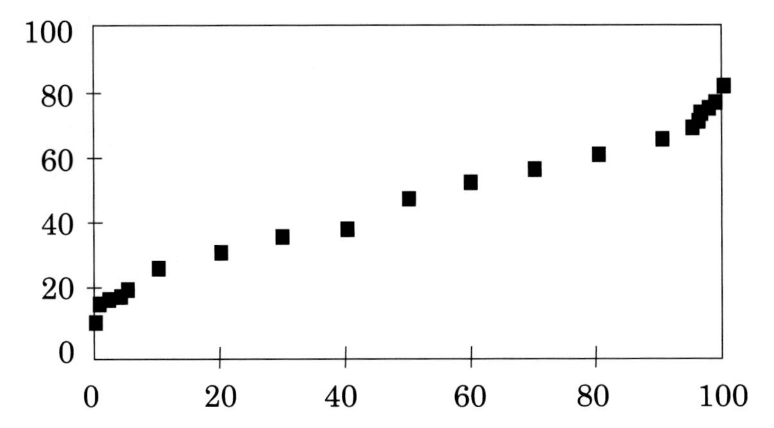

Fig. 5.4 **a** Mean rating values in Frame 1, Frame 1, e.g., 10% discount from original price. **b** Mean rating values in Frame 2, Frame 2, e.g., discount of 880 yen from original price. **c** Mean rating values in Frame 3, Frame 3, e.g., you may purchase the item at 90% of the original price. **d** Mean rating values in Frame 4, Frame 4, e.g., you may purchase the item at 7920 yen after the discount

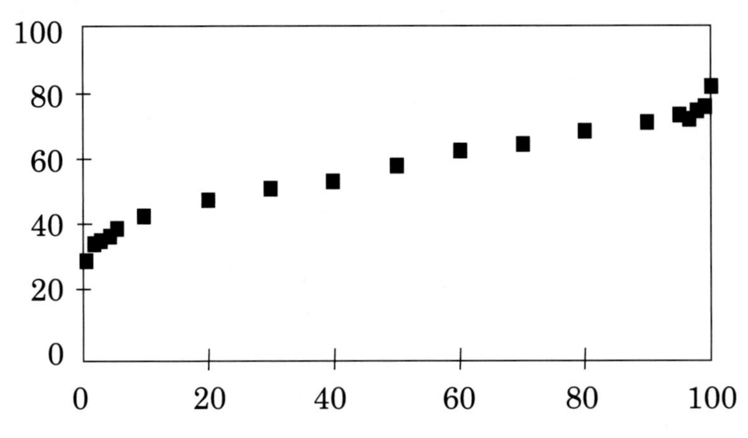

Fig. 5.4 (continued)

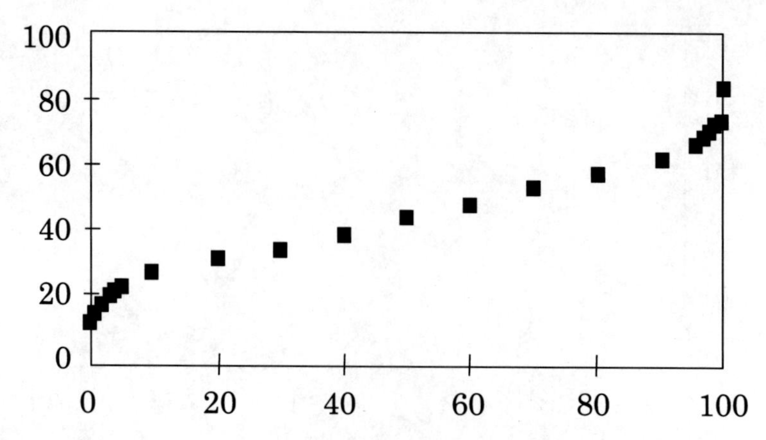

Fig. 5.4 (continued)

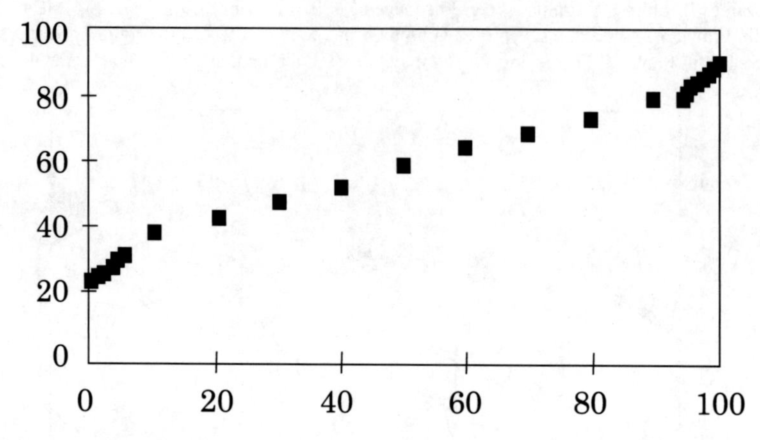

Fig. 5.4 (continued)

Figure 5.4a–d Mean rating values for a mobile phone in Frame 4 versions of price reduction description.

Box 1: Gustav Theodor Fechner
Born in 1801; deceased in 1887. He is the founder of psychophysics and demonstrated that psychology could be studied using quantitative models and experiments. His psychophysical studies not only strongly affected modern psychology, but also left a substantial impact on such philosophers as E. Mach, E. Husserl, and H. Bergson.

Photograph: Wikipedia

5.3 Context Effect of Psychology on Price

5.3.1 *Priming Effect*

In recent years, product purchases have been active on retail websites. For example, Rakuten and Amazon are sites that have a strong influence on product purchase. Normally, information on the products themselves is considered to influence the purchasing decisions of consumers. However, previous studies have shown that information that is unrelated to the product itself, such as the color or background design of the Web page, also affects purchase decision making. This phenomenon can be considered psychologically to be a kind of priming effect. This effect makes the judgment of the next task faster, slower, accurate, or inaccurate by presenting information to people without making them aware of it. Background images and banner advertisements on product purchase screens on the website may become a visual prime and may cause a priming effect.

To change the visual prime and examine the priming effect on judgment and decision making, Mandel and Johnson (2002) changed the image of the background of the product presentation screen of a website, as shown in Fig. 5.5. Their study used two product categories, cars and sofas, and presented two items for each product category. They prepared the website design that becomes the prime showing "comfort" ("safety" in case of cars), as shown in Fig. 5.5a, and the website design that becomes the prime showing "money," as shown in Fig. 5.5b, and conducted choice experiments under each condition. Their experimental hypothesis was that in the experiment where participants were asked to choose on the screen (a) showing comfort, more participants were likely to choose comfortable sofas (or safe cars) even if the price was high, and the participants who chose on the screen (b) were more likely to choose cheap and economical sofas (or cheap and economical cars). This experimental result supported the hypothesis.

5.3.2 *Context-dependent Effect*

Consumer decision making can change depending on the context in which options are presented. The occurrence of such a context effect suggests that promotion can promote consumer's purchase decision making by controlling the presented brand information and configurations. Context effects of consumer decision making include the following (Okuda, 2003):

(1) Attraction effect: Attraction effect is a phenomenon reported by Huber et al. (1982); due to the existence of an unappealing brand that no one chooses, the attraction of the brand rises and the choice rate changes. Figure 5.6 shows brand X with high quality but with high price, and brand Y with medium quality and price. A has the same quality as Y and inferior to Y in price, B is inferior to Y

Fig. 5.5 **a** Website background of "comfort" condition used in priming experiments. *Source* Mandel and Johnson (2002). **b** Website background of "money" condition used in priming experiments. *Source* Mandel and Johnson (2002)

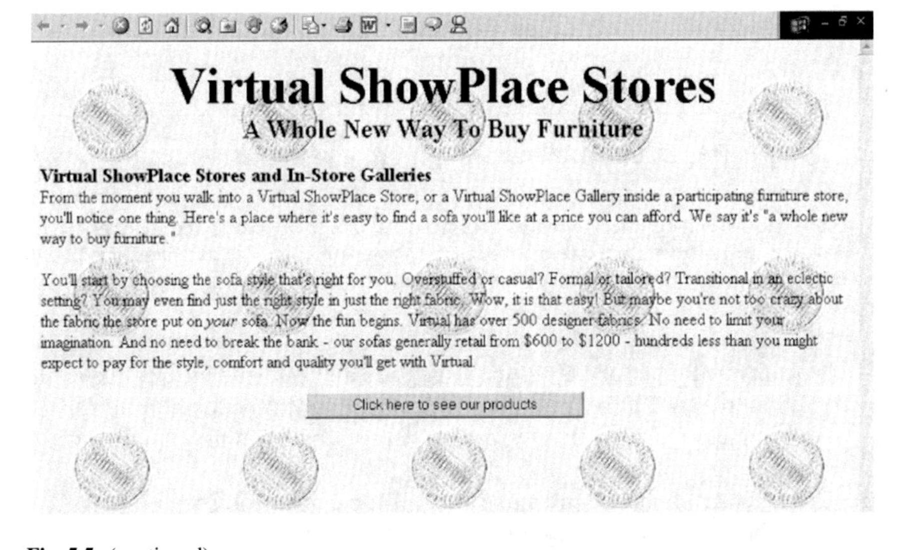

Fig. 5.5 (continued)

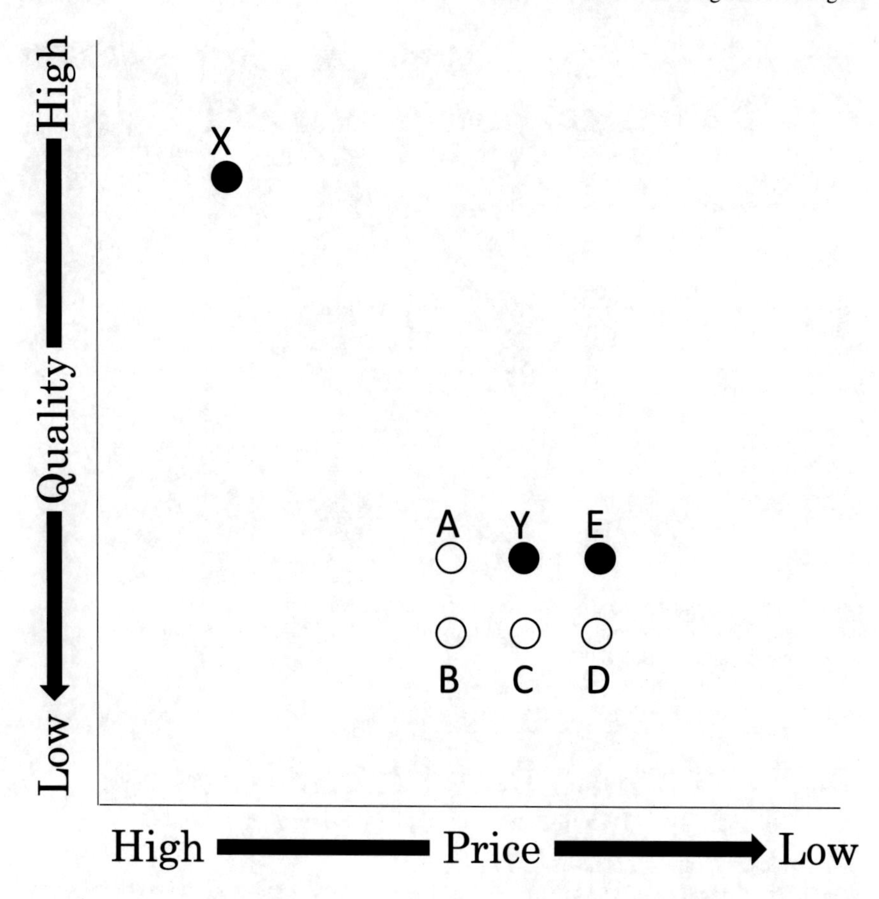

Fig. 5.6 Price and context effect. Adopted from Okuda (2003). *Note* Vertical line indicates quality of goods, and the horizontal line indicates price, *X* is a competitive good; *Y* is a target good. *A*, *B*, and *C* are alternatives that were dominated by *Y*, where *E* dominates *Y*. *D* is an extreme alternative

in quality and price, and *C* has the same price as *Y* but with inferior quality. In addition to *X* and *Y*, by presenting *A*, *B*, or *C* as not superior to *Y*, *Y* is made more attractive than when *A*, *B*, or *C* is not presented and the choice rate of *Y* is raised.

(2) Compromise effect: Compromise effect is a phenomenon shown in Fig. 5.6, where brand *D* is neither superior nor inferior to *Y*, but because of the existence of the brand that is extreme in terms of advantage and disadvantage compared with *Y*, the choice rate of *Y* rises (Huber & Puto, 1983).
(3) Phantom effect: The phantom effect is a phenomenon illustrated by *E* in Fig. 5.6, where even when brands superior to *Y* are difficult to obtain, *Y* is also more likely to be chosen. For example, if battery *E*, which is superior to *Y*, is reported to be out of stock, then the choice rate of *Y* rises more than when it is chosen between the two (Pratkanis & Farquhar, 1992).

Okuda (2003) organized these three types of context effects and conducted psychological experiments on the appearance of these effects by using products such as personal computers, recreational vehicles, mobile phones, and dry batteries as subjects. The results of the psychological experiment showed that relatively inferior brands increased the choice rate of target brands while superior brands and brands at the same level reduced the choice rate of target brands but increased the choice rate of the target brand when they were out of stock. In any case, the choice rate of competing brands declined.

5.4 Mental Accounting

5.4.1 What Is Mental Accounting?

To explain decision making related to money, Tversky and Kahneman (1981) used the concept of mental accounting. Mental accounting refers to a way for people to handle monetary decision making problems mentally and refers to the function of psychological purses that Kojima (1986) pointed out.

Tversky and Kahneman (1981) examined mental accounting by asking 383 experiment participants several questions. The following question on the ticket loss condition was posed to 200 experiment participants: "Imagine the following scene: You decide to see a movie, and after purchasing a ticket for 10 dollars, go to a movie theater. When you enter the theater, you notice that you have lost the ticket. Will you buy another ticket?"

In addition, the remaining 183 people were asked the following question on cash loss condition: "Imagine the following scene: You decide to see a movie and go to a movie theater. The price of the ticket is 10 dollars. When you enter the movie theater, you notice that you have lost 10 dollars in cash. Will you buy the ticket?"

According to the results, in the ticket loss condition, 46% of the experiment participants answered that they would buy the ticket, while in the cash loss condition, 88% of the experiment participants answered that they would buy the ticket.

The point to note here is that in both conditions, the participants suffered a loss equivalent to 10 dollars and were asked to decide whether to buy a ticket equivalent to that amount of money. Tversky and Kahneman (1981) explain that the results are different because the way of mental accounting differs between the ticket loss condition and the cash loss condition. In other words, in the ticket loss condition, they have to buy another ticket from the account of the ticket expenditure (a type of psychological purse), whereas in the cash loss condition, the ticket expenditures are in different accounts, so the participants do not feel the pain of buying the ticket twice, which can be interpreted as the reason for a higher intention to purchase a ticket. When buying a ticket, the loss of cash did not have a strong influence because only the account of tickets is used. Thus, Tversky and Kahneman (1981) explain that mental accounting is not performed based on a comprehensive assessment of money, but more likely based on the topic.

Tversky and Kahneman (1981) also asked a total of 192 experiment participants several questions about the mental accounting of consumers. The following question about a 15-dollar calculator condition was posed to 88 participants: "Imagine the following situation: When you try to buy a $125 jacket and a $15 calculator, you hear from a clerk that at a store branch 20 min away by car, a $15 calculator is sold for $10. Will you go to that branch to buy the calculator?"

In addition, they asked the remaining 93 experiment participants the following questions about the $125 calculator condition: "Imagine the following situation. When you try to buy a $125 calculator and a $15 jacket, you hear from a clerk that at a store branch 20 min away by car, a $125 calculator is sold for $120. Will you go to that branch to buy the calculator?"

Both conditions are common in the purchase decision making of buying a calculator and a jacket. Furthermore, these conditions are completely the same as for the shopping with a total of $140 and going to the branch with the cost of driving for 20 min to have a five-dollar gain. According to the results, in the $15 calculator condition, 68% of the experiment participants answered that they would go to the store branch, while in the $125 calculator condition, only 29% of the participants answered that they would go to the branch.

Thus, the experiment participants performed framing by separating two decision making problems without considering the calculator and jacket shopping in an integrated sense. This result also indicates that the mental accounting process is not comprehensive but may be executed separately based on the topic. If your problem recognition is whether to spend a total of $140 or a total of $135, then the evaluation results of both conditions should be the same. However, in the $10 calculator condition, the part where $15 of the calculator's fixed price becomes $10 attracted attention, and in the $125 calculator condition, the part where $125 of the calculator's fixed price becomes $120 attracted attention. If a downward convex negative utility function, as employed in the prospect theory, is assumed here, then the decline in cost where $15 of the calculator's fixed price becomes $10 is valued largely compared with the cost decline where $125 becomes $120.

5.4.2 Psychological Purse

The task performed by consumers in a budget or closing is called mental accounting, and the concept of psychological purse is similar to this mental accounting. Psychological purse deals with a broader range of phenomena than mental accounting.

Kojima (1959, 1994) points out that the consumer's price judgment is relative and that purchase behavior and satisfaction after purchasing are greatly influenced by how consumers grasp purchase decision making. Kojima (1986) also explains what kind of situation-dependent problem recognition exists by using the construct of psychological purses. According to this concept, consumers behave as if they

Fig. 5.7 Image of psychological purses. *Note* The character may have various psychological purses, while real purse is only one

owned several purses and spent money from these purses depending on the type of purchased goods and services and the situation at the time of purchase (see Fig. 5.7). Since each of these psychological purses has a value scale of a different dimension, even if the same amount is paid for the same item, if the purse from which the amount paid is different, then the satisfaction obtained from it and the psychological pain associated with the expense are different. From the viewpoint of having different psychological purses, consumer psychology can show that while cutting down on food expenses in daily life, consumers can have dinner at a high-class restaurant while traveling without feeling any psychological pain.

Regarding psychological purses, Kojima (1986) reported the following findings. First, due to a temporary economic income such as bonuses and lottery prizes, psychological purses generally expand, and purchase behavior is promoted. Moreover, due to this temporary economic income, the psychological purses do not shrink and purchase behavior is not suppressed. However, depending on losses, such as dropping money, psychological purses generally tend to shrink, and purchase behavior tends to be suppressed. On the contrary, some findings show that psychological purses expand due to a large loss. For example, if a person loses a large amount of money in gambling, or spends a large amount on purchasing a condominium, one may think that "it is no use saving 10,000 yen or 20,000 yen now" or "since I am purchasing a condominium for 40 million yen, I am not worried about purchasing a carpet for 200,000 yen."

In addition, Kojima (1986) pointed out that using credit and loans in product purchase has a strong influence on the expansion of psychological purses. In credit and loans, (1) initial payment requires only a downpayment, (2) money is automatically transferred from accounts without paying directly for the purchase, (3) a time lag (or time gap) exists between purchase and payment, (4) the consumer need not be strongly conscious of payment at the time of purchase, and (5) the purchase is not restricted by the amount of money on hand. Thus, psychological pain accompanying expenditure is reduced, psychological purses are expanded, and purchase behavior is promoted by credit and loans.

5.5 Reference Price and Prospect Theory

5.5.1 What is a Reference Price?

In price judgment, consumers do not evaluate the price absolutely but rather judge relatively. For example, as shown by Kojima (1986), the consumer considers a tie worth 4000 yen "expensive" among products worth 2000–3000 yen, but among products worth 5000–6000 yen or more, the consumer considers it "cheap." An important concept in explaining such a relative judgment is the reference price. The reference price is the standard by which consumers evaluate the price of products. An internal reference price exists, which consumers perceive to store, along with an external reference price, which is presented externally as an information stimulus at a store. According to research (Mazumdar & Papatla, 1995), consumers with high loyalty to a specific brand are likely to use the external reference price for judgment, and consumers with low brand loyalty are likely to use the internal reference price for judgment because they value price. One considers the 4000-yen tie "expensive" among the products worth 2000–3000 yen because the reference price is low and feels that the tie is "cheap" among the products worth 5000–6000 yen or more because the reference price is high.

Hsee (1998) found that the internal reference price influenced the price judgment. According to the study, when a consumer was presented a $55 wool coat (the price range is from at least $50 to a maximum of $500) or a $45 wool scarf (the price range is from at least $5 to a maximum of $50) as a gift from a friend, the consumer felt the $45 wool scarf as a more valuable gift. The result shows the idea that the internal reference price changes according to the price range, and sometimes a person would rather appreciate less expensive gifts.

Shirai (2005) examined the internal reference price and asked consumers what internal reference price they use for the product categories of personal computers, mobile phones, and shampoos. Nine types of internal reference prices were shown to consumers, and they were asked which internal reference price was used at the time of purchase decision making: "The price considered fair when considering the cost of the manufacturer (fair price)," "the price considered too high at a price higher than this (reservation price)," "the price that quality is considered inferior at a price lower than this (lowest acceptable price)," "the cheapest price among the prices observed in the past (lowest observation price)," "the highest price among the prices observed in the past (highest observation price)," "the average price of the given product category observed in the past" (average observation price), "the standard price that you think the product is sold at around this level" (usual price), "the price you estimate that the product will be sold at around this price now" (expected price), and "the price

1. Price considered as being fair (Fair price)
2. Highest acceptable price (Reservation price)
3. Lowest acceptable price that quality is considered as being not inferior (Lowest acceptable price)
4. Cheapest price among the prices in the past (Lowest observation price)
5. Highest price among the prices in the past (Highest observation price)
6. Average of various prices in the past (Average observation price)
7. Standard price that the product is usually sold at around this level (Usual price)
8. Price estimated that the product will be sold at around this price (Expected price)
9. Price paid at the time of purchase in the past (purchase price)
10. Price shown in advertisements such as folding flyers
11. Price heard from family members and acquaintances
12. Usual selling price shown at the store
13. Special sale prices shown at the store
14. Lowest price shown in price comparison sites, shopping sites, and others.
15. Standard price shown in price comparison sites, shopping sites, manufacturer sites, and others.

Fig. 5.8 Reference prices used in the survey. *Source* Okuse (2014).

paid when purchasing a personal computer in the past" (purchase price). As a result of this survey, for high-price image products and high-involvement products, the reservation and fair prices were regarded as important and the lowest and highest observation prices were not considered important. Furthermore, the evaluation of the nine types of internal reference prices showed that the usual price had the highest importance and the highest observation price had the lowest importance.

Okuse (2014) conducted a survey considering not only the internal reference price reported by Shirai (2005) but also the external reference price (see Fig. 5.8). According to the result, many respondents answered that "the price is considered fair when considering the quality (fair price)" and "the standard price that you think the product is sold at this level (usual price)." Regarding the importance of each price, the responses showed that external reference prices were considered important such as "the price to be seen in advertisements such as folding flyers," "the price at the time of special sales to be seen at the store front," "the lowest price to be seen in price comparison sites and shopping sites," and so on in addition to the fair price, the usual price, and others listed as the price to be used at the time of purchase decision making. These survey results also showed that as the reference price that consumers use at the time of purchase decision making, the internal reference prices such as the fair price and usual price are easy to use, but with regard to importance, the external reference prices such as the price to be seen on flyers, store fronts, and websites, and the lowest price are found to be high as well.

(1) Editing Phase
By being influenced by the context and linguistic expression, the decision-maker performs a mental configuration of the problem (framing) and the reference point is determined. ⇒
(2) Evaluation Phase
A reference point is determined without depending on the situation.
Evaluation based on evaluation functions is performed.

Fig. 5.9 Editing and evaluation phases assumed in prospect theory

5.5.2 *Reference Price and Prospect Theory*

The concept of reference price corresponds to the reference point in the prospect theory proposed by Kahnenman and Tversky (1979). The prospect theory, as shown in Chap. 4, integrated the findings of the psychological research on decision making and nonlinear utility theory (or generalized expected utility theory). The prospect theory was initially proposed as a descriptive theory dealing with decision making under risk (Kahneman & Tversky, 1979) but was later developed into a theory that can explain decision making under uncertainty (Tversky & Kahneman, 1992).

In the prospect theory, the decision making process is divided into an editing phase, where the problem is perceived and the decision making framework is determined, and an evaluation phase, where options are evaluated according to the problem perception (Kahneman & Tversky, 1979). As shown in Fig. 5.9, the former stage is subject to the circumstances and changes according to context information and differences in some linguistic expressions. However, at the latter stage, once the problem is identified, evaluation not dependent on the circumstances and decision making are performed.

The special characteristic of the prospect theory is that the reference point corresponding to the origin of the value evaluation is assumed to move easily according to the way of editing the decision making problems. In the prospect theory, the result is evaluated on the basis of the deviation from the reference point, which is the psychological origin, and the decision maker evaluates the result as either gain or loss. The reference price, especially the internal reference price, can be regarded as changeable depending on how one grasps the situation, in the same manner as the reference point in the prospect theory.

In prospect theory, as explained in Chap. 4, the slope of the value function (evaluation function: $v(x)$) is generally larger in the loss region than in the gain region. This result shows that the psychological impact when one loses 100 yen is larger than that when one gains 100 yen. That is, in the prospect theory, at $x > 0$, $v'(x) < v'(-x)$, where $v'(x)$ is a derivative function of $v(x)$. This condition indicates that losses have a stronger effect than gains; this property is called loss aversion.

5.5.3 *Price Judgment Phenomenon Derived from Loss Aversion*

As a phenomenon derived from the property of loss aversion, the strength of psychological resistance to price increases can be emphasized. Once consumers form a low internal reference price, they regard higher prices as losses and lower prices as gains; however, due to the loss aversion property, these perceptions do not have a strong impact. Therefore, once a low internal reference price is formed due to a price cut, the desirability is significantly impaired if the price becomes higher than this reference price, and one has to lower the price considerably to increase the discount effect (Moriguchi & Tsurumi, 2004). For example, loss aversion is considered to be the reason why the low price concept of McDonald's Japan was not so successful, where the internal reference price of consumers was set low. Frozen foods sold at 40–50% discount in special sales at supermarkets are similar examples. If frozen foods are discounted as much as half the original price, those prices become internal reference prices, and they will cause consumers to feel that usual selling prices are too high, thereby preventing the consumers from purchasing the products.

As a phenomenon derived from the property of loss aversion, the endowment effect can be pointed out as well (Kahneman, Knetch, & Thaler, 1990, 1991). In this condition, when a good is given and a person is holding it, the selling price of the good becomes higher than the buying price in the case where the good is not given. Simply put, this phenomenon causes difficulty in parting with the goods that one has initially owned. Sometimes, this phenomenon is interpreted as representing the status quo bias.

Kahneman et al. (1990) conducted a series of experiments to confirm the endowment effect. In one such experiment, the researchers first randomly divided 77 students at Simon Fraser University into three groups: "selling," "buying," and "choosing" conditions. The experiment participants with the condition to sell gave coffee mugs and investigated the price at which the other participants were willing to part with their mugs. The experiment participants with the condition to buy investigated the price that the other participants were willing to pay. The experiment participants with the condition to choose presented various prices and asked the other participants to choose between the mug and the cash. As a result, the medians of the pricing were $7.12 for the selling condition, $2.87 for the buying condition, and $3.12 for the choosing condition. In the selling condition, the reference point was the condition of possessing the mug. In the buying and choosing conditions, the reference point was the condition of not having the mug, which caused the pricing differences.

The endowment effect is also considered to work easily in the second-hand housing market. Sellers of houses tend to set higher prices because of the status quo bias, but on the contrary, buyers can tolerate only lower prices and make strict price judgments. Particularly in the situation where land and house prices are decreasing, the price judgment of the seller and the price judgment of the buyer differ considerably, and the seller tends to continue owning the second-hand house without selling it.

5.5.4 *Hedonic Framing*

According to prospect theory, in consumers' price judgment, a reference point changes depending on the mental configuration (framing) of a problem, and different judgments and decision making are made. "Framing effect" is the phenomenon in which the decision making result varies depending on the framing. This effect is observed, for example, in purchase decision making for car insurance. For example, when comparing an insurance with an enrollment premium of $1000 and payment for damage of $600 or less being exempted, and an insurance that requires an enrollment premium of $1600 and payment for the damage of $600 or less, as well as a cashback in the absence of any accident, we know that people prefer the latter option although the result is the same in both cases (Johnson, Hershey, Meszaros, & Kunreuther, 1993). Recently, foreign insurance companies advertising insurance with a cashback of the latter type have also used the framing effect on consumers.

Thaler (1985, 1999) assumes that mental accounting is based on the principle of hedonic framing that integrates and separates various elements of the decision making problem so that the comprehensive evaluation value would be high. This concept assumes that when considering two elements x, y based on the assumption that $x \circ y$ is a combination of x and y, the hedonic framing is performed by the rule

$$v(x \circ y) = \mathrm{Max}(v(x + y), v(x) + v(y)).$$

Considering the assumption of the value function of the prospect theory in relation to hedonic framing, Thaler (1985, 1999) points out the following features:

(1) Gains are framed separately for each topic (the value function of gains is concave downward, so the comprehensive evaluation value is higher with the separation).
(2) Losses are framed by integrating various topics (the value function of losses is convex downward, so the comprehensive evaluation value is higher with the integration).
(3) Small losses and large gains are framed together (loss aversion is deducted in accounting).
(4) Small gains and large losses are separated and framed (as the value function of the gain region has a steep gradient near the origin, increasing gains have a slightly stronger impact than that of reducing a large loss slightly).

According to the principle of hedonic framing by Thaler (1985, 1999), the discount in the form of reward points and rebates gains are especially separated by consumers, so it is easy to frame. For example, according to Thaler and Johnson

(1990), regarding the question on which makes a person happier, the case where one has won the lottery twice ($50 and $25) separately in two weeks or the case where one has won a total of $75 on the same day, 63% of the experiment participants said that the case with two tickets on different days was a happier event. However, regarding $20 and $25 traffic violation fines and $100 and $50 additional tax collections, relatively more people (75 and 57%, respectively) disliked these incidents happening on the same day rather than happening on different days, which is a reversed result expected from the prediction that various losses are separated and framed. Thaler and Johnson (1990) modified the hedonic framing and proposed quasi-hedonic framing. According to this concept, a person becomes risk-seeking when he/she obtains a profit (house money effect), becomes risk-averse after suffering a loss and becomes risk-seeking when losses can be canceled out (break-even point effect).

As shown by the calculator question result by Tversky and Kahneman (1981), if discounts are framed separately for each product, when consumers are trying to purchase more than one product, one can predict that increasing the amount of discount for low-priced products is more advantageous than increasing the amount of discount for high-priced products in terms of marketing. For example, in the supermarket, an egg pack that is usually priced at 200 yen is significantly discounted to 100 yen to attract customers, which is a strategy often adopted to increase the overall purchase unit price without considerably lowering the prices of other expensive items (in addition, the discount is easy for consumers to know because the price of eggs is comparatively stable). If the supermarket made the discount equal to or greater than that of the total price of the egg packs, then the half-price effect of the 100-yen discount is larger than that of the 500-yen or more discounts of high-priced items such as television sets. In this way, if one understands the characteristics of mental accounting and utilize them, one can consider a meaningful strategy in terms of marketing, and from the standpoint of consumers, one can be careful enough not to be controlled by companies.

Box 2: Ernst H. Weber

Born in 1795; deceased in 1878. He invented the method of measuring the concept of threshold and studied the minimum distinguishable value of the distance between two points on skin. He also established Weber's law that the minimum distinguishable difference between the amounts of physical stimulus is approximately proportional to the amount of physical stimulus.

Photograph: Wikipedia

5.6 Consumers' Price Judgment and Model of Mental Rulers

5.6.1 Existing Qualitative Models to Explain Context Effects in Price Judgment

How should we deal with the situation-dependent decision making where mathematical systematization is difficult? One approach is to understand the complex decision making phenomena by qualitatively or metaphorically describing the situation dependence of decision making. Such an approach may be effective because it helps conceptualize the immediate problems when thinking about marketing and others. The decision frame and psychological purse models are representative examples of such an approach.

As mentioned, the decision frame (Tversky & Kahneman, 1981) is a psychological framework for recognizing decision making problems. The decision making process could be divided into an editing phase, in which the problem is recognized, and an evaluation phase, in which options are evaluated according to the problem recognition; according to the way of the decision frame in the former phase, even the same purchase decision making problems might generate different decision results. Tversky and Kahneman (1981) did not describe the property and function of the decision frame in detail, but they explained the situation dependence by using the "frame," a concept which is intuitively easy to understand, and they showed in experiments that various purchase choice reactions occur when different frames are given to consumers.

The psychological purse model points out that purchase behavior and the satisfaction after purchasing are greatly influenced by situation-dependent problem recognition, and explains what kind of situation-dependent problem recognition exists. Kojima (1959, 1994) argued that consumers behave as if they own several psychological purses and paid from these different purses depending on the type of purchased goods and services and the situation at the time of purchase. Thus, even if the same amount is paid for the same item, if the purse from which the amount is paid is different, then the satisfaction obtained from it and the psychological pain associated with the expense are different (Kojima, 1959, 1994).

5.6.2 Problems of Existing Qualitative Models

In the qualitative model of the decision frame, they showed only the positive frame corresponding to the value function in the gain region in the prospect theory (Kahneman & Tversky, 1979; Tversky & Kahneman, 1992) and the negative frame corresponding to the value function in the loss region. However, various frames exist in actual decision making scenes. We can understand that decision making can be broadly divided into positive and negative frames, but certain phenomena may be difficult to classify into either frame. For example, judgment and decision making on which object is beautiful, large, and generous are difficult to classify and how to translate such judgments to the prospect theory corresponding to the model of the decision frame is unknown. Similar to the work of Hsee (1998) shown earlier, explaining the judgment on the degree of generosity only by positive and negative frames is difficult.

On the other hand, in the psychological purse model, factor analysis studies have clarified what kind of psychological purse exists. We highly evaluate that point, but from a strict viewpoint, it is limited to the classification of the situation. Furthermore, the psychological purse model is a model of consumer purchase behavior. Although this model may be applicable to decision making phenomena other than purchase behavior, such as organizational behavior and accounting behavior of companies in the future, it cannot explain other decision making phenomena in daily life.

In addition, a common feature of both models is that we do not clearly understand how the construct works to guide decision making both in the decision frame and in the psychological purse model. In other words, both models deal with the mental configuration of decision making problems and point out that the mental configuration has an important effect on decision making, but they do not sufficiently describe how decision makers psychologically construct the situation, what kind of property and function the decision frame or the psychological purse has, or whether the property and function lead to judgment and decision making.

5.6.3 Assumption of Mental Ruler Model

The basic assumption of this model is that people make decisions as if they have a ruler. The fact that people make decisions by using rulers with different values is often figuratively mentioned, but this metaphor is limited to the domain of everyday conversation. Looking into this metaphor closely, we can understand that it is surprisingly effective in explaining situation-dependent decision making. Here, what is subject to mental rulers is basically divided into the region of gains and losses similar to the model of the decision frame. In addition, interpersonal impression judgment and probability judgment for generosity, quietness, and other phenomena that are not necessarily divided into losses and gains are included.

Let us first consider the basic meaning of a "ruler." A ruler measures length. People use rulers to observe variations and distortions in judging the length of an object, and this task may be more inconvenient than the commonplace fact of seeing the object without using a ruler. Psychologically, people cannot make judgments confidently without measuring them with a ruler. People use a ruler as a criterion of judgment and can judge the length at ease if a physical ruler exists. What do people do in the absence of a physical ruler? In such a situation, a person constructs a ruler in his/her mind. This condition is considered as a type of creative process in the problem recognition in decision making.

Let us consider the features of decision making metaphorically by thinking further about the characteristics of the ruler.

(1) Basic property 1: A ruler is characterized by gradations. We can assume that people make decisions based on the fine and large gradations on their mental ruler. For example, a mental ruler when considered as a physical ruler may have a gradation of 1 mm unit and cm unit. Let us think about price judgment. When the gradation is fine, the consumer is sensitive even to a difference of 1 yen. However, when the gradation is large, the consumer becomes insensitive to the difference in tens of thousands of yen. The difference in sensitivity with respect to such prices can be expressed by the largeness and fineness of the gradation of the consumer's mental ruler. As will be described later, the largeness of the gradation of this ruler may change even within the same individual depending on circumstances.

(2) Basic property 2: The length of a ruler is finite (this characteristic is called "boundedness"). This feature seems extremely common, but the metaphor has an important meaning. For example, in price judgment, it is almost impossible to easily judge targets that exceed the range of a ruler such as extremely high and extremely low. The reason is that the evaluation target is beyond the range of the ruler. Consumers might sometimes use two rulers if one is too short, but in such a case, the variation in judgment becomes quite large.
(3) Basic property 3: A ruler is one-dimensional. A physical ruler measures the one-dimensional property of length. We believe that people make decisions based on multi-dimensional information, but one-dimensional judgments are also highly possible in the end. Many people stick to the deviation value education, even though they say that this approach should not be used, and stick to rankings such as for restaurant gourmet classification, which might show the property of human one-dimensional judgment.

5.6.4 Basic Functions of Mental Rulers

Takemura (1998) made the following theoretical predictions on the basic functions based on the basic properties of the mental ruler described.

(1) Basic function 1: People construct an appropriate mental ruler according to the situation. This condition means that people make the gradation and size of a ruler according to the situation. Perhaps people are nearly unaware because they perform such tasks in an extremely natural manner. However, such situations happen often when comparing purchase situations. For example, consumers who are considering buying a car make the gradation by 10,000 yen for car discount negotiations and purchasing optional items because the price of the car is over 1 million yen. In such a case, the price range of several hundred yen units is handled as if it was an error and is rarely considered. However, after visiting dealers, the same consumer is pleased to find that a pack of eggs is cheaper than usual by 20 yen at the supermarket, or conversely stops buying when the price is higher by 30 yen, where she is making a decision with the gradation of 10 yen. This condition shows that people focus on the immediate situation, construct the situation subjectively, and construct a mental ruler for that situation.
(2) Basic function 2: The positions to which the reference and end points of the ruler are applied vary depending on the situation. For example, the reference point of the ruler varies depending on the groups to be compared when making a price judgment. In other words, depending on the comparative targets, whether a product is cheaper than that in other stores or cheaper compared with the retail price of the same store so far, the price judgment and purchase decision making change depending on where to place the reference point where the mental ruler is at zero. In addition, the end point of the ruler also changes according to the difference in the situation such as the group to be compared.

(3) Basic function 3: The gradation of a ruler narrows particularly in the vicinity of the end and reference points (nonlinearity of a ruler). This property is different from that of a physical ruler. For example, consumers who are considering buying a product with a budget of 10,000 yen are more sensitive to the price difference between the 10,000-yen and 9500-yen products than the price difference between the 5000-yen and 5500-yen products. However, as described in the basic properties, evaluating beyond the end point is extremely difficult. For example, consumers considering a product with a budget of 10,000 yen become considerably insensitive to the price difference between 15,000 yen and 15,500 yen, and their evaluation is considered to become unstable.
(4) Basic function 4: High knowledge and high involvement narrows the gradation of a ruler. For example, when consumers have considerable knowledge about products or their involvements are high, the ruler gradation becomes finer, and the consumers become sensitive to a slight difference such that even a similar product is clearly distinguished. Therefore, consumers with high knowledge and high involvement are likely to buy products with slightly better performance at high prices, while consumers with low knowledge and low involvement are hardly aware of the difference.
(5) Basic function 5: Even if information is given in a multi-dimensional manner, one-dimensional judgment is made by the mental ruler. This condition not only means that humans are only trying to simplify the problem by avoiding the information processing load. Through a type of creative process in the problem recognition in decision making, consumers sometimes create new rulers according to the situation. For example, while reading fashion magazines and repeating their shopping behavior, consumers may construct such a ruler as "good taste" from complex information on clothes and make purchase decisions by using the ruler. In such cases, a one-dimensional ruler is used.
(6) Basic function 6: Comparison between different mental rulers is difficult. People construct various mental rulers according to the situation, but comparing and evaluating one's own ruler is difficult. Contradictory judgments and decision making between situations, such as examples of car purchasing and egg purchasing situations, are less noticeable among consumers themselves. In general, people focus on the situation, construct the situation subjectively, and construct one ruler for that situation. Thus, constructing two or more rulers for one situation from the viewpoint of cognitive load is difficult. An assumption is that in some cases, when people have to perform the same evaluation from the viewpoint of economic rationality, they may have evaluated objects by using different rulers, and on the contrary, even in the situation where they should have measured with different rulers, they may have evaluated objects by using the same ruler.

5.6.5 Explanation of Existing Experiment Results

We qualitatively explain several experimental results using the model of mental rulers proposed in this study.

5.6.5.1 Interpretation of Mental Accounting Experiments (Tversky and Kahneman, 1981)

We explain experimental results on purchase decision making conducted by Tversky and Kahneman (1981) by using the model of mental rulers. Tversky and Kahneman (1981) interpret the results by using the decision frame concept as the indication that when experiment participants should consider the total expenditure, they consider decisions by using the frame for each product. We interpret the process from which this result came by using the model of mental rulers. First, the situation of this problem can be expressed as follows:

Version 1 $S'_1 = \{$($125 jacket, $15 calculator, no travel), ($125 jacket, $10 calculator, 20 min of travel)$\}$
Version 2 $S'_2 = \{$($15 jacket, $125 calculator, no travel), ($15 jacket, $120 calculator, 20 min of travel)$\}$.

Then, in the subjective situation, the common information is canceled out and erased. Also, the information "no travel," which was not mentioned, is not considered. That is,

Version 1 $S_1 = \{$($15 calculator), ($10 calculator, 20 min of travel)$\}$
Version 2 $S_2 = \{$($125 calculator), ($120 calculator, 20 min of travel)$\}$.

Here, in each version, the calculator is compatible and only the price is different, so we can compare them. We assume that to evaluate the discount, the experiment participants construct the evaluation function v of a mental ruler to make comparisons with the travel by car. Here, when the evaluation function F of the discount amount is expressed by the functions v_1, v_2 of the mental ruler, the function becomes the following:

Version 1 F ($5 as a discount from $15) = v_1 ($5)
Version 2 F ($5 as a discount from $125) = v_2 ($5).

Here, considering the evaluation function F in association with the evaluation function v, in v_1 in version 1 and v_2 in version 2, $m(x)$ is $5 in either case, but the maximum value of the evaluated target is $15 in v_1 of version 1 and $125 in v_2 of version 2. According to the property of the evaluation function v in the positive region defined earlier, in both evaluation formulas,

$v_1(\$5) > v_2(\$5)$ is given.

Therefore, F ($5 as a discount from $5) > F ($5 as a discount from $125) is obtained.

5.6.5.2 Interpretation of Context Effect Experiment on Price (Hsee, 1998)

When we interpret these experimental results, the subjective situation is diagrammatically described as follows:

Version 1 {\$55 wool coat, minimum \$50, maximum \$500}
Version 2 {\$45 wool scarf, minimum \$5, maximum \$50}.

The experiment participants construct mental rulers in the given information. That is, mental rulers are assumed to be made between \$5 and \$500 in version 1 and between \$5 and \$50 in version 2.

Assuming that the experiment participants are using the highest price as a comparison target, we can consider the following relationship. Here, when the evaluation function F is expressed by functions v_1, v_2 of mental rulers with the maximum values of different evaluation targets, the following are obtained:

Version 1 F (\$55 wool coat) = v_1 (\$55)
Version 2 F (\$45 wool scarf) = v_2 (\$45).

Here, given the evaluation function F in association with the above evaluation function v, m (x) is \$55 in v_1 of version 1 and $m(x)$ is \$45 in v_2 of version 2, and m (x^*) is \$500 in v_1 of version 1 and \$50 in v_2 of version 2. Also, when a psychological reference point is applied to the lowest and highest prices, it is \$5 (\$55–\$50) in v_1 of version 1 and \$40 (\$45–\$5) in v_2 of version 2, and the maximum values of evaluated targets are \$450 (\$500–\$50) in v_1 of version 1 and \$45 (\$50–\$5) in v_2 of version 2.

Both in the former case of not considering the reference point of the lower limit and in the latter case of considering the reference point of the lower limit, that is, in either case, according to the property of the evaluation function v in the positive region defined earlier, in both evaluation formulas,

v_1 (\$55) < v_2 (\$45) is given.

Therefore, F (\$55 wool coat) > F (\$45 wool scarf) is obtained.

Hsee (1998) conducted an experiment asking people how much they pay for 8 oz of Haagen-Dazs ice cream contained in a large 10-ounce cup and for 7 oz of the same ice cream brand contained in a small 5-ounce cup (experiment 2, see Fig. 5.10). The results revealed that when experiments were conducted with factors between experiment participants, they would pay significantly more money for the 7-ounce than for the 8-ounce ice cream. The average of the former was \$1.66 and that of the latter was \$2.26. These results suggest that people make a base set from the subjective situation, construct a mental ruler for it, and make judgments based on this ruler, so if they consider the subjective situation widely, they make an inconsistent decision. Applying the mental ruler model to the case of the 10-ounce cup, we find that the value to be evaluated corresponds to 8 oz and the maximum value that can be taken corresponds to 10 oz. Furthermore, in the case of the 5-ounce cup, the value to be evaluated corresponds to 7 oz and the maximum value that can be taken

Fig. 5.10 Evaluated ice cream. *Source* Hsee (1998). *Note* Left figure indicates 8 oz ice cream in a 10 oz cup. Right figure indicates 7 oz icecream in a 5 oz cup

corresponds to 5 oz. This condition leads to the evaluation that the 7-ounce ice cream in a 5-ounce cup is more desirable than the 8-ounce ice cream in the 10-ounce cup.

Hsee (1998) conducted this experiment with factors within the experiment participants. When the payment amount was reversed, the experiment participants presented higher payment amounts for the 8-ounce ice cream than for the 7-ounce ice cream. According to the model of mental rulers, this phenomenon is also interpreted as follows. In the factors among the experiment participants, both versions are presented at the same time. Thus, in the subjective situation, the participants focus on either 7 or 8 oz, and mental rulers are formed on this basis. As Hsee (1998) points out, such experimental results are difficult to interpret through the decision making theory explaining the conventional situation dependence. For example, even if we attempt to explain the results by using the decision frame model and the prospect theory, both versions can be regarded as frames of the gain region. Thus, satisfactorily explaining the experiment result is impossible. Moreover, applying the decision frame model and prospect theory to such evaluation of generosity of experiment 1 is difficult. Such a phenomenon cannot be sufficiently explained from the psychological purse model either. Even if we try to explain it as an ex-post fact, we can only argue that the psychological purse and decision frame would have been different.

Hsee (1998) referred to the evaluability hypothesis to explain such experiment results. This hypothesis assumes that in judgment and decision making, attributes that are easy to evaluate, such as the size of the ice cream container and the amount of ice cream, are connected and have an influence on decision making as a result. According to this hypothesis, in factors among the experiment participants, the comparison between the amounts of ice cream is easier than that between the ice cream containers, so the size of the container does not have a strong influence on decision making. This explanation is consistent with the previous explanation by the mental ruler model. However, although the evaluability hypothesis by Hsee (1998) is meaningful to understand the configuration of the subjective situation in the mental ruler model, it only describes the type of attributes related to each other and how decisions are influenced by them. Furthermore, although the mental ruler model explains these attributes, it does not explain how subjective situations are constructed and evaluated, and what types of judgment and decision making occur.

In this chapter, we have discussed the one-dimensional property of evaluation in consumers' judgment. However, these discussions have limitations. In people's judgment, multi-dimensional attributes and information may be evaluated multi-dimensionally. For example, we can sometimes observe the manner of judgment and decision making consciously by considering the multi-dimensional information that is assumed in multi-attribute attitude theory and multi-attribute decision making theory. In the future, we have to consider the situation where the one-dimensional evaluation assumed in the mental ruler model is easy to make and the situation where the evaluation in which the multi-dimensional information is considered is easy to make.

The mental ruler model is essentially a qualitative one and has many ambiguous aspects that require strict formulation in the future, but it can further bring forth predictions and interpretations, thereby enabling empirical tests. For example, we can conduct studies with various aspects, such as the prediction of perceptual judgment by the evaluation functions proposed in this chapter, prediction of evaluation functions in social judgment and decision making, reinterpretation of probability weighting functions, prediction of risk evaluation, and prediction of consumer behavior. From now on, we need to empirically test these predictions and refine the models.

Box 3: Sotohiro Kojima

Born in 1925; deceased in 2004. He studied economics and experimental psychology at Nagoya University, and then teaching economic psychology at Kagoshima University, Doshisha University, and Aichi Gakuin University. He was working on the application of perceptual psychology to marketing, developing measures of psychological purses. Kojima wrote many books of consumer psychology and marketing.

Summary

- The role of price is important in economic activities. Particularly with regard to revealed preference, consumer preference and utility functions can be estimated based on price information and choices. However, for this estimation, we have to assume the rationality of consumers.
- Consumers know that price information functions as a clue for quality as "how good this brand's quality is" and as a clue for saving as "how cheap this brand is."
- Considering these clues, we obtain "reaction to price," which is the consumers' comprehensive evaluation of the price. As a basic process of consumer price judgment, psychophysical laws such as Weber's and Fechner's laws are involved. These laws mean that even in price judgment, the evaluation value is not absolute but relative with respect to the amount of money.
- An important concept in explaining such relative judgment is reference price. This price is the standard by which consumers evaluate the price of products. The reference price has an internal reference price, which consumers perceive to store, and an external reference price, which is presented externally as an information stimulus at a store. The reference price corresponds to the reference point in prospect theory.
- In prospect theory, the slope of the value function is generally larger in the loss region than in the gain region. This condition indicates loss aversion. Considering this property of loss aversion, we can explain consumers' psychological resistance to price increases and endowment effect.
- In comprehensive decision making on price, the concepts of psychological purses, mental accounting, hedonic framing, and mental rulers are relevant. Each concept explains that the manner of mental editing of problem situations where consumers are placed affects their price judgment and purchase decisions.

Recommended Books and Reading Guides for Further Learning

– Poundstone, W. (2010). *Priceless: The myth of fair value (and how to take advantage of it)*. New York, NY: Hill and Wang.

This book provides psychological and behavioral explanation of pricing using many practical examples. This book focuses on phenomenon related to pricing and includes detailed descriptions of consumers' price judgment. The theoretical contents of consumer behavior are also described.

– Belsky, G., & Gilovich T. (2010). *Why smart people make big money mistakes—and how to correct them: Lessons from the life-changing science of behavioral economics* (revised ed.). New York, NY: Simon Schuster.

This book explains the psychological patterns of thinking and decision making behind seemingly irrational economic behavior. They explain why so many people make irrational financial and economic choices using the psychological concepts such as mental accounting.

– Kahneman, D., & Tversky, A. (Eds.). (2000). *Choices, values, and frames*. New York, NY: Cambridge University Press.

This book is a classical book edited by Kahneman and Tversky who proposed prospect theory and framing. This book provides psychological explanation of choice in risky and riskless contexts. In line with prospect theory, this book provides the psychophysics of value induce risk aversion in the domain of gains and risk-seeking in the domain of losses.

References

Aoki, M. (2004). Kakaku to shōhisha shinri [Price and consumer psychology]. In T. Ueda & T. Moriguchi. (Eds.), *Kakaku puromōshon senryaku [Pricing and promotion strategy]* (pp. 37–58). Tokyo, JP: Yuhikaku Publishing.

Dickson, P., & Sawyer, A. (1990). The price knowledge and search of supermarket shoppers. *Journal of Marketing, 54*(3), 42–53.

Fechner, G. T. (1860). *Elemente der psychophysik*. Leibzing, DE: Breitkopf & Härtel.

Hsee, C. K. (1998). Less is better: When low-value options are valued more highly than high-value options. *Journal of Behavioral Decision Making, 11*(2), 107–121.

Huber, J., Payne, J. W., & Puto, C. (1982). Adding asymmetrically dominated alternatives: Violations of regularity and the similarity hypothesis. *Journal of Consumer Research, 9*, 90–98.

Huber, J., & Puto, C. (1983). Market boundaries and product choice: Illustrating attraction and substitution effects. *Journal of Consumer Research, 10*(1), 31–44.

Indo, T. (Ed.). (1977). *Shinri sokutei, gakusyū riron [Psychological measurement, learning theory]*. Tokyo, JP: Morikita Publishing.

Johnson, E. J., Hershey, J., Meszaros, J., & Kunreuther, H. (1993). Framing, probability distortions, and insurance decisions. *Journal of Risk and Uncertainty, 7*, 35–51.

Kahneman, D., Knetsch, J. L., & Thaler, R. H. (1990). Experiment tests of the endowment effect and coase theorem. *Journal of Political Economy, 98*, 1325–1348.

Kahneman, D., Knetsch, J. L., & Thaler, R. H. (1991). The endowment effect, loss aversion, and status quo bias. *Journal of Economic Perspectives, 5,* 193–206.

Kahneman, D., & Tversky, A. (1979). Prospect theory: An analysis of decision under risk. *Econometrica, 47,* 263–292.

Kojima, S. (1959). *Shōhisha shinri no kenkyū [Investigation of consumer psychology].* Tokyo, JP: Nihon seisansei honbu.

Kojima, S. (1986). *Kakaku no shinri: Shōhisya ha nani wo kōnyū kettei no monosashi ni surunoka [Price psychology: What measure do consumers use to make purchasing decisions?].* Tokyo, JP: Diamond Inc.

Kojima, S. (1994). Psychological approach to consumer buying decisions: Analysis of the psychological purse and psychology of price. *Japanese Psychological Research, 36,* 10–19.

Mandel, N., & Johnson, E. J. (2002). When Web pages influence choice: Effects of visual primes on experts and novice. *Journal of Consumer Research, 29*(2), 235–245.

Mazumdar, T., & Papatla, P. (1995). Loyalty differences in the use of internal and external reference prices. *Marketing Letters, 6*(2), 111–122.

Moriguchi, T., & Tsurumi, H. (2004). Burando ikusei to puromōshon [Brand development and promotion]. In T. Ueda & T. Moriguchi (Eds.), *Kakaku puromōshon senryaku [Pricing and promotion strategy]* (pp. 109–134). Tokyo, JP: Yuhikaku Publishing.

Okuda, H. (2003). Ishikettei ni okeru bunmyaku kōka: Miryoku kōka, maboroshi kōka, oyobi tasū kōka [Context effects in decision making: Attraction, phantom, and plurality effects]. *Shakai shinrigaku kenkyū [Japanese Journal of Social Psychology], 18*(3), 147–155.

Okuse, Y. (2014). Shōhisha ga koubai ishikettei ji ni shiyō suru sanshō kakaku ni kansuru jisshōteki kōsatsu [An empirical study on the use of reference prices for consumer decision making process]. *Senshu daigaku shogaku kenkyu shohou [Bulletin of the Research Institute of Commerce], 45*(5), 1–23.

Pratkanis, A. R., & Farquhar, P. H. (1992). A brief history of research on phantom alternatives: Evidence for seven empirical generalizations about phantoms. *Basic and Applied Social Psychology, 13*(1), 103–122.

Shirai, M. (2005). *Shōhisha no kakaku handan no mekanizumu: Naiteki sanshō kakaku no yakuwari [Mechanism of consumers' price judgment: Role of internal reference price].* Tokyo, JP: Chikura Shobo.

Stevens, S. S. (1975). *Psychophysics: Introduction to its perceptual, neural and social prospects.* New York, NY: Wiley.

Takemura, K. (1996). Ishikettei to sono shien [Decision-making and support for decision-making]. In S. Ichikawa (Ed.), *Ninchi shinrigaku 4kan shiko [Cognitive psychology thoughts]* (Vol. 4, pp. 81–105). Tokyo, JP: University of Tokyo Press.

Takemura, K. (1998). Jōkyō izonteki isikettei no teiseiteki moderu: Shinteki monosashi riron niyoru setsumei [Qualitative model of contingent decision-making: An explanation of using the mental ruler theory]. *Ninchi Kagaku [Cognitive Studies], 5*(4), 17–34.

Takemura, K. (2001). Contingent decision making in the social world. In C. M. Allwood & M. Selart (Eds.), *Decision making: Social and creative dimensions* (pp. 153–173). Dordrecht, NL: Kluwer Academic.

Thaler, R. H. (1985). Mental accounting and consumer choice. *Marketing Science, 4,* 199–214.

Thaler, R. H. (1999). Mental accounting matters. *Journal of Behavioral Decision Marketing, 12,* 183–206.

Thaler, R. H., & Johnson, E. J. (1990). Gambling with the house money and trying to break even: The effects of prior outcomes on risky choice. *Management Science, 36,* 643–660.

Tversky, A., & Kahneman, D. (1981). The framing of decisions and the psychology of choice. *Science, 211,* 453–458.

Tversky, A., & Kahneman, D. (1992). Advances in prospect theory: Cumulative representation of uncertainty. *Journal of Risk and Uncertainty, 5,* 297–323.

Wada, Y., Oyama, T., & Imai, S. (Eds.). (1969). *Kankaku chikaku shinrigaku handobukku [A handbook for sensory and perception psychology].* Tokyo, JP: Seishin Shobo.

Wakayama, D. (2014). Saitei kakaku hoshō tempo ni okeru kokyaku no taten chirashi riyōgata nesage yōsē kōi [Price-matching guarantees for customers' buying behavior using flyers of competing retailers]. *Komazawa daigaku keiei gakubu kenkyu kiyou [Journal of Komazawa University], 43,* 19–39.

Chapter 6
Multi-attribute Decision Making Process in Economic Behavior: Process Tracking of Decision Making and Computer Simulation

Keywords Multi-attribute decision making process · Process-tracing technique

Most consumers make purchase decisions on the basis of various considerations such as price, design, and brand performance. Thus, the behavior of consumers can be considered as multi-attribute decision making. Knowing the process of multi-attribute decision making is important in relation to management policies and marketing. In this chapter, we outline the theoretical framework that describes the process of consumers' multi-attribute decision making and the method for its analysis. First, we show the qualitative conditions that allow mathematical expression to analyze multi-attribute decision making by consumers. Second, we outline conjoint measurement in the case where the utility of each attribute can be expressed by additive forms. Furthermore, the qualitative condition of decision making whose additive expression is impossible and statistical analysis is difficult is theoretically examined. Finally, we illustrate the computer simulation of multi-attribute decision making.

6.1 Expression of Multi-attribute Decision Making by Consumers and Measurement Theory

6.1.1 Definition of Multi-attribute Decision Making

In multi-attribute decision making that considers multiple aspects such as product performance, price, and design, a multi-attribute option x is considered as an option expressed by q-dimension attributes and is considered as an element of the subspace of the direct product of the set $X_1, X_2, \ldots, X_q$ of attributes expressing various values (Takemura, 2011). That is,

K. Takemura, *Foundations of Economic Psychology*,
https://doi.org/10.1007/978-981-13-9049-4_6

$$x \in X_1 \times X_2 \times \cdots \times X_q$$

In addition, we consider the direct product $X_k \times X_k$ for any attribute k and that the ordered pair of this element expresses the preference relation with respect to the attribute value. Based on the assumption that this preference relation is R_k, R_k is a subset of $X_k \times X_k$. For each attribute, R_k is assumed to satisfy the properties of transitivity and completeness. When

$$R^q = R_1 \times R_2 \times \cdots \times R_q$$

is given, we call a function that associates preference relation R by multi-attribute decision making with an element of R^q as a multi-attribute value function. In other words, the multi-attribute value function U can be expressed as

$$U : R^q \to R.$$

For the preference relation that can be expressed by the multi-attribute value function, weak order assumption is often requested when considered from the perspective of the possibility of quantitative measurement and rationality described above.

6.1.2 Weak Order Property of Multi-attribute Decision Making and Conjoint Measurement

Among analyses assuming the weak order property of preference, conjoint measurement assumes the additivity of the operation that sums utility values for each attribute. Conjoint measurement is often used, particularly to understand consumers' preferences in marketing. For instance, in product development, the conjoint measurement may be used to determine which value of the attributes of existing products should be changed to create a new product that consumers prefer the most and calculate the market share of the new products by simulation. Conjoint measurement is often applied to marketing, but it is also applicable to research on preference judgment, such as preference surveys for university entrance applications. Furthermore, it has high potential applicability for research on risk assessment by civil engineering experts.

As acknowledged in the pioneering research by Luce and Tukey (1964), conjoint measurement is an analytical technique originally developed in mathematical psychology and was formulated to construct additive real-valued functions (additive utility functions) that are equivalent to the interval scale from the preference data at the ordinal scale level (specifically, the scale satisfying the weak order property). A requirement for constructing such an additive real-valued function is that the preference relation must satisfy a group of axioms.

In utility estimation in the original conjoint measurement, under the influence of research from such an axiomatic point of view and assuming the ordinal scale for judging the preference of experiment participants, most researchers conducted estimation by using monotone conversion methods such as monotone analysis of variance (MONANOVA) (e.g., Shepard, Romney, & Nerlove, 1972). However, in recent years, conjoint measurement through ordinary least squares (OLS) method based on dummy variables has been used frequently (Cattin & Wittink, 1989; Louviere, 1988). Conjoint measurement using the OLS method requires preference judgment to be not less than the interval scale level in a precise sense, but simulation research shows that this approach produces highly similar results to MONANOVA, which performs monotone conversion on the assumption of ordinal scale (Carmone, Green, & Jain, 1978).

We briefly explain conjoint measurement using the least squares method incorporated in ordinary statistical packages. In conjoint measurement, the response (evaluation result) r_i of an experiment participant to the evaluation target i is expressed by the following linear model:

$$r_i = \beta_0 + \sum_{i=1}^{p} u_j(k_{ji})$$

However, $u_j\ (k_{ji})$ is the utility (partial utility) of the k_{ji} level of the factor (attribute) j in the evaluation target i (hereafter referred to as u_{jk} for simplicity).

Regarding the estimation of the evaluation result r_i, partial utility function u_j employs different calculation methods in accordance with the following cases: (1) the case of discrete factors in which relationships, such as a line format and quadratic expression, cannot be necessarily assumed between the levels of the factor; (2) the case of a linear factor that can assume a linear relationship between the levels; and (3) a quadratic function factor that can assume the relationship of a quadratic function between the levels (the case where an ideal point exists [ideal factor] and the case where an anti-ideal point exists [anti-ideal factor]). For example, in the case of a linear factor, the estimated evaluation value changes with a linear function of the level value, whereas in the case of a quadratic function factor, the estimated evaluation value changes with a quadratic function of the level value. We estimate u_{jk} under the specification of these functions.

In actual data acquisition, all profiles to be evaluated are presented to experiment participants, and evaluation scores and ranking data are collected. However, as the number of attributes and attribute levels to be picked up increases, the respondents experience difficulty in evaluating them by rank and other features. Various efforts are necessary to reduce the burden. Reducing the number of profiles presented to experiment participants using orthogonal design is a common practice.

6.1.3 Additive Conjoint Structure and Measurement

From the theoretical viewpoint of measurement, the conjoint measurement shown above cannot be expressed by the additive structure from the theoretical viewpoint if it does not have a property called additive conjoint system. In the following, we explain the viewpoint of the additive conjoint system from the perspective of the axiomatic measurement theory by Krantz, Luce, Suppes, and Tversky (1971) and the multi-objective decision theory by Ichikawa (1980).

Definition 1: Independence When the relationship $\succsim$ on the set $X_1 \times X_2$ is independent, the following is a necessary and sufficient condition: approximately $a, b \in X_1$, for a $p \in X_2$, if $(a, p) \succsim (b, p)$, for any $q \in X_2$, $(a, q) \succsim (b, q)$ is given, and approximately $p, q \in X_2$, for an $a \in X_1$, if $(a, p) \succsim (a, q)$, for any $b \in X_1$, $(b, p) \succsim (b, q)$ is given.

Then, independence in each attribute is similarly defined.

Definition 2: Independence within Attributes We assume that the relationship $\succsim$ on the set $X_1 \times X_2$ is independent.

$\succsim_1$ on X_1 is approximately $a, b \in X_1$ if $a \succsim_1 b$, and only then, for a $p \in X_2$, $(a, p) \succsim (b, p)$.

$\succsim_2$ on X_2 is approximately $p, q \in X_2$ if $p \succsim_2 q$, and only then, for an $a \in X_1$, $(a, p) \succsim (a, q)$.

Definition 3: Double Cancellation and Thomsen Condition A double cancellation relationship $\succsim$ on the set $X_1 \times X_2$ means that approximately any $a, b, f \in X_1$ and any $p, q, x \in X_2$, if $(a, x) \succsim (f, q)$ and $(f, p) \succsim (b, x)$, then $(a, p) \succsim (b, q)$ is given. Additionally, the condition that replaced $\succsim$ in this weak order relationship with the indifferent relationship ~ is called Thomsen condition. In the Thomsen condition, approximately any $a, b, f \in X_1$ and any $p, q, x \in X_2$, if $(a, x) \sim (f, q)$ and $(f, p) \sim (b, x)$, $(a, p) \sim (b, q)$ is given.

Figure 6.1 shows the Thomsen condition, which states that if points A and B are indifferent and points E and F are indifferent, points C and E are indifferent.

Definition 4: Archimedean Property Real numbers have an Archimedean property, in which an integer n satisfies $nx \geq y$ for any positive number x, regardless of how small it is and for any number y, regardless of how large it is. In other words, even if any other real number $\beta > 0$ is taken, by adding to a positive real number α, $\alpha < 2\alpha < 3\alpha < \cdots < (n-1)\alpha \leq \beta \leq n\alpha$, the numerical sequence 1, 2, 3, ..., n of natural numbers such as above is finite.

For a set N of consecutive integers (which can be either positive or negative, finite or infinite), the set $\{a_i | a_i \in X_i, i \in N\}$ is said to be a standard sequence for attribute X_1 only when the following holds; that is, there is $p, q \in X_2$, which is not $p \sim_2 q$, and for any $i, i+1 \in N$, $(a_i, p) \sim (a_{i+1}, q)$ is given. The standard sequence $\{a_i | a_i \in X_2, i \in N\}$ is that for any $i \in N$, if $b, c \in X_1$ satisfying $c \succ_1 a_i \succ_1 b$ exists

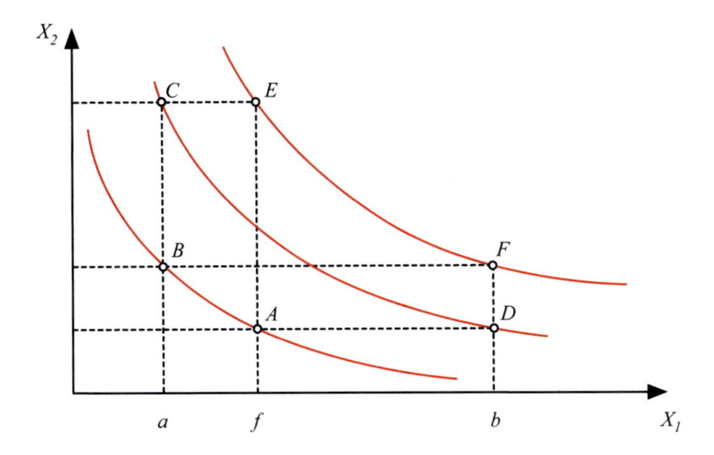

Fig. 6.1 Thomsen condition (adapted from Krantz et al., 1971)

and only then bounded. For X_2 as well, a similar standard sequence can be defined. About any $a, b \in X_1$ and any $p, q \in X_2$, when the bounded standard sequence is finite, the set $(X_1 \times X_2, \succsim)$ indicates that the weak order relationship $\succsim$ on the set $X_1 \times X_2$ is Archimedean (Fig. 6.2).

Definition 5: Unrestricted Solvability That the relationship $\succsim$ on the set $X_1 \times X_2$ satisfies unrestricted solvability means that for $a, b \in X_1$, $p, q \in X_2$, when three elements are given, the remaining one that satisfies $(a, p) \sim (b, q)$ exists.

Definition 6: Restricted Solvability That the relationship $\succsim$ on the set $X_1 \times X_2$ satisfies restricted solvability means that for any $a, b^*, b_* \in X_1$, $p, q \in X_2$, when $(b^*, q) \succsim (a, p) \succsim (b_*, q)$, a $b \in X_1$ exists and $(b, q) \sim (a, p)$ is satisfied. Furthermore, for any $a, b \in X_1, p, q^*, q_* \in X_2$, when $(b, q^*) \succsim (a, p) \succsim (b, q_*)$ is gained, $p, q \in X_2$ a $q \in X_2$ exists and $(b, q) \sim (a, p)$ is satisfied.

Figure 6.3 shows the diagram of restricted solvability on X_1. According to this diagram, if point *B* exists between the indifference curve passing through the point *A* and the indifference curve passing through point *C*, then point *D* always exists and an indifference curve passing through *B* and *D* exists.

Definition 7: Essentiality We assume the relationship $\succsim$ on the set $X_1 \times X_2$. That X_1 is essential means that for any $a, b \in X_1$, $p \in X_2$, $(a, p) \sim (b, p)$ is not given. That X_2 is essential means that for any $a \in X_1$, $p, q \in X_2$, $(a, p) \sim (a, q)$ is not given. This condition means that the indifference curves are not parallel to the X_1 or X_2 axis.

Definition 8: Additive Conjoint Structure We assume that sets X_1 and X_2 are non-empty sets and have a relationship $\succsim$ on $X_1 \times X_2$. The set of triads $(X1, X2, \succsim)$ is called an additive conjoint structure when it satisfies the following conditions:

(1) Weak order,

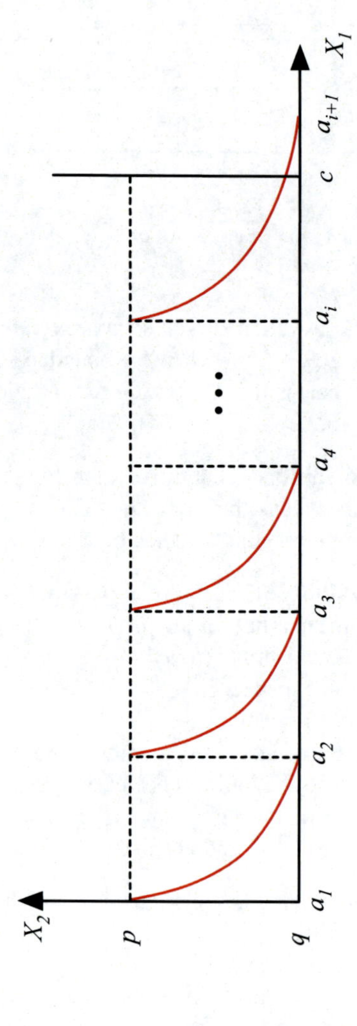

Fig. 6.2 Archimedean property with two attributes. *Source* Takemura and Fujii (2015)

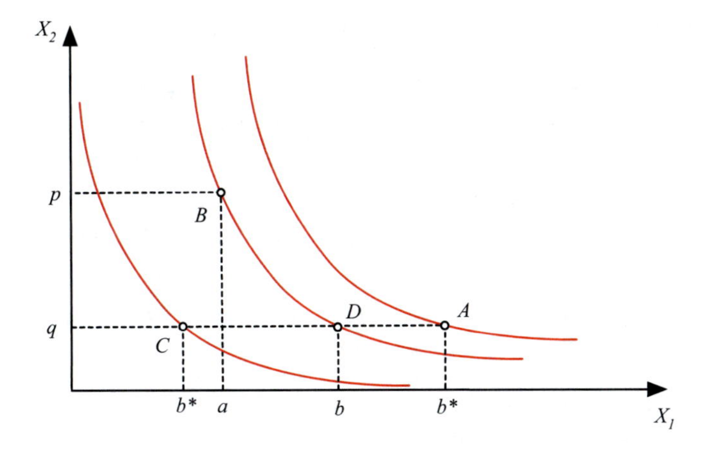

Fig. 6.3 Restricted solvability on X_1. *Source* Takemura and Fujii (2015)

(2) Independence (Definition 1),
(3) Thomsen condition (Definition 3),
(4) Archimedean property (Definition 4),
(5) Restricted solvability (Definition 6), and
(6) Each attribute is essential (Definition 7).

Expression Theorem of Additive Conjoint Structure (Krantz et al., 1971)

We assume that the set of triads $(X1, X2, \succsim)$ is an additive conjoint structure. From each of X_1, X_2 to real numbers, functions φ_1 and φ_2 exist, respectively, and for $a, b \in X_1$, $p, q \in X_2$, $(a, p) \succsim (b, q) \Leftrightarrow \varphi_1(a) + \varphi_2(p) \geq \varphi_1(b) + \varphi_2(q)$ is obtained.

Additionally, each function is unique in the range of φ'_1, φ'_2 of the positive linear transformation.

That is, $\varphi'_1 = \alpha\varphi_1 + \beta_1$, $\varphi'_2 = \alpha\varphi_2 + \beta_2$, where $\alpha > 0$.

This theorem shows that when the preference relation has an additive conjoint structure, the preference relation can be expressed by the utility of the additive system, and even if this utility function makes a positive linear transformation, its essential characteristic is not changed. This condition means that utility measured by additive conjoint measurement is an interval scale.

6.2 Multi-attribute Decision Making Not Satisfying Weak Order

As consumers' multi-attribute decision making is assumed in conjoint measurement, the utility of an attribute can be expressed by the additive system, and if the preference is in weak order, then quantitative analysis is also easy. However, consumers' multi-attribute decision making does not necessarily have a weakly ordered structure (Takemura, 2011).

First, regarding the property of completeness in weak order, we consider, for example, the preference based on the superiority principle as shown in the following.

6.2.1 Preference According to the Superiority Principle

For the preference relation of all attributes, only when "x is more indifferent than or preferred to y" that "x is more indifferent than or preferred to y" is given as an overall preference. That is, certain preferences are based on the superiority principle that $x \succsim y$ is obtained only when "$x \succsim_i y$ for all attributes i." The theorem on the principle of superiority holds from the following properties:

(1) Completeness: For any element x, y of the option set A expressed by multiple attributes ($\forall x, y \in A$), the relationship of $x \succsim y$ or $y \succsim x$ holds. Additionally, the preference relation $\succsim_i$ with respect to different values of a certain attribute i also satisfies the completeness.
(2) Transitivity: For any element x, y, z ($\forall x, y, z \in A$) of the option set A expressed by multiple attributes, if $x \succsim y \succsim z$, the $x \succsim z$ relationship holds. Additionally, the preference relationship $\succsim_i$ with respect to different values of an attribute i also satisfies the transitivity.
(3) Non-limiting property of the problem space of multi-attribute decision making: In multi-attribute decision making, one can have any preference regarding the value of each attribute as far as completeness and transitivity are satisfied, and any combination of preferences can also occur for multiple attributes.

Decision theory based on superiority principle: Under the preference by the superiority principle based on conditions (1), (2), and (3), that is, only when "for all attributes i, $x \succsim y$" does the preference that gives $x \succsim y$ not satisfy completeness or is not in weak order, or no multi-attribute value function expresses preference relationships.

Proof Let us consider that y is in weak preference with x for other attributes if $x \succsim_k y$ for an attribute k with the preference by the superiority principle. In this case, neither $x \succsim y$ nor $y \succsim x$ holds, so completeness does not hold. Therefore, the above theorem holds.

Conducting ordinary quantitative analysis is also difficult for decision making by the superiority principle assumed in many psychological models (see Takemura, 2009a), and we can understand that it is different from the so-called principle of consumer utility maximization. However, regarding superiority, when it is in strict preference in all the attributes, the case that the option is strictly preferred (weak Pareto rule) or when at least one attribute is in strict preference and the other is indifferent, the option is strictly preferred (strong Pareto rule), and the impossibility of multi-attribute value functions does not occur.

Preference by the principle of highest superior attribute number: Regarding the preference relation of attributes, only when there is a large number of attributes that "x is more indifferent than or preferred to y," that "x is more indifferent than or preferred to y" is given as an overall preference. The following theorems hold in this regard as well.

6.2.2 *Theorem of Decision Making by Highest Superior Attribute Number*

Under (1) completeness, (2) transitivity, and (3) the non-limiting property of the problem space, the preference by the principle of the highest superior attribute number is not in weak order, and the multi-attribute value function expressing the preference relation does not exist.

Proof When the number of attributes is an even number and the number of superior attributes is the same, neither of them is preferred, comparison is impossible, and completeness is not satisfied. In addition, when the number of attributes is an odd number, creating a cyclic order that does not satisfy transitivity is possible. Therefore, the above preference relation is not in weak order, so no multi-attribute value function exists either.

6.2.3 *General Possibility Theorem on Multi-attribute Decision Making*

As a social choice theory for thinking about decision making in a social group, the general possibility theorem of Arrow (1951) focuses on democratic group decision making. Takemura (2011) shows that if the situation pointed out by this theorem is changed to a multi-attribute decision making problem and reinterpreted, then the following general possibility theorem on multi-attribute decision making can be derived. This theorem shows that constructing a multi-attribute value function that satisfies all of the following conditions is impossible, which means that conditions satisfying the completeness and transitivity, which are the conditions of rationality, and conditions assumed to be appropriate for the following rational decision making do not hold at the same time.

Let A be the direct product of the set of attributes $X_1, X_2, \ldots, X_q$ expressing various values, and let the elements (ordered set expressing the q term relationship) x, y, z be options described with multiple attributes. When three or more multi-attribute options are described by two or more attributes, no multi-attribute value function satisfies the following condition, and concurrently satisfying these conditions leads to contradiction. In addition, the value function that satisfies conditions (1), (2), (4), and (5) under the condition of three or more options with two or more attributes represents the preference relation that is one-dimensional or imposing. Here being one-dimensional means that preference can be expressed only by the preference relation of a certain attribute, and imposing means that for a pair of options, the preference relation is decided regardless of what attribute value it is.

(1) Completeness: As shown above.
(2) Transitivity: As shown above.
(3) Non-limiting property of the problem space of multi-attribute decision making: In multi-attribute decision making, one can have any preference regarding the value of each attribute as far as completeness and transitivity are satisfied, and any combination of preferences can also occur for multiple attributes.
(4) Monotonicity (weak Pareto rule): When the preference relation of all attributes is consistent with "x is preferable to y," overall preference is also "x is preferable to y" (i.e., $x \succsim y$ is obtained when "$x \succsim_i y$ for all attributes i").
(5) Independence from an unrelated choice object (IIA characteristic): The preference related to the options x and y is determined only by attribute ordering related to these two options. That is, it is not affected by the attribute preference of the other option z (i.e., it shows that to know whether $x \succsim y$ holds or not, it is sufficient if a profile exists for all attributes i describing whether either or both $x \succsim_i y$ or/and $y \succsim_i x$ hold(s) for those particular x, y).
(6) Multipurpose property (multidimensionality): Only a single attribute exists, and when that attribute determines that x is preferable to y, there is no such attribute that x is always preferable to y in the overall preference (i.e., the attribute i that "for any preference profile, if $x \succsim_i y$, then $x \succsim y$ is obtained" does not exist). This condition requires that humans make decisions within multiple purposes and that they always make decisions not in only one dimension.

General possibility theorem for multi-attribute decision making: When three or more options are available, the preference described that satisfies all the conditions of the six axioms on the overall preference of the decision making subject are not in weak order, and no multi-attribute value function exists.

In other words, the two conditions for consumers to be able to select options rationally (preferences are comparable and transitive) are not compatible with the four conditions that suggest rationality for multi-attribute decision making. This fact suggests not only that it is extremely difficult for consumers to have optimal and rational decisions in multi-attribute decision making, but also that the quantitative analysis of consumers as conditions for such decision making is extremely difficult. Most studies on consumer behavior and quantitative analysis of marketing, including linear regression analysis, structural equation modeling, and conjoint measurement, are included in a series of models of additive conjoint systems, so they have a high probability of differing from the properties of actual consumer decision making.

6.3 Process Analysis of Multi-attribute Decision Making

6.3.1 Process-Tracing Technique

The multi-attribute decision analysis of consumers has been suggested as difficult to measure quantitatively and express. How, then, should we analyze consumer decision making?

One way is through the process-tracing technique. Kühberger, Schulte-Mecklenbeck, and Ranyard (2011) divided the methods of analyzing the multi-attribute decision making process into the following three methods. The first is the method for tracing information acquisition; the second is the method of tracing information integration; and the third is the method for tracing physiological, neurological, and other concomitants of cognitive processes.

The first method for tracing information acquisition is the method of letting experiment participants freely search information on brands and analyzing the attributes of the available options and the order it was searched in. This method involves using the information presentation board to examine the situation of the attribute information (e.g., prices) of brands shown on the card being sequentially acquired and the method analyzing the gaze pattern in decision making by using the eye movement measurement apparatus (Bettman, 1979; Bettman, Johnson, & Payne, 1991; Okubo, Morogami, Takemura, & Fujii, 2006; Takemura, 2009a). When analyzing experimental data using the method for tracing information acquisition, one infers the decision strategy, for example, if information on all the options is being reviewed by the option-type information search, then the strategy is additive type, or if information is searched based on attributes to reduce the reviewed options sequentially, then the strategy is elimination by aspects (EBA) type. Figure 6.4 shows an example of how information is presented in the method of tracing information acquisition.

The second method of tracing information integration includes the so-called verbal protocol method and the evaluation method of a self-description formula. In the verbal protocol method, experiment participants are allowed to speak or write about the decision making process and then record the content and examine the decision strategy adopted from the record. The verbal protocol method involves simultaneously making participants speak during the decision making process and taking records based on memory immediately after decision making. As a variation of this method, presenting a list of decision strategies is another way to test participants beforehand and ask them to report subsequently on which strategy was used.

The third is the method of tracing physiological, neurological, and other concomitants of cognitive processes, including skin potential resistance, electroencephalogram, pupillary response, etc. In recent years, functional brain image analysis and other methods have entered the mainstream. In particular, consumer behavior studies using functional brain image analysis are also called neuromarketing and neuroeconomics (Takemura, 2009b).

Manufacturer	Toshiba	Mitsubishi Electric	Casio calculator	Sanyo	Motorola
Price	¥12,800	¥17,900	¥21,680	¥9,240	¥19,800
# of colors that can be displayed simultaneously	Approx. 260,000	Approx. 260,000	Approx. 260,000	Approx. 260,000	65,535
Weight	142 g	128 g	144 g	117 g	168 g
Display size	2.4 inches	2.4 inches	2.6 inches	2.2 inches	2.9 inches
Camera pixels	1,310,000	2,000,000	3,200,000	1,330,000	1,310,000
Other function	TV reception	None	None	FM radio reception	Web browsing

Fig. 6.4 Mobile phone selection task. *Source* Okubo et al. (2006). *Note* Five cell phones and seven attribute information were presented

6.3.2 Types of Consumers' Information Search

Let us assume a scene where one purchases certain products at a store. For example, we consider the purchase scenes of digital audio players. Consumers make decisions by checking various attributes such as price, number of recordable songs, acoustic performance, and designs in stores and catalogs. In multi-attribute decision making, consumers make decisions by searching for multiple pieces of information.

As shown in Fig. 6.5, the consumer's information search starts from an internal information search that retrieves relevant information in the memory, and if information is insufficient in the memory, an external information search is made to the information source outside (Aoki, 1992; Engel, Blackwell, & Miniard, 1993; Mowen, 1990). For example, in one study, when deciding on a car repair service, many consumers rely mostly on an internal information search from memory and only 40% perform an external information search (Biehal, 1983). Furthermore, when satisfied with the previous purchasing, consumers make decisions only based on the internal information search (Engel et al., 1993).

Whether or not consumers make decisions only through an internal information search depends on their existing knowledge, for one thing. For example, consumers who visit an electronics store for the first time have limited knowledge about prod-

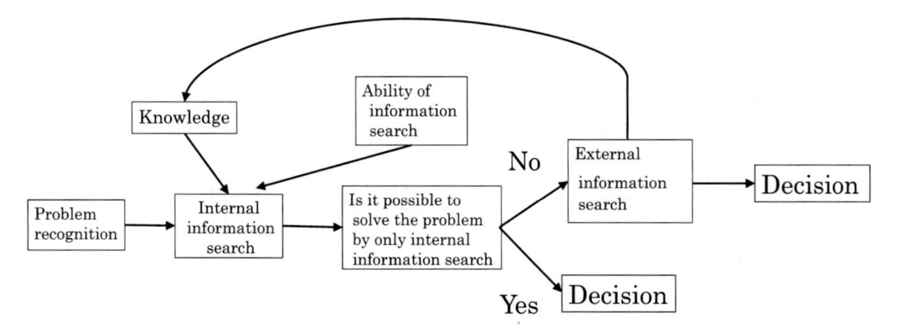

Fig. 6.5 Consumers' problem recognition and decision making process. *Source* Takemura (1997)

ucts, so they cannot search the information necessary for decision making by an internal information search. Therefore, consumers search for necessary information through external means, such as catalogs, or by asking store clerks. Even if a consumer has knowledge about the product, when he/she feels that the knowledge on the product category and search capability is insufficient, the external information search is promoted (see Fig. 6.5). The external information search can be divided into a prepurchase search and an ongoing search. The former is an information search made before buying a specific product, and the latter is conducted regularly for knowledge acquisition and not necessarily for purchasing. For example, if a person who operates computers for pleasure regularly reads magazines related to personal computers (PCs), that act is called an ongoing search. However, a prepurchase search is when a person who wants to buy a PC reads a magazine related to a PC for information gathering.

6.3.3 Situation of Consumers' Information Search

The way of consumers' information search depends on the type of purchase situations, which can be categorized into three cases (Howard, 1989).

The first case is the situation of repetitive purchase behavior or routine problem solving. For example, the problem is recognized in an extremely simple manner when the product used at home is out of stock, as in the case where printer ink has run out.

The second case is the situation of limited problem solving, where consumers are familiar with information on various options and multiple options are available, which consumers can recall for decision making. This condition corresponds, for example, to cases where a consumer who is familiar with PCs compares the computers of several manufacturers on the basis of hard disk capacity, CPU type, DVD-R, and monitor type before making a purchase decision.

In the third case, consumers have not learned the product attributes as in the second case, so they are still in the situation where they start comparing and considering a number of options after forming concepts on what attributes to use in comparison, that is, a situation of extensive problem solving. This situation corresponds to the case of low product knowledge, such as purchasing a PC for the first time. In this case, consumers have to learn the features of the hard disk or CPU.

As shown in Table 6.1, in an extensive problem-solving situation, consumers generally search a large amount of information and the time required for searching is long. In other words, consumers search for information on various brands and many attributes such as prices and manufacturer names are found in many stores. When searching for external information, consumers also try to obtain information from many sources such as store clerks, advertisements, and acquaintances. By contrast, in routine problem solving, consumers do not search for information extensively but only look into one attribute for each brand; the time required for searches is also extremely short. Furthermore, in limited problem solving, the amount of an information search is an intermediate amount between extensive and routine problem solving.

6.4 Consumers' Various Decision Strategies

6.4.1 Information Search and Option Evaluation

The ways of information search and option evaluation are closely related. For example, in purchase decision making for PCs, evaluating various PCs is necessary, but the overall evaluation of a PC is much influenced by information search such as the attributes in the order searched by the consumer. Even if each attribute of the brand has been evaluated as shown in Fig. 6.6, the evaluation of options is clearly different between the case of performing information searches for all the attributes and the case of performing information searches

Table 6.1 Situation of decision making and way of information searches (partly modified from Engel et al., 1993)

Nature of information search	Extensive problem solving	Limited problem solving	Limited problem solving
Number of searched brands	Many	Small	One
Number of searched stores	Many	Small	Unknown
Number of searched attributes	Many	Small	One
Number of search sources	Many	Small	No
Search time	Long	Short	Extremely short

	Brand A	Brand B	Brand C	Brand D
Price	1,098 dollar (60 points)	798 dollar (80 points)	798 dollar (80 points)	598 dollar (90 points)
Design	Not very good (90 points)	Reasonably good (70 points)	Fairly good (9[illegible] points)	Very good (90 points)
Function	Very good (90 points)	Reasonably good (70 points)	Fairly good (80 points)	Not very good (50 points)

Fig. 6.6 Patterns of information search by the method of monitoring information acquisition

only for some attributes (Takemura, 1996a, b, 1997). Furthermore, when searching the brand with the most appropriate price in the most important attribute such as price, the results of evaluation of options and decision making are often different between the case where, first, the information search for prices of all the brands is made and, then, the information search for the second most important attribute is made and a decision is made (an information search of the attribute type) and the case where the information search is made for each brand and a decision is made after the comprehensive evaluation of each brand (an information search of the option type) (Bettman, 1979; Bettman et al., 1991; Takemura, 1997). Thus, the way of information search and evaluation of options are closely related.

6.4.2 *What is a Decision Strategy?*

A key concept in understanding the decision making process from the viewpoint of information search is decision strategy. The decision strategy is how a consumer evaluates and adopts options in the kind of sequence of mental operations. The decision strategy is also called decision heuristics. The concept of heuristics is contrasted with an algorithm, which is an execution strategy that always brings an optimal solution. The use of heuristics, compared with the use of the algorithm, often solves problems quickly and efficiently; however, in some cases, it may bring inappropriate solutions and lead the consumer to an inconsistent and situation-dependent decision. Decision strategies in human decision making are mostly heuristic, so they are often called decision heuristics. The decision strategy is conceptually distinguished from the information search, but in reality the two are almost similar. Indeed, with regard to decision strategies, researchers often analyze the information search pattern of decision makers (Klayman, 1983; Bettman et al., 1991).

Research on the decision making process has revealed that human decision strategies hardly go through the process of maximizing the utility (the degree of satisfaction expressing a preference relation) as assumed in the utility theory (Abelson & Levi, 1985; Gigerrenzer & Selten, 2001; Simon, 1957). Simon (1957) pointed out that humans do not make decisions based on the principle of maximization or optimization where humans select the best choice from as many available options as possible, but because of the limit of the information-processing capabilities, they make decisions based on the principle of satisfaction that they seek for options that can satisfy them at a certain level. Since then, a number of decision strategies that are mainly attributed to the limitation of human information-processing capabilities have been found (Beach & Mitchell, 1978; Payne, 1976; Payne & Bettman, 2004; Takemura, 1985, 1996a, b, 1997).

6.4.3 Types of Consumers' Decision Strategies

The decision strategies found so far include the following:
Additive type: In this decision strategy, each option is examined from all dimensions, an overall evaluation of each option is made, and the option with the best overall evaluation is chosen. The additive type has different weights placed on each attribute (weighted additive type) and others (equal weight type). This strategy is illustrated in Fig. 6.6. In this strategy, we first consider brand A, with a price of 1098 dollar (60 points). Design is not very good (50 points). The function is very good (90 points), and the overall quality is judged as reasonably good (total 200 points). Similarly, brand B is very good (total 250 points), brand C is fairly good (240 points), and brand D is reasonably good (total 190 points). In this manner, we evaluate all options and select brand B (250 points in total) with the highest evaluation through the decision method. Here, the overall evaluation of one option is additively decided.
Additive difference type: In this decision strategy, the evaluation values of each attribute are compared for an arbitrary pair of options X and Y. If the number of options is three or more, then those winning the comparisons by pair are successively compared in the so-called tournament method, and the last remaining option is adopted. For example, in Table 6.2, first, brands A and B are compared. Brand B is superior in price, with the same design evaluation, and brand A is excellent in function. If the weights of the attributes are equal, then the difference between brands A and B is negative, i.e., (60 points $-$ 80 points) $+$ (50 points $-$ 70 points) $+$ (90 points $-$ 70 points) $= -20$ points), and brand B is selected for the moment. Then, brands C and D are compared, and brand C is selected for the moment. Finally, the remaining brands B and C are compared, and brand B is selected.

Conjunctive type: In this decision strategy, necessary conditions are set for each attribute, and if even one attribute does not satisfy the necessary condition, then the information processing for the option is cut off regardless of the values of other attributes, and the option is rejected. If only one option is chosen with this decision

Table 6.2 Evaluation of options in brand purchase decision making

	Brand A	Brand B	Brand C	Brand D
Price	1098 dollars (60 points)	798 dollars (80 points)	798 dollars (80 points)	598 dollars (90 points)
Design	Not very good (50 points)	Reasonably good (70 points)	Fairly good (80 points)	Very good (90 points)
Function	Very good (90 points)	Reasonably good (70 points)	Fairly good (80 points)	Not very good (50 points)

strategy, then the first option that cleared the necessary conditions over all attributes is selected. For example, in Table 6.2, assuming that the necessary condition of all the attributes is 80 points or more, if the evaluation of options is sequentially made from brand A, then we select brand C, which cleared the condition first. In this case, the remaining brands C and B are not considered.

Disjunctive type: In this decision strategy, sufficient conditions are set for each attribute, and if at least one attribute satisfies the sufficient condition, then the option is adopted regardless of the values of the other attributes. For example, in Table 6.2, we assume that sufficient conditions for all attributes are 80 points or more. When the evaluation of options is sequentially performed from brand A, the brand does not satisfy this condition for price and design, but since the function is 70 points or more, brand A is selected immediately. In this case, remaining brands B, C, and D are not considered.

Lexicographic type: In this decision strategy, the option with the highest evaluation value for the most important attribute is selected. If options of the same rank for the most important attribute have appeared, then the judgment is made based on the next most important attribute. However, a difference within a small range is regarded as being in the same rank, and if judgment is made with the next important attribute, it is called a lexicographic semi-order type. For example, in Table 6.2, assuming that price is the most important attribute, we select brand B, which has the lowest price. In this case, function and design are not considered.

Elimination by aspects (EBA) type: This decision strategy examines whether or not the necessary condition is satisfied for each attribute, and the option that does not satisfy the necessary condition is rejected. This decision strategy is similar to the conjunctive type except that the former adopts an attribute-type strategy that scans multiple options for one attribute. For example, in Table 6.2, we assume the same standard as the conjunctive type as the necessary condition (70 points or more) and also assume that attributes are sequentially examined in the order of function, price, and design. Then, brands A, B, and C remain based on the function. Furthermore, among the remaining three brands, B and C remain based on the price. Finally, brand C remains based on the design, so this brand is selected.

Majority of confirming dimensions (MCD) type: In this decision strategy, the evaluation values are compared for each attribute for an arbitrary pair of options by the brute force method. Unlike the additive difference type, the evaluation method makes comparison on whether the number of more superior attributes is large or not and adopts the option with the highest winning percentage in brute force comparison. For example, when comparing PCs, brand B is superior in two attributes between brands A and B, and brand A is superior in one attribute, so we evaluate that brand B is winning. The option with the highest number of winnings is chosen in such a procedure.

Effect referral type: This strategy involves habitually selecting a brand so that the consumer forms the most favorable attitude from past purchase and use experience. This strategy is often used in purchase decision making by consumers with high preference and loyalty to specific brands, such that they rarely search for new information. For example, as shown in Table 6.2, if a consumer always uses brand D and feels familiar with it, he/she chooses brand D almost without considering other brands.

6.4.4 Classification of Consumers' Decision Strategies

Various decision strategies have been identified and are commonly classified as compensatory and non-compensatory. In the compensatory type, even if the evaluation value of an attribute is low, if the evaluation value of other attributes is high, then the overall evaluation is compensated. The additive type, additive difference type, and MCD are included in this category. For the compensatory type, information on all options is considered. Furthermore, in the non-compensatory type, no compensatory relation exists between the attributes. The conjunctive, disjunctive, lexicographic, EBA, and effect referral types are included in this category.

Under the decision strategies of the non-compensatory type, the decision result may differ depending on the order of considering options and attributes; thus, it may cause inconsistent decision making. Consider, for example, a situation in which consumers select a brand of television based on the conjunctive type. In the conjunctive type, the option that has cleared the necessary conditions first is adopted, so the order in which one examines the brands is crucial. Even if the brand that the consumer prefers most is placed in another store, if a certain brand satisfies the necessary condition in the store he/she has visited first, then that brand is purchased. Therefore, whether the consumer purchases the brand he/she prefers most or not is likely to be influenced by situational factors such as store locations and product arrangement at stores.

Moreover, in actual decision making situations, both types are often mixed depending on the decision stage. Alternatively, to reduce cognitive strain, after consumers first narrow down the number of options to a few with the strategy rejecting options such as the EBA type, they often use compensatory-type strategies such as the additive type (Bettman, 1979; Takemura, 1996a, 1997). Thus, the decision strategy itself may change depending on the progress of the decision making process.

Such decision making is sometimes called a multistage decision strategy (Takemura, 1993).

6.4.5 How to Identify Consumers' Decision Strategies

As shown in the preceding subsections, several methods can be applied in identifying decision strategies. Here, we outline two representative methods: verbal protocol and monitoring information acquisition (Bettman et al., 1991; Payne & Bettman, 2004; Takemura, 1997) .

In the verbal protocol method, experiment participants are allowed to speak or write about the decision making process and record the content, and they can examine the type of decision strategy adopted from the record. The verbal protocol method involves simultaneously making participants speak during the decision making process and taking a record based on memory immediately after decision making. A variation of this method involves presenting a list of several decision strategies to test participants beforehand and allow them to report on which strategy was used.

For example, in studying the verbal protocol method, Takemura (1996b) asked Japanese consumers planning to purchase various products to write diaries. In the study, protocols presumed to have used conjunctive-type decision strategies at the final stage of the decision were reported as shown in Table 6.3. The diary in Table 6.3 was written by a 22-year-old college student who reported the process until she purchased a half coat in preparation for her trip to Canada.

The method of monitoring information acquisition involves letting experiment participants freely search for information on brands and analyzing the attributes of the options as well as the order they were searched in. This method includes the task of using an information presentation board on which one examines the situation of attribute information (e.g., prices) of brands shown on the card being sequentially acquired, and the method analyzing the gaze pattern in decision making by using measurement apparatus such as eye cameras (Bettman et al., 1991; Takemura, 1996a, b, 1997). In the analysis of experimental data using the method of monitoring infor-

Table 6.3 Example of verbal protocols

Verbal protocols of buying decision by a female college student
On January 6: "I heard that the temperature in Canada and the eastern United States was below zero. I am going to buy a waterproof coat with excellent cold protection."
On January 12: "I went to a sports store in Umeda. When I went to the ski wear corner, I found half coats. Among the half coats, I found one with a hood. The edge of the hood [was] adorned with faux fur and it [was] cute. The outer material [was] not waterproof, but it [was] processed so that it [would] repel water, so it [would] be fine even if it [got a little wet]. I was [slightly] dissatisfied with the color, but the cold protection function seemed good. I decided to compromise on color and bought that half coat."

Source Takemura (1996b)

mation acquisition, one may estimate in the following manner: If information on all the options has been reviewed via the option-type information search, it should be an additive type, or if information has been searched based on attributes to reduce the reviewed options sequentially, then it should be an EBA type.

Figure 6.6 shows the information search pattern of a fictional experiment participant in the method of monitoring information acquisition with the same brand as shown in Fig. 6.6. According to this figure, the experiment participant first searched for attribute-type information on prices, examined relatively cheap brands B and C, and selected brand C, which excelled in design and function. The design and function of brand D were not examined. In this case, we presume that the lexicographic semi-order type was used in the first half of the decision making process, and the conjunctive type was used in the latter half. Such an information review process is sometimes examined by monitoring information acquisition using, for example, an eye camera as shown in Figs. 6.7 and 6.8.

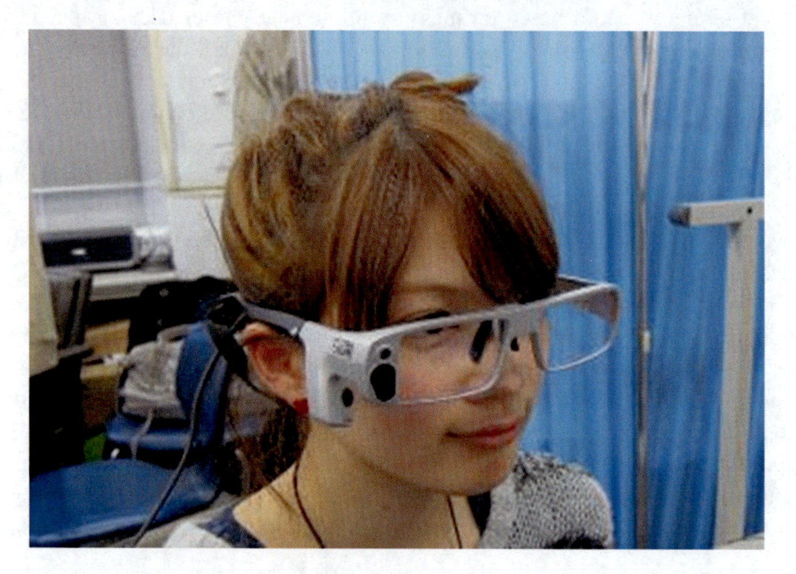

Fig. 6.7 Contact-type eye movement measurement device (Tobii glass eye-tracker manufactured by Tobii). *Source* Okubo and Takemura (2011)

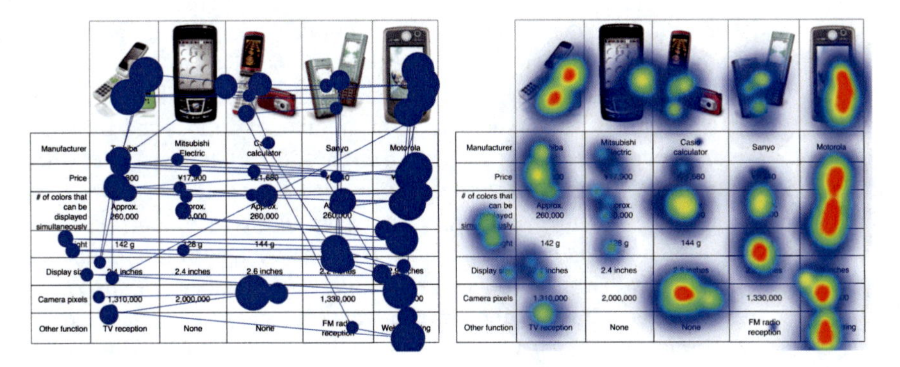

Fig. 6.8 Example of visualization of eyesight data on cell phone specifications table measured using an eye movement measurement system. *Source* Okubo and Takemura (2011). *Note* Figure on the right presents sections that attracted greater attention in order of red, yellow, green, and blue, similarly to thermography

6.5 Situation Dependence of Consumers' Decision Strategies

6.5.1 Decision Strategies and Situational Factors

Various decision strategies in decision making processes have been identified and are commonly classified into two: compensatory and non-compensatory. The first type is the strategy in which compensatory overall evaluation is performed if the evaluation values of other attributes are high even if the evaluation value of an attribute is low as a linear compensatory type. The second type is the decision strategy without such a compensatory relationship between attributes and includes the attitude-dependent, conjunctive, disjunctive, and lexicographic types. Under the decision strategies of the non-compensatory type, the decision result may differ depending on the order of considering options and attributes, so it may cause inconsistent decision making, which is dependent on the order of an information search. Consider, for example, a situation in which a consumer makes decisions based on the conjunctive type. In the conjunctive type, the brand that has cleared the necessary conditions first is adopted, so an important consideration is the order in which the consumer examines brands. Even if the brand that the consumer prefers most is placed in another store, if there is a product that satisfies the necessary conditions in the store he/she has visited first, then that brand is purchased. Therefore, whether the consumer purchases the brand he/she prefers most or not is likely to be influenced by situational factors such as the location of stores and the product arrangements in them.

Moreover, in actual decision making processes by consumers, various decision strategies are often mixed depending on the decision stage. To reduce cognitive strain, after consumers first narrow down the number of options to a few by deciding, such as in the lexicographic type, they often use the way of decision making such as the linear compensatory type (Bettman, 1979; Takemura, 1996b, 2009a).

Consumers' decision making research to date shows that these decision strategies vary according to the properties of tasks such as the number of options and attributes (Bettman, 1979; Bettman et al., 1991; Takemura, 1996b, 2009a). For example, when the number of options is small, the compensatory type is adopted, and when the number of options increases, the non-compensatory type is easy to adopt.

The reason why the decision strategy varies with the change in the number of options and attributes is that under conditions with a large number of options and attributes, large amounts of information must be processed and an information overload occurs, so a simple decision strategy with a low information-processing load is adopted to avoid cognitive strain.

The decision strategy is influenced not only by the number of attributes and options but also by the operation of decision making task variables such as the form of the information presentation of decision making problems and the reaction mode of decision making, the operation of motivation variables such as the involvement of decision makers, emotional manipulation such as moods, and situational factors (Abelson & Levi, 1985; Bettman et al., 1991; Cohen & Areni, 1991; Engel et al., 1993; Takemura, 1996b, 2009a) . For example, Isen and Means (1983) examined the effect of positive emotions (good mood) on the decision strategy. They found that when experiment participants received a fake feedback that they succeeded in perceptual movement tasks (positive emotion group), the time they spent on choosing a fictitious car was short and they did not thoroughly explore the information when deciding, compared with experiment participants who did not receive a feedback (controlled group). This condition suggests that a positive emotion promotes non-compensatory decision making.

6.5.2 Theory Explaining Situation Dependence of Multi-attribute Decision Making Process by Consumers

Thus, because the decision strategy is adopted depending on various situation factors, the result of the decision also depends on the situation. Therefore, the purchase behavior of consumers is greatly influenced by the situation. Why do the variation phenomena of situation-dependent decision strategies occur? Research based on a computational framework provides a representative theory to explain such situation-dependent variation phenomena. In the computational approach, an assumption is that the decision maker calculates the cost (expenditure) and benefit (profit) of using the relevant decision strategy to adapt to the situation and thus adopts an appropriate decision strategy. What can be considered in calculating the cost and benefit is the magnitude of cognitive effort required in decision making and the optimality of decision making.

The first model based on this computational approach is the situation-ready model by Beach and Mitchell (1978), and the model that has been refined so that computer simulation can be performed by expanding the basic idea of this model is the adaptive decision making model by Payne, Bettman, and Johnson (1993). Payne et al. consider that certain decision strategies are adopted in certain situations as a result of decision makers trading off the magnitude of cognitive effort required in deciding for the optimality (accuracy) of decision making. The researchers varied the number of options and attributes, performed a computer simulation, and calculated the cognitive effort (operationally defined by the number of operations of basic information processing) accompanying the execution of each strategy and the relative accuracy (operationally defined by the index that takes a value of 1 when the result is exactly the same as the weighted additive type and takes a value of 0 when the reaction is completely random) of the decision result under each condition. Although the accurate decision is possible with the weighted additive type, the type requires much cognitive effort as the number of options and attributes increase. In addition, the lexicographic type requires almost no cognitive effort by increasing the number of options and attributes and also maintains accuracy to a certain extent. Furthermore, in the non-compensatory type, cognitive efforts are not required as much as in the weighted additive type, even if the number of options and attributes are increased. Experimental studies to date show that the adoption rate of non-compensatory-type decision strategies increases along with the increasing number of options and attributes, and we can interpret this phenomenon consistently from the results of this simulation. That is, we can understand that in situations where the number of options and attributes are small, complementary strategies such as the weighted additive type with high accuracy are easy to use because cognitive effort is not strictly required. However, in situations with a large number of options and attributes, complementary-type decision strategies that require a great deal of cognitive effort are not adopted and non-complementary decision strategies that do not require much cognitive effort are easy to adopt.

Payne et al. (1993) correlate the results of this simulation with the results of many psychological experiments and conclude that decision makers trade off the accuracy of choice and cognitive effort and adaptively choose the decision strategy. The model of Payne et al. can quantitatively predict what kind of decision strategy is adopted and what kind of decision is likely to be made in which situations, so this model is considered useful in predicting consumer behavior in marketing and the support of consumers' decision making. Such a viewpoint has also been adopted in the decision making theory of consumer behavior in recent years (Takemura, 2009a).

However, one cannot always explain situation-dependent option evaluation and decision making only by using the model of Payne et al. At present, the framing effect that appears in the editing phase of decision making processes and the effects, such as emotions and motivations, as shown in the prospect theory cannot be fully explained from the computational point of view. Moreover, their model does not explain how trade-offs are made between the accuracy of choice and cognitive effort and decision makers adaptively choose decision strategies. To elucidate such a problem, Takemura (1996b) proposes a model that assumes metacognition to monitor and control the decision making process under given processing resources, but, currently, its empirical evidence is not sufficiently obtained. In addition, the allocation of attention is related to such metacognition (Takemura, 2009a), and research that elucidates the overall structure of the consumer's decision making process is necessary.

Box 1: Richard H. Thaler

Born in 1945. He received Ph.D. degree in 1974 from the University of Rochester. He was awarded the Nobel Prize in 2017 for the contributions have built a bridge between the economic and psychological analyses of individual decision making. He is now working as the Charles R. Walgreen Distinguished Service Professor of Behavioral Science and Economics at the University of Chicago Booth School of Business.

Photograph: The Nobel Prize Organization

https://www.nobelprize.org/prizes/economic-sciences/2017/thaler/facts/.

6.6 Consumers' Multistage Decision Making and Computer Simulation of Decision Strategies

As a result of analyzing the decision strategy by the method of monitoring information acquisition, several decision strategies are often mixed according to the decision stage. For example, in experimental studies to date, decision makers first narrow down the number of options by the strategy of rejecting options such as the LEX and EBA types to reduce cognitive effort and then using compensatory-type strategies such as the additive type (Takemura, Haraguchi, & Tamari, 2015). Thus, the decision strategy itself may change depending on the progress of the decision making process. Such decision making is called a multistage decision strategy and is approximately expressed by a variation of the two-stage decision making strategy such as the first and second halves.

Takemura et al. (2015), using the decision strategy presented earlier, and considering that the decision strategy is changed in the course of decision making by setting two stages in the decision making process, performed computer simulation based on the assumption that the decision strategy changes in the decision making process. In this computer simulation, they first considered what kind of decision strategy and what combination of decision strategies require less cognitive effort and are relatively accurate. Furthermore, they considered what ways of the combination of decision strategies changing with two stages required relatively less cognitive effort and were accurate, and examined their psychological functions. All decision strategies examined here are classified into either the compensatory type or the non-compensatory type. In the decision strategies of the compensatory type, even if the evaluation value of an attribute is low, if the evaluation value of other attributes is high, then the overall evaluation is compensated. The additive type (WAD and EQW), additive difference type (DIF), and MCD are included in this category. For the compensatory type, information on all options is considered. On the other hand, the decision strategies of the non-compensatory type, which are different from those of the compensatory type, have no compensatory relation between the attributes. The conjunctive (CON), disjunctive (DIS), lexicographic (LEX), EBA, and MCD types are included in this category.

In this study, the cognitive effort (elementary information processes [EIP], which are operationally defined by the number of operations of basic information processing) and the relative accuracy of decision results (relative accuracy [RA], which is operationally defined by the index that takes a value of 1 when the result is exactly the same as the weighted additive type and takes a value of 0 when the reaction is completely random) are calculated under each condition.

As shown in Fig. 6.9, as a result of this computer simulation, cognitive loads are relatively low and accurate when narrowing down the options to two using LEX or EBA and then making comparisons by WAD. When thinking about this in a realistic situation, decision makers narrow down the options with the most important attribute to themselves. If they cannot decide, they narrow down the options with the next important attribute and further narrow down the final candidates to two to make a

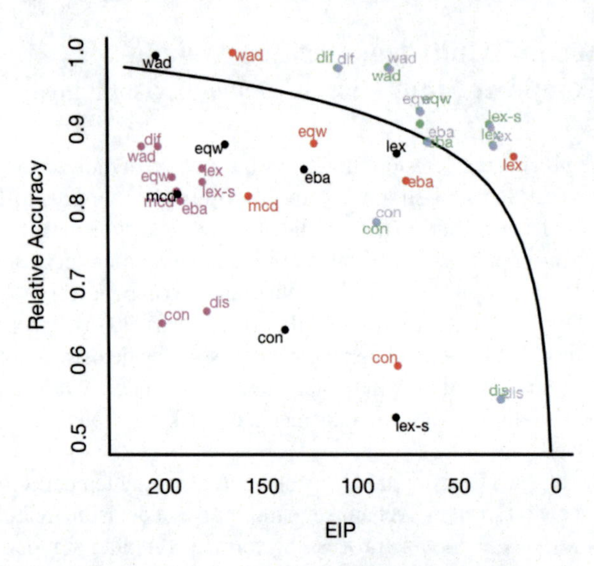

Fig. 6.9 Relationship between cognitive effort and relative accuracy in two-stage decision making (partial enlargement). *Source* Takemura et al. (2015). *Note* The pink plot uses CON, the black plot uses DIS, the blue plot uses EBA, the green plot uses LEX, and the red plot uses NONE in the first stage

decision. Thus, decision making with relatively less cognitive effort that is close to the result of the weighted additive type from the beginning can be performed. Previous research on multi-attribute decision making shows that many people exhibit such decision making processes (Bettman, 1979; Takemura, 1996b, 2014). Thus, many consumers who show multistage decision making strategies are considered to make relatively rational decisions.

Additionally, in this study, for each number of options left in the second stage, the relationship of cognitive effort is shown in Fig. 6.10. According to this figure, the effect on the cognitive effort of the number of options left in the second stage is smaller for the combination whose options were narrowed down by using EBA compared with the combination using LEX. The RA for each number of options left in the second stage is shown in Fig. 6.11. As shown in Fig. 6.11, in decision making by combining the same strategies, the RA is hardly influenced by the options left in the second stage. Thus, if we combine the same strategies with attribute-type strategies, we will observe the options again from exactly the same point of view, so no difference of results has been generated regardless of how many options are left in the second stage. In addition, in the combination using WAD in the second stage, the RA increases when the number of options left in the second stage is large. This tendency can be considered as the result of the lower probability of excluding options with higher RA by narrowing down the options before adopting those with higher RA in terms of WAD.

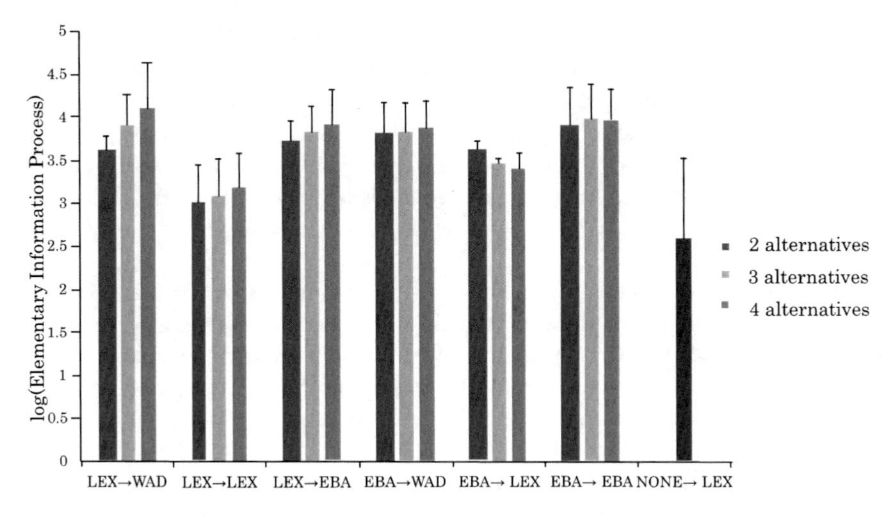

Fig. 6.10 Relationship between cognitive effort and number of options Left in second stage. *Source* Takemura et al. (2015)

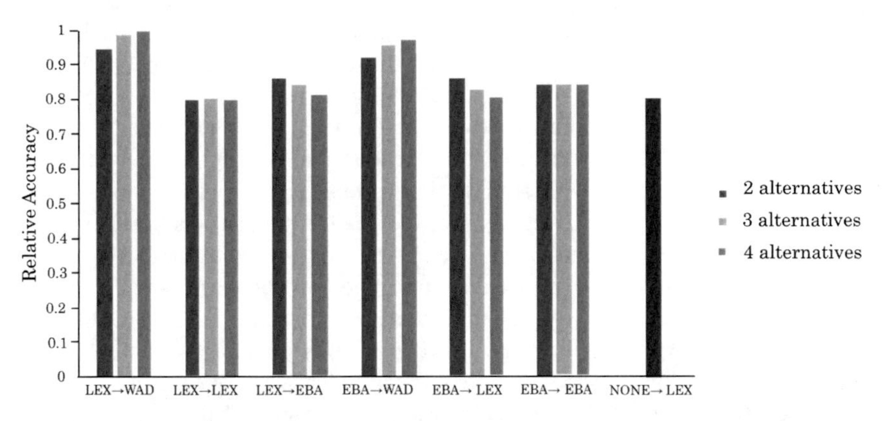

Fig. 6.11 Relationship between relative accuracy and number of options left in second stage. *Source* Takemura et al. (2015)

However, we have to understand that RA was measured by the weighted additive type in this study in the same way as in the work of Payne et al.. From the normative decision theory, the question remains as to whether the weighted additive type is truly a rational and accurate decision strategy. From the viewpoint of multipurpose optimization, restrictions exist on the conditions under which the weighted additive type becomes Pareto optimal, and in this sense also, it is not necessarily true that the weighted additive type is optimal and rational. Furthermore, when considering the axiomatic condition of the weighted additive type, we have to closely examine the assumption that all attribute values can be weighted and added numerically.

Takemura (2018) also examined two-stage multi-attribute decision strategies to avoid bad decisions in different conditions, wherein the numbers of alternatives and attributes vary. A Monte-Carlo computer simulation was used to assess the accuracy and cognitive effort of each of the two-stage decision strategies. The simulation result suggested that comparatively effortless and accurate strategies to avoid bad decisions comprised the two-phased strategy that used the lexicographic strategy to eliminate up to a few alternatives in the first stage and used weighted additive strategy in the second stage.

Nevertheless, we cannot necessarily say that non-compensatory strategy, such as the lexicographic strategy, leads to irrational decision making, as shown by the results of computer simulation. On the contrary, in multi-attribute decision making, trying to make a compensatory decision by integrating information on many attributes for making the best decision may lead to irrational decision making owing to the limitation of information-processing capability. The more important decision making is, the more important attributes have to be considered; thus, one has to process a large amount of information. To this end, as shown by the simulation result, one can perform decision making in a slightly better manner and avoid making bad decisions by thoroughly examining the last two options after narrowing down the range of options by employing methods of the attribute base, such as LEX and EBA. In sum, in this case, one should first consider "what is the most important" and narrow down the options and then perform the multifaceted consideration.

Such kind of decision making is easy to perform because it requires a relatively low cognitive effort. However, in critical decision making situations, some cases are not decided by using such a strategy. For instance, with regard to the problem caused by a massive earthquake, we have to wait for the analysis based on what kinds of measures and social policies were appropriate, but we should remember that the life and health of people are the most important. However, we need to take seriously the fact that the more important the decision making and the situation are, the easier it is to forget these matters. This condition is similar to the psychological phenomena of having a reckless war despite knowing in advance that many lives would be lost yet people do not take measures in advance. This condition seems to be related to the psychological properties of the focal point. Attention has been paid to specific attributes based on linguistic messages or imagery, and one is likely to make decisions based on the evaluation of such attributes. Fujii and Takemura (2001) propose a contingent focus model to explain the decision making based on such a focus of attention (Fujii & Takemura, 2001; Takemura, 1994; Takemura & Fujii, 2015). Given the assumption in this model, if attention plays an important role in decision making, in the case of emergency or serious decision making, processing resources will be used more frequently to interpret the situation and the capacity of attention will be reduced. Thus, the focus lies on attributes that are not important but easy to understand. This fact is considered as the reason why decisions only from a one-sided perspective are frequently made in emergencies such as war and disasters, when viewed in hindsight.

6.7 Purchase Environment and Consumers' Decision Making Process

6.7.1 *Information Overload and Consumers' Decision Strategies*

The research on consumers' decision making process to date shows that decision strategies vary according to the properties of tasks such as the number of options and attributes. When the number of options is small, the compensatory type is adopted, and when the number of options increases, the decision strategies of the non-compensatory type are easy to adopt. For example, Takemura (1993) conducted an experiment using the verbal protocol method on purchase decision making about radio cassette recorders. Takemura changed the number of options and attributes presented in the problem over two levels (4 and 10) and compared the decision strategies adopted between experimental conditions. As a result, the decision strategy recognized under each condition is shown in Table 6.2. As this table indicates, many experiment participants adopted a form that confuses multiple decision strategies in multiple stages. In general, however, under conditions where the numbers of options and attributes are large, the non-compensatory-type decision strategy is easy to adopt. Furthermore, the analysis of the data elucidates that decision strategies based on attributes are easy to adopt when the number of options is large.

The reason why the decision strategy varies with the change in the number of options and the number of attributes as described above is that under conditions with a large number of options and attributes, a large amount of information must be processed and an information overload occurs. Thus, to avoid cognitive strain, a simple decision strategy with a low information-processing load was adopted. In this state of information overload, the adopted decision strategy becomes simple, and consequently, the result of decision making changes. Moreover, when information is overloaded, not only is the decision strategy simplified but the avoidance of the decision making situation is also likely to occur. The avoidance of the situation without making a decision is also a type of simplification of information processing. Takemura (1996a) conducted consumer behavior observation and in-store interview surveys at supermarkets and reported that consumers who were wavering in the state of information overload were likely to escape from the sales floor to avoid conflicts.

6.7.2 *Consumers' Decision Strategies, Involvement, and Emotion*

Consumers' decision strategies are affected not only by the number of attributes and options but also by the extent to which the decision maker is psychologically involved in the purchase decision (Abelson & Levi, 1985; Bettman et al., 1991; Engel et al., 1993). Consumers are more involved when they select products of interest, and compensatory-type decision strategies are easy to adopt. However, for products with low interest and low-involvement, non-compensatory decision strategies are easy to adopt. Thus, in the case of high-involvement products such as houses and cars, consumers tend to examine all kinds of information and make purchase decisions, but in the case of low-involvement products such as soft drinks or disposable lighters, consumers do not consider information and are likely to make purchase decisions influenced by the display methods at the storefront.

Decision strategies are also influenced by the feelings of decision makers as detailed in Chap. 8 (Cohen & Areni, 1991; Luce, Bettman, & Payne, 2001; Payne & Betteman, 2004; Takemura, 1996b, 1997). For example, Isen and Means (1983) examined the effect of positive emotions (good mood) on the decision strategy. They showed that when experiment participants received a fake feedback that they succeeded in perceptual movement tasks (positive emotion group), the time they spent on choosing a fictitious car was short and they did not thoroughly explore the information when deciding, compared with experiment participants who did not receive a feedback (controlled group). Isen and Means (1983) also revealed that by using the verbal protocol method, experiment participants with positive emotion tended to make decisions by using the EBA-type decision strategies of eliminating options sequentially according to the attributes of interest as compared with experiment participants in the controlled group. The EBA-type decision strategy does not necessarily lead to the optimal decision but involves a low cognitive load, so positive emotions are thought to promote the use of this strategy. By feeling good in the store environment and hearing background music (BGM), consumers are likely to apply non-complementary decision strategies and make purchase decisions without fully examining products.

6.7.3 Consumers' Decision Making in Stores

When consumers actually make purchase decisions in stores, what options and attributes do they consider? In-store shopping behavior has been studied to examine the decision making process in real situations (see Figs. 6.12 and 6.13). Dickson and Sawyer (1990) studied in-store consumer behavior toward four products: cereals (grain food), coffee, margarine, and toothpaste from 802 shopper's data. The study showed that consumers made decisions within 12 s on average, and approximately half the number of consumers decided within five seconds. In addition, only about 25% of consumers made price comparisons with other brands, and over 40% did not even examine the price of brands in their carts. Takagi et al. (see Takemura, 1996b) studied in-store consumer behavior at supermarkets toward curry, hashed stew, retort food, and microwavable food; the researchers found that nearly 60% of the 201 consumers examined only one brand. In addition, the researchers investigated the purchase behavior toward products such as soft drinks and functional drinks, as shown in Fig. 6.14, and found that most of the consumers considered only

Fig. 6.12 An example of in-store shopping study by Takemura' laboratory

Fig. 6.13 An example of in-store shopping study using eye-gaze recorder by Takemura' laboratory

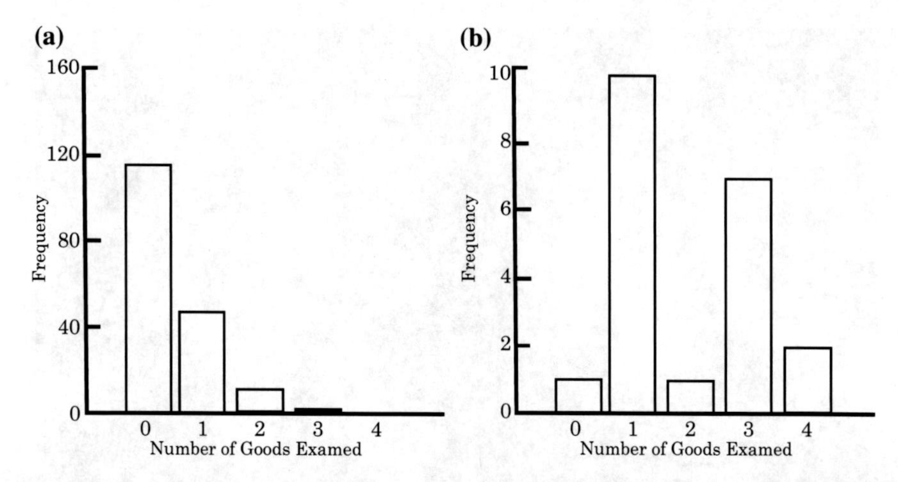

Fig. 6.14 Number of goods examined by purchasers and non-purchasers. *Source* Takemura (1996b). *Note* The horizontal axis indicates the number of goods examined by purchasers (left) and non-purchasers (right), and the vertical axis indicates the frequency

one brand, and the number of brands considered increased to approximately three at most. Otsuki (1991) reported that consumers' attention tended to focus on the right side of supermarket store shelves and this side was selected with a probability of 1.5–2 times more than that of the left side. These studies show that consumers use non-compensatory decision strategies when making decisions without considering all information.

6.7.4 Factors Influencing In-store Consumer Behavior

In-store consumer behavior is also influenced by context information such as consumer emotions and BGM. These factors can be considered to be more prominent when consumers are using non-compensatory decision strategies as examined in the preceding subsection.

Donovan and Rossiter (1982) conducted an interview survey to assess the emotional value of stores and found that when consumers evaluated the store environment as comfortable, their purchasing intention for products was stronger. Furthermore, when consumers evaluated the environment in the store as comfortable, purchasing intention increased with the arousal level; however, when consumers evaluated the environment as uncomfortable, or neither uncomfortable nor comfortable, the purchasing intention was not related to the arousal level.

Milliman (1986) applied BGM in a store and examined its effect on consumer behavior. In this two-month study, the researcher created conditions in the supermarket such as the following: consumers cannot listen to music, they can listen to music with 60 beats per minute, and they can listen to music with 108 beats per minute. Further, the relevance of these conditions to in-store behavior was analyzed. Various kinds of music were used, but all of them were BGM, of which no one would become aware. With regard to the walking speed in the store, the non-music condition was an intermediate result between the other two conditions, but the speed with fast music (108 beats) condition was 17% faster than that with slow music (60 beats) condition. Furthermore, with respect to the total purchase price, the fast speed music condition was 38% more influential on the purchase decision than the slow speed music condition. Milliman (1986) also found that the slow-tempo BGM at the restaurant caused customers to stay for a longer time and consume a larger amount of alcoholic beverages.

6.7.5 Consumers' Unplanned Purchase

As shown above, consumers are likely to use non-compensatory-type decision strategies in stores, but in the first place, if consumers are sufficiently considering what to buy before coming to the store to make purchase plans, then they will be able to make compensatory-type decisions. However, research on in-store consumer behavior shows that many consumers come to stores without having clear purchase plans. An unplanned purchase occurs when the brand that has not been intended to be bought is bought in the store.

Research on unplanned purchase can be traced back to Dupont company's study on consumer buying habits in 1935. Dupont conducted interview surveys with consumers entering and leaving the store, and examined the degree of non-purchase rate continuously. Subsequent research on unplanned purchase has been promoted in Japan since the 1980s, mainly at the Distribution Economics Institute, and the results are reported by Otsuki (1991). Aoki (1989) reviewed the previous survey studies on unplanned purchase; according to a collaborative study between Dupont and the Point-of-Purchase (POP) Advertising Institute conducted in the United States in 1977, an unplanned purchase rate in a broad sense (in-store decision ratio) was 64.8%, and a series of survey studies conducted in Japan by the Distribution Economics Institute in the 1980s showed an unplanned purchase rate of 87.0–96.8% (Aoki, 1989; Nagano, 1997; Saido, 2000). These results suggest that in promotion in supermarkets and other locations, not only the word-of-mouth and advertisements are important, but also POP advertisements and the technique of sales staff in the store. Aoki (1989) has compiled studies on unplanned purchase and emphasizes that unplanned purchase exists in a narrow sense as described in the following:

Recall purchase: When consumers come to stores or have seen POP advertisements at the sales floor, they recall the potential necessity of products and decide on their purchase.

Related purchase: The case where necessity is recognized from recalling the relevance to other purchased products and consumers decide on purchasing.

Conditional purchase: Consumers do not have a clear purchasing intention at the time of visiting the store, but they vaguely recognize the necessity of a specific product, and they decide to purchase when they find that prices and other conditions are satisfying.

Impulse purchase: This case is different from the aforementioned three; here, the purchase is due to product novelty or complete impulsiveness.

Furthermore, Aoki (1989) points out that although sometimes classified as planned purchase rather than unplanned purchase in a narrow sense, the following is considered as purchase becoming unplanned purchase in a broad sense:

Brand choice: The case where commodity categories are planned, but brands are decided in stores.

Brand change: The case where purchasing a specific brand has been planned before the consumer comes to the store, but a different brand is bought as a result of decision making in the store.

When the conditional purchase wherein unplanned purchase occurs because of improvements (such as price conditions) is excluded from these unplanned purchases; other purchases can be assumed to be decision strategic and many non-compensatory decision strategies exist. Even in conditional purchase, the preference changes according to the relevant conditions; thus, a non-compensatory-type decision strategy is highly likely. In particular, impulse purchase is made by emotion-dependent non-compensatory-type decision strategies.

6.8 Context Factors of Consumer Behavior

6.8.1 Priming Effect

In recent years, consumers have been able to purchase products from Internet-based sites. Amazon is an example of an influential retail website. Normally, information on products themselves is considered as a factor that affects purchase decision making. However, previous studies have shown that information that is unrelated to the product itself, such as the color or background design of the Web page, affects purchase decision making. This phenomenon can be considered psychologically as a kind of priming effect. The priming effect makes the judgment of the next task faster, slower, accurate, or inaccurate by presenting information to people without having them be aware of it. Background images and banner advertisements on product purchase screens on the website may become a visual prime and cause a priming effect.

Mandel and Johnson (2002) changed the image of the background of the product presentation screen of the website, as shown in Fig. 5.6, to change the visual prime and examined the priming effect on judgment and decision making. Their research used two product categories, namely cars and sofas and presented two items under each product category. Mandel and Johnson (2002) prepared a web design that becomes the prime showing "comfort" (safety in the case of cars) in Fig. 5.6a and a web design that becomes the prime showing "money" in Fig. 5.6b; then, the researchers conducted choice experiments under each condition. Their experimental hypothesis was that in the experiment to select on the screen of (a) showing comfort, more subjects were likely to choose comfortable sofas (or safe cars) even if these products were high in price, and the subjects who selected on the screen of (b) were more likely to choose cheap and economical sofas (or cheap and economical cars). This experimental result supported the hypothesis.

6.8.2 *Context-Dependent Effect*

Consumer decision making can change depending on the context in which options are presented. The occurrence of such a context effect suggests that promotion can influence consumers' purchase decision making by controlling the presented brand information and configurations. The context effects of consumer decision making include the following (Okuda, 2003):

(1) Attraction effect: This phenomenon has been reported by Huber, Payne, and Puto (1982); due to the existence of an unappealing brand that no one will choose, the attraction of a certain brand increases and the choice rate changes. Figure 5.7 shows brand X with high quality but with high price and brand Y with medium quality and price. Compared with Y, A has a lower price but the same quality, B is inferior in quality and price, and C has the same price and inferior quality. In addition to X and Y, by presenting A, B, or C as not superior to Y, Y is made more attractive than when A, B, or C is not presented and the choice rate of Y is raised.
(2) Compromise effect: This phenomenon is shown in Fig. 5.7, where brand D is neither superior nor inferior to Y, but the choice rate of Y rises because of the existence of a brand that is relatively extreme in terms of advantage and disadvantage (Huber & Puto, 1983).
(3) Phantom effect: The phenomenon is illustrated by brand E in Fig. 5.7; when brands superior to Y are difficult to obtain, Y is also more likely to be chosen. For example, if battery E is reported to be superior to Y and is out of stock, the choice rate of Y rises more than when it is chosen between the two brands (Pratkanis & Farquhar, 1992).

Okuda (2003) organized these three types of context effects and conducted psychological experiments by using products such as PCs, recreational vehicles, mobile phones, and dry batteries as subjects. The experiment results showed that relatively inferior brands increased the choice rate of target brands, while superior brands and those at the same level reduced the choice rate of the target brands but increased the choice rate of the target brand when they were out of stock. In any case, the choice rate of competing brands declined.

6.9 Process After Consumers' Decision Making

6.9.1 Cognitive Dissonance Theory

Consumers may regret a purchase or change cognition after their purchase decision making process. Festinger (1957) proposed the cognitive dissonance theory and tried to explain the characteristics of psychological changes after decision making. He summarized every piece of knowledge, opinion, belief, or feeling about the self or one's own environment with the term of cognition, and assumed that when dissonance between these cognitive units occurred, people would try to resolve it. The state of dissonance is unpleasant, so changes in cognition, behavior, addition of new cognition, and selective contacts to new information, occur to reduce the discomfort. According to Festinger, in the situation after decision making, since action has been done, dissonance occurs easily and actions to resolve it are easy to take. Thus, the attractiveness of options after choice is predicted to increase more than that before choosing.

Festinger and Carlsmith (1959) reported that people who received a one-dollar reward by telling another experiment participant a lie, specifically, that boring work tasks were interesting, were more likely to believe that work tasks were interesting than those who obtained a 20-dollar reward due to the justification of one's action. This result is consistent with the cognitive dissonance theory. Despite a cheap reward of one dollar, reporting that boring work tasks are interesting is in a relationship that is more dissonant than doing the same thing with a reward of 20 dollars. Thus, a change in cognition is thought to occur easily.

6.9.2 Cognitive Dissonance and Consumer Behavior

Assuming that cognitive dissonance works, one can expect that the items one has purchased will be evaluated higher after purchase than before purchase. In fact, in one study, people were asked right before buying and right after buying betting tickets at a racetrack about the approximate winning probability of the horse they were betting on. The results showed that the group that had already bought betting tickets more strongly believed that the horse they were betting on would win than the group that was asked right before buying the tickets (Knox & Inkster, 1968). Moreover, an interview survey asked people who have bought a new car recently and people who have a car of more than three years old about the kind of car advertisements they saw recently; they found that people who had bought a new car recently were more frequently watching advertisements for cars they had purchased (Ehrlich, Guttman, Schönbach, & Mills, 1957).

Thus, even after purchase decision making, consumers change the evaluation of the selected brand and watch the advertisement to justify their decision making. In this sense as well, promotional strategies such as advertisements become important for consumers who have already purchased products.

The cognitive dissonance theory covers a wide range of human social behaviors and research using the cognitive dissonance concept continues (Harmon-Jones & Mills, 1999; Matz & Wood, 2005). Furthermore, the cognitive dissonance theory has focused on the process of justification after decision making. However, as Montgomery (1983, 1993) and Luce et al. (2001) have done, some researchers point out influencing decision strategies before decision making to justify certain decisions.

6.10 Multi-attribute Decision Making Process and "Good Decision"

In this chapter, we aim to suggest the normative contents of multi-attribute decision making. In this type of decision making, the choice strategy of the non-compensatory type is often used, but we cannot necessarily say that this strategy leads to irrational decision making as shown by the results of computer simulation. Rather, in multi-attribute decision making, trying to make a compensatory decision by integrating information on many attributes for the purpose of making the best decision may lead to irrational decision making due to the limit of information-processing capability.

The more important decision making is, the more attributes have to be considered, so one has to process a large amount of information. To this end, as shown by the result of the computer simulation shown earlier, one can perform slightly better decision making by thoroughly examining the last two options after narrowing down the range of options by employing methods of the attribute base such as LEX and EBA. Briefly, in this scene, one should first consider "what is the most important" and narrow down the options and then perform multifaceted consideration.

Such a way of decision making is easy to perform because it requires a relatively low cognitive effort. However, in critical decision making situations, some cases are not decided by such a strategy. For instance, consider the discussions on the reconstruction policy of the East Japan earthquake that occurred in 2011. In this situation, besides the earthquake, a large number of other factors have to be considered such as damage by tsunami, radioactive contamination due to nuclear power plant accidents, residents' life and health, and economic reconstruction, which caused confusion in the country. Here, the most important consideration is to protect the life and health of the residents, on which many people share similar views in normal times. For this purpose, promptly developing projects is necessary by issuing government bonds, and so on. However, in the discussion immediately after the great earthquake, few experts and media practitioners were pointing out such issues. Rather, many economists emphasized that issuing deficit bonds would cause a collapse of the Japanese economy and people who needed help should rely on volunteers; hardly any discussions raised different opinions on these matters. Immediately after the natural disaster, I myself discussed the need for policies that emphasize life and health rather than economic policies, such as finance, with graduate students and experts, but most people did not understand my opinion.

With regard to the problem caused by the massive earthquake, we have to wait for judgment by history on what kinds of measures and social policies were appropriate, but we should remember that the life and health of people are the most important. However, we need to take seriously the fact that the more important the decision making and the situation are, the easier these matters can be forgotten. This condition is similar to the psychological phenomena of having a reckless war despite knowing in advance that many lives would be lost and of people not worrying about the issue initially.

Why are people likely to make decisions based on non-essential attributes that are not important from the intrinsic point of view? If non-essential attributes stand out in an easy-to-understand manner, they may seem important. This author believes that this condition is related to psychological properties of the focus of attention. In this case, attention focuses on specific attributes based on linguistic messages or imagery expression, and one is likely to make decisions based on the evaluation of attributes on which the attention has been focused.

Takemura and Fujii propose a psychological model called situation-dependent focus model to explain decision making based on such a focus of attention (Fujii & Takemura, 2001; Takemura, 1994; Takemura & Fujii, 2015). This model explains, for example, that the risk attitude changes according to the degree of the focus of attention of the result and also the probability in the decision making under risk. In the psychological experiment of decision making, the authors reveal that the result of decision making may differ from the prospect theory by manipulating the attention to attribute information, and show that the manipulation of attention has a relatively strong impact on decision making. As the assumption in this model, if attention plays an important role in decision making, then in the case of emergency or serious decision making, processing resources will be used more to interpret the situation and the capacity of attention is reduced. Thus, attention focuses on attributes that are not important but easy to understand. This fact is considered as the reason why decisions only from a one-sided perspective are made frequently in emergencies, such as war and disasters, when viewed in hindsight.

In contemporary society, decision makers have to bear heavy mental burdens such as serious matters developing in various aspects, as well as a large amount of information to be processed. Thus, people increasingly prefer decision making that is psychologically simple and formally rational. For example, people tend to prefer decision making by excessive formalism, such as excessively applying compliance and accountability and excessively seeking procedural rationality. By conducting a questionnaire survey, Takahashi et al. (2010) suggest that the psychological tendency to pursue such formalism stems not from a sense of justice but from the psychology of responsibility avoidance, and that individuals with such strong tendencies are rather low in altruism. Furthermore, according to social psychological studies on decision making, people who pursue rationality tend to have a higher sense of depression and a lower sense of subjective well-being (Schwartz, 2004; Schwartz et al., 2002). Decision making from a pluralistic perspective is difficult psychologically as discussed in this paper and also from a formal perspective. However, we argue that "good decision making" does not necessarily mean psychological easiness or rationality from a formal perspective, and that the attitude of thinking first about essentially important matters is important.

Box 2: Henry Montgomery

Born in 1943. He received his Ph.D. in 1975 from the Department of Psychology, Stockholm University. He was working as a professor at the Department of Psychology, Stockholm University. He made important contributions to decision process theory and economic psychology. He started the process-tracing study of decision making in the early 70s with Ola Svenson and Tommy Gärling.

Summary

- Consumers often make multi-attribute decisions by searching multiple attributes such as prices and functions from various pieces of information.
- Purchase situations can be categorized into three: extensive problem-solving situations with a large amount of information searching, limited problem-solving situations with a moderate amount, and routine problem-solving situations with the smallest amount.
- Consumers make decisions by searching for information with the decision strategy of how to decide in any situation.
- The decision strategies can be divided into two major categories: compensatory type, where even if the evaluation value of an attribute is low, the overall evaluation is compensated if the evaluation value of other attributes is high, and non-compensatory type, where the aforementioned relationship does not exist.

- In the non-compensatory-type decision strategy, since all the information is not always examined, decision making depending on the context, such as product displays and emotion, is likely to occur. In-store purchases are unplanned in many cases, and non-compensatory decision strategies are used, so the in-store environment, such as product displays, strongly influences the decision making process of consumers.
- Computer simulation may be used in theoretical considerations on the influence of decision strategies and multistage decision making.
- Consumers reevaluate the selected brand due to cognitive dissonance after purchase decision making, so promotion, such as through advertisements, is also effective.

Recommended Books and Reading Guides for Advanced Learning

- Payne, J. W., Bettman, J. R., & Johnson, E. J. (1993). *The adaptive decision maker*. New York, NY: Cambridge University Press.

This book provides a compelling overview of the empirical results and conceptual framework that the authors have been pursuing in their research program of multi-attribute decision making process. This book introduces a model that shows how decision makers balance effort and accuracy considerations and predicts which strategy a person will use in a given situation. A series of experiments testing the model were introduced.

- Ogaki, M., & Tanaka, S. (2017). *Behavioral economics: Toward a new economics by integration with traditional economics*. Tokyo, JP: Springer.

This book explains basic concepts of behavioral economics in an easy-to-understand manner, clarifying relationships with traditional economics, and also explains the findings of recent neuroeconomic studies of decision making process. This book is intended as a textbook for a course in behavioral economics for advanced undergraduate and graduate students.

- Schulte-Mecklenbeck, M., Kühberger, A., & Ranyard, R. (Eds.). (2011). *A handbook of process tracing methods for decision research: A critical review and user's guide*. New York, NY: Psychology Press.

This book provides a critical review and user's guide to conducting and analyzing decision process using process-tracing techniques. Each chapter covers a specific process-tracing method of multi-attribute decision making process. This book also illustrates a multi-method approach to decision making process study.

References

Abelson, R. P., & Levi, A. (1985). Decision making and decision theory. In G. Lindzey & E. Aronson (Eds.), *The handbook of social psychology* (3rd ed., Vol. 1, pp. 231–309). New York, NY: Random House.

Aoki, Y. (1989). Tentō kenkyū no tenkai hōkō to tenponai shōhisha kōdō bunseki [Direction of in-store research and in-store consumer behavior]. In Y. Tajima & Y. Aoki (Eds.), *Tentō kenkyū to shōhisha kōdō bunseki: Tenpo nai kōbai kōdō bunseki to sono shūhen [In-store research and consumer behavior analysis: Analysis of buying behavior and related issues]* (pp. 49–81). Tokyo, JP: Seibundō shinkōsha.

Aoki, Y. (1992). Shōhisha jōhō shori no riron [Theory of consumer information processing]. In Y. Osawa (Ed.), *Māketingu to shōhisha kōdō: Māketingu saiensu no shin tenkai [Marketing and consumer behavior: New development in marketing science]* (pp. 129–154). Tokyo, JP: Yuhikaku Publishing.

Arrow, K. J. (1951). *Social choice and individual values*. New York, NY: Wiley.

Beach, L. R., & Mitchell, T. R. (1978). A contingency model for the selection of decision strategies. *Academy of Management Review, 3,* 439–449.

Bettman, J. (1979). *An information processing theory of consumer choice*. Reading, MA: Addison Wesley.

Bettman, J., Johnson, E. J., & Payne, J. W. (1991). Consumer decision making. In T. S. Robertson & H. H. Kassarjian (Eds.), *Handbook of consumer behavior* (pp. 50–79). Englewood Cliffs, NJ: Prentice Hall.

Biehal, G. J. (1983). Consumers' prior experiences and perceptions in auto repair choice. *Journal of Marketing, 47,* 87–91.

Carmone, F. J., Green, P. E., & Jain, A. K. (1978). Robustness of conjoint analysis: Some Monte Carlo results. *Journal of Marketing Research, 15,* 300–303.

Cattin, P., & Wittink, D. R. (1989). Commercial use of conjoint analysis: An update. *Journal of Marketing, 53,* 91–96.

Cohen, J. B., & Areni, C. S. (1991). Affect and consumer behavior. In T. S. Robertson & H. H. Kassarjian (Eds.), *Handbook of consumer behavior* (pp. 188–240). Englewood Cliffs, NJ: Prentice Hall.

Dickson, P., & Sawyer, A. (1990). The price knowledge and search of supermarket shoppers. *Journal of Marketing, 54*(3), 42–53.

Donovan, R. J., & Rossiter, J. R. (1982). Store atmosphere: An environmental psychology approach. *Journal of Retailing, 58*, 34–57.

Ehrlich, D., Guttman, I., Schönbach, P., & Mills, J. (1957). Postdecision exposure to relevant information. *The Journal of Abnormal and Social Psychology, 54*(1), 98–102.

Engel, J. F., Blackwell, R. D., & Miniard, P. W. (1993). *Consumer behavior* (7th ed.). New York, NY: Dryden.

Festinger, L. (1957). *A theory of cognitive dissonance*. Stanford, CA: Stanford University Press.

Festinger, L., & Carlsmith, J. M. (1959). Cognitive consequences of forced compliance. *Journal of Abnormal and Social Psychology, 58,* 203–210.

Fujii, S., & Takemura, K. (2001). Risuku taido to chūi: Jōkyō izonteki shōten moderu ni yoru furēmingu kōka no keiryō bunseki [Risk attitude and attention: A psychometric analysis of framing effect by contingent focus model]. *Kodo keiryogaku [The Japanese Journal of Behaviormetrics], 28,* 9–17.

Gigerenzer, G., & Selten, R. (Eds.). (2001). *Bounded rationality: The adaptive toolbox*. Cambridge, MA: MIT.

Harmon-Jones, E., & Mills, J. (Eds.). (1999). *Cognitive dissonance: Progress on a pivotal theory in social psychology*. Washington, DC: American Psychological Association.

Howard, J. A. (1989). *Consumer behavior in marketing strategy*. Englewood Cliffs, NJ: Prentice Hall.

Huber, J., Payne, J. W., & Puto, C. (1982). Adding asymmetrically dominated alternatives: Violations of regularity and the similarity hypothesis. *Journal of Consumer Research, 9,* 90–98.

Huber, J., & Puto, C. (1983). Market boundaries and product choice: Illustrating attraction and substitution effects. *Journal of Consumer Research, 10*(1), 31–44.

Ichikawa, A. (Ed.). (1980). *Tamokuteki kettei no riron to hōhō [Theory and method of multi-objective decision].* Tokyo, JP: Keisoku jidō seigyo gakkai.

Isen, A. M., & Means, B. (1983). The influence of positive affect on decision making strategy. *Social Cognition, 2,* 18–31.

Klayman, J. (1983). Analysis of predecisional information search patterns. In P. Humphreys, O. Svenson, & A. Vari (Eds.), *Analyzing and aiding decision processes* (pp. 401–414). Amsterdam, NL: North-Holland.

Knox, R. E., & Inkster, J. A. (1968). Postdecision dissonance at post time. *Journal of Personality and Social Psychology, 8,* 319–323.

Krantz, D. H., Luce, R. D., Suppes, P., & Tversky, A. (1971). *Foundations of measurement Vol. 1: Additive and polynomial representations*. New York, NY: Academic.

Kühberger, A., Schulte-Mecklenbeck, M., & Ranyard, R. (2011). Introduction: Windows for understanding the min. In M. Schulte-Mecklenbeck, A. Kühberger, & R. Ranyard (Eds.), *A handbook of process tracing methods for decision research* (pp. 3–17). New York, NY: Psychology Press.

Louviere, J. J. (1988). Analyzing decision making: Metric conjoint analysis. *Journal of Marketing Research, 26*. https://doi.org/10.2307/3172612.

Luce, M. F., Bettman, J. R., & Payne, J. W. (2001). Emotional decisions: Tradeoff difficulty and coping in consumer choice. *Monographs of Journal of Consumer Research, Conflict and Tradeoffs in Decision Making, 1,* 86–109.

Luce, R. D., & Tukey, J. W. (1964). Simultaneous conjoint measurement: A new type of fundamental measurement. *Journal of Mathematical Psychology, 1,* 1–27.

Mandel, N., & Johnson, E. J. (2002). When web pages influence choice: Effects of visual primes on experts and novice. *Journal of Consumer Research, 29*(2), 235–245.

Matz, D., & Wood, W. (2005). Cognitive dissonance in groups: The consequences of disagreement. *Journal of Personality and Social Psychology, 88,* 22–37.

Milliman, R. E. (1986). The influence of background music on the behavior of restaurant patrons. *Journal of Consumer Research, 13,* 286–289.

Montgomery, H. (1983). Decision rules and the search for a dominance structure: Towards a process model of decision making. In P. C. Humphreys, O. Svenson, & A. Vari (Eds.), *Analyzing and aiding decision processes* (pp. 343–369). Amsterdam, NL: North-Holland.

Montgomery, H. (1993). The search for a dominance structure in decision making: Examining the evidence. In G. A. Klein, J. Orasanu, R. Calderwood, & C. E. Zsambok (Eds.), *Decision making in action: Models and methods* (pp. 182–187). Norwood, NJ: Ablex.

Mowen, J. C. (1990). *Consumer behavior* (2nd ed.). New York, NY: Macmillian.

Nagano, M. (1997). Shōhisha kōdō ni okeru jōkyō yōin [Situational factors in consumer behavior]. In T. Sugimoto (Ed.), *Shōhisha rikai no tame no shinrigaku [Psychology for understanding consumer]* (pp. 192–205). Tokyo, JP: Fukumura Shuppan.

Okubo, S., Morogami, S., Takemura, K., & Fujii, S. (2006). Kanjō ga jōhō kensaku ni ataeru eikyō no jikken kenkyū: Aikamera ni yoru katei tsuiseki o mochīte [An experimental study of the effects of emotions on information search: Using eye camera-based process tracking]. In *Collection of Lecture Papers for 2004 Entaateimento Kansei Waakushoppu [Entertainment Sensitivity Workshop]* (pp. 14–19).

Okubo, S., & Takemura, K. (2011). Shirīzu shōhisha kōdō to māketingu 2 gankyū undō sokutei to shōhisha kōdō [Consumer behavior and marketing (2) measurement of the eye movement and consumer behavior]. *Seni seihin shohi kagaku [Journal of the Japan Research Association for Textile End-uses], 52*(12), 744–750.

Okuda, H. (2003). Ishikettei ni okeru bunmyaku kōka: Miryoku kōka, maboroshi kōka, oyobi tasū kōka [Context effects in decision making: Attraction, phantom, and plurality effects]. *Shakai shinrigaku kenkyū [Japanese Journal of Social Psychology], 18*(3), 147–155.

Otsuki, H. (1991). *Tentō māketingu no jissai [In-store marketing practice]*. Tokyo, JP: Nihonkeizaishinbunsha.

Payne, J. W. (1976). Task complexity and contingent processing in decision making: An information search and protocol analysis. *Organizational Behavior and Human Performance, 16,* 366–387.

Payne, J. W., & Bettman, J. R. (2004). Walking with the scarecrow: The information-processing approach to decision research. In D. J. Koehler & N. Harvey (Eds.), *Blackwell handbook of judgment and decision* (pp. 110–132). Malden, MA: Blackwell.

Payne, J. W., Bettman, J. R., & Johnson, E. J. (1993). *The adaptive decision maker*. New York, NY: Cambridge University Press.

Pratkanis, A. R., & Farquhar, P. H. (1992). A brief history of research on phantom alternatives: Evidences for seven empirical generalizations about phantoms. *Basic and Applied Social Psychology, 13,* 103–122.

Saido, M. (2000). Shōhisha no hikeikaku kōbai katei [Consumer's non-planning buying decision]. In K. Takemura (Ed.), *Shōhi kōdō no shakai shinrigaku: Shōhisuru ningen no kokoro to kōdō [Social psychology of consumer behavior: Mind and action in human consumption]* (pp. 40–50). Tokyo, JP: Fukumura Shuppan.

Schwartz, B. (2004). *The paradox of choice: Why more is less*. New York, NY: Harper Collins.

Schwartz, B., Ward, A., Monterosso, J., Lyubomirsky, S., White, K., & Lehman, D. R. (2002). Maximizing versus satisficing: Happiness is a matter of choice. *Journal of Personality and Social Psychology, 83,* 1178–1197.

Shepard, R. N., Romney, A. K., & Nerlove, S. (1972). *Multidimensional scaling* (Vol. 1). New York, NY: Seminor Press.

Simon, H. A. (1957). *Administrative behavior: A study of decision making process in administrative organization*. New York, NY: McMillan.

Takahashi, H., Matsui, H., Camerer, C., Takano, H., Kodaka, F., Ideno, T., et al. (2010). Dopamine D1 receptors and nonlinear probability weighting in risky choice. *Journal of Neuroscience, 30,* 16567–16572.

Takemura, K. (1985). Ishikettei sutoratejī jikkō ni okeru meta ninchi katei moderu [A metacognitive model for the implementation of decision strategies]. *Doshisha shinri [Doshisha Psychological Review], 32,* 16–22.

Takemura, K. (1993). The effect of decision frame and decision justification on risky choice. *Japanese Psychological Research, 35,* 36–40.

Takemura, K. (1994). Furēmingu kōka no rironteki setsumei: Risukuka deno ishikettei no jōkyō izonteki shōten moderu [A theoretical explanation of the framing effect: The contingent focus model of decision making under risk]. *Japanese Psychological Review, 37*(3), 270–293.

Takemura, K. (1996a). Ishikettei to Sono Shien [Decision-making and support for decision-making]. In S. Ichikawa (Ed.), *Ninchi Shinrigaku 4kan Shiko [Cognitive psychology vol. 4 thoughts]* (pp. 81–105). Tokyo, JP: University of Tokyo Press.

Takemura, K. (1996b). *Ishikettei no shinri: Sono katei no tankyū [Psychology of decision making: Its process and investigation]*. Tokyo, JP: Fukumura Shuppan.

Takemura, K. (1997). Shōhisha no jōhō tansaku to sentakushi hyōka [Alternative evaluation and consumer buying decision]. In T. Sugimoto (Ed.), *Shohisha rikai no tame no shinrigaku [Psychology for understanding consumer]* (pp. 56–72). Tokyo, JP: Fukumura Shuppan.

Takemura, K. (2009a). *Kōdō ishiketteiron: Keizai kōdō no shinrigaku [Behavioral decision theory: Psychology of economic behavior]*. Tokyo, JP: Nippon hyoron sha Co. Ltd.

Takemura, K. (2009b). Ishikettei to shinkei keizaigaku [Decision making and neuroeconomics]. *Rinshō seishin igaku [Japanese Journal of Clinical Psychiatry], 38,* 35–42.

Takemura, K. (2011). Tazokusei ishikettei no shinri moderu to "yoi ishikettei" [Psychological model of multi-attribute decision making and good decision]. *Operēshonzu risāchi [Journal of the Operations Research Society of Japan], 56,* 583–590.

Takemura, K. (2014). *Behavioral decision theory: Psychological and mathematical representations of human choice behavior.* Tokyo, JP: Springer.

Takemura, K. (2018). *Avoiding bad decisions: From the perspective of behavioral economics*. Keynote Paper presented at the International Congress of Applied Psychology, Montreal, Canada.
Takemura, K., & Fujii, S. (2015). *Ishikettei no shohō [Prescription for decision making]*. Tokyo, JP: Asakura Shoten.
Takemura, K., Haraguchi, R., & Tamari, Y. (2015). Tazokusei ishikettei katei ni okeru kettei houryaku no ninchiteki doryoku to seikakukusa: Keisanki simyurēshon ni yoru kōdō ishiketteiron teki kentō [Effort and accuracy in multi-attribute decision making process: A behavioral decision theoretic approach using computer simulation technique]. *Ninchi kagaku [Cognitive Studies], 22*, 368–388.

Chapter 7
Deployment on the Consumer's Interaction Research: Behavioral Game Theory and Problems of Happiness

Keywords Decision making process · Interaction · Diffusion process · Game theory · Happiness

Nowadays, people are communicating more efficiently in an economic society. Communication is typically a two-way interaction, but as the number of members increases from tripartite to quadripartite, the combination of interactions increases dramatically and the means of communication have become complicated as well. Communication can also be done strategically in situations, wherein the interests of the parties involved are in conflict and they are assumed to behave strategically as in game theory. Nonetheless, mutual communication can increase trust and cooperation as well as suppress the emergence of social dilemmas to a certain extent. This chapter explains game theory as a theoretical viewpoint to examine consumer interaction and discusses the research on happiness in relation to people's social welfare.

7.1 Interaction in Economic Psychology

An economic society presents plenty of situations, wherein people are required to communicate. Communication between two people is basically a two-way interaction. As the number of members increases from tripartite to quadripartite, the combination of interactions increases dramatically and the way of communication becomes complicated. Communication can also be done strategically in situations, wherein the interests of each party are in conflict and they are assumed to behave strategically as in game theory. With mutual communication, increased trust and cooperation can be achieved and the emergence of social dilemmas can be suppressed to a certain extent.

Communicating within an economic society involves verbal communication; large-scale information transmission by newspapers, magazines, broadcast media,

K. Takemura, *Foundations of Economic Psychology*,
https://doi.org/10.1007/978-981-13-9049-4_7

and so on (also called "mass communication"); and information transmission by other forms of media such as the Internet. Regarding the influence of mass communication, a two-step theory indicates that communication passes through two steps from opinion leaders to followers. Regarding the diffusion process of ideas and technologies, the sociologist Rogers (2003) proposed the occurrence of decision making processes involving knowledge, attitudes, decision, execution, and beliefs. Various groups of people, such as innovators, adopt new innovations early, while laggards only introduce innovations and popular trends at a later time.

Meanwhile, communication within a society involves interactions among people and communication with organizations, such as national and local governments and organizations, especially with regard to dangerous events and social decision making. In people's decision making, there is the distortion of probability judgment and the presence of biases that demand zero risks. In this kind of communication, "trust" also plays an important role, and problems of communication from an organization losing its trust value have no effect. In this chapter, we explain these problems.

7.2 Interaction and Communication in Society

As shown in Chap. 5, people are conducting interpersonal communication among various social relationships, but the process is not just limited to two- or three-people relationships. Nowadays, we are communicating with many more people. In addition, information is conveyed through large-scale information transmission (i.e., mass communication) through newspapers, magazines, television programs, etc., and through other platforms such as the Internet (information transmission by the Internet may be included in mass communication).

Humans live in a society containing various human relations, groups, and organizations. "Society" is a collective term for the relationship between individuals and groups of individuals when abstractly described. In a society, people live with complicated relationships.

When we refer to "society," it usually means many organized groups in which there is mutual communication of people. For simplicity, we first consider the relationship of only two people to show complex social relations. As shown in Fig. 7.1, suppose that there are two people, A and B, and that Mr. A likes Miss B. Suppose that one day he asked her out on a date, saying "Would you like to go see a movie together next

Fig. 7.1 Interaction and social relations between two persons

week?" This communication from Mr. A to Miss B is indicated by →. Suppose that Miss B did not dislike Mr. A, and she was a little pleased with the invitation, but she did not know well that Mr. A could really be a love interest, so then she declined it by saying, "I cannot go because I have to work next week." The communication of this refusal is indicated by ←. In this way, the communication between Mr. A and Miss B can be expressed by the relationship Mr. A → Miss B, and the communication from Miss B to Mr. A can be expressed by the relationship Miss B → Mr. A.

In this interpersonal relationship comprising only two people, there is at least a set of Mr. A, Miss B, the relationship from Mr. A to Miss B, and the relationship from Miss B to Mr. A. When it is written with a set, {Mr. A, Miss B, (Mr. A, Miss B), (Miss B, Mr. A)}; however, (Mr. A, Miss B) is an ordered pair expressing the relationship from Mr. A to Miss B, and (Miss B, Mr. A) is an ordered pair expressing the relationship from Miss B to Mr. A. An ordered pair is a set consisting of a pair in which the order of two elements is considered and is generally (Mr. A, Miss B) ≠ (Miss B, Mr. A). Here, the ordered pairs of (Mr. A, Miss B), (Miss B, Mr. A) are the elements of a set of ordered pairs (direct product: {(Mr. A, Mr. A), (Mr. A, Miss B), (Miss B, Mr. A), (Miss B, Miss B)}) of all combinations made with Mr. A and Miss B as elements.

Furthermore, in the relationship between the two, there is a dynamic interdependent relationship wherein actions from Mr. A to Miss B define actions from Miss B to Mr. A and vice versa. For example, Miss B's behavior changes in accordance with the case whether Mr. A has asked her out on a date. Furthermore, Mr. A himself defines his next action. For example, the fact that Miss B declined a date once may affect Miss B's own action this time. Furthermore, among the relationships between the two, each person may behave strategically. For example, Mr. A may think that "If I asked Miss B out, what would Miss B think? I don't want to be disliked or refused, so I should refrain from asking her out." Meanwhile, Miss B may think that "Although I was asked out by Mr. A, I should decline it once because if I accept it at once I will be thought of as being too receptive." In this way, they may decide their actions according to the actions of the other party. Communication between only two persons is complicated, and this becomes even more complicated when three or more persons are involved. For example, Miss C may also observe the communication of Mr. A and Miss B, and Mr. A and Miss B may also communicate strategically by anticipating the actions of the observation result of Miss C.

In this way, just considering the relationship of communication between two or three persons is complicated, and it is difficult to predict the behavior of each person. What would happen if we are to consider the relationships among five persons, as shown in Fig. 7.2. There would be 20 units of ordered pairs in any two-person relationship among the five persons, excluding the self-relationships (${}_5C_2 \cdot 2! = 20$). Furthermore, considering a three-person relationship, such as Mr. A → Miss B → Miss C, it makes a three-term relationship in terms of set theory, and when self-relationships are excluded, 60 units (${}_5C_3 \cdot 3! = 60$) are given. Similarly, considering four- and five-person relationships, 120 units (${}_5C_4 \cdot 4! = 120$) and 120 units (${}_5C_5 \cdot 5! = 120$) are given, respectively. In this way, mutual relationships are complicated even in a five-person society. To understand the human behaviors within a society, it is

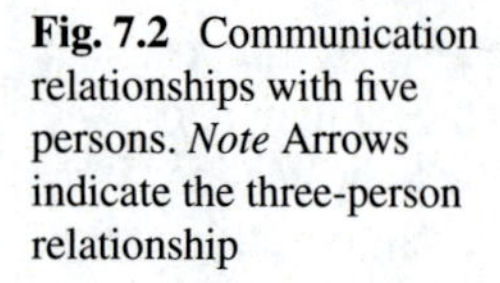

Fig. 7.2 Communication relationships with five persons. *Note* Arrows indicate the three-person relationship

important to know that individuals are also taking actions while communicating in such a complicated relationship.

Dunbar (1998) advocated the social brain hypothesis (Machiavelli intelligence hypothesis), which states that the human brain rapidly evolved to cope with such complex social relationships. He then examined the validity of this hypothesis from the viewpoint of "the prediction of the brain size" among primates. With respect to the size of the brain, Dunbar analyzed the correlation with various indicators in various species using the ratio of the neocortex to the total brain size. As a result, he found a high correlation of the ratio of neocortex only in the group size, thus supporting the social brain hypothesis. This analysis is based only on correlation, and causality is not sufficiently examined. However, this correlation suggests that highly developed brain processing capabilities are required to facilitate communication in human relations within a group.

7.3 Strategic Characteristic of Communication and Game Theory

Due to the interdependent relationships of communication and strategic relationships in human social relations, communication and various consequences accompanying it may become complicated. For example, despite the fact that individuals in society are acting rationally, uncertain results may ensue due to the irrational behaviors of some members of a social group. To exemplify this, first consider the simpler two-person relationship and discuss the problem of the prisoner's dilemma.

Table 7.1 Interdependent relationships in the prisoners' dilemma

		Prisoner 2	
		Not confess	Confess
Prisoner 1	Not confess	1 year each	10 years for 1 and 3 months for 2
	Confess	3 months for 1 and 10 years for 2	8 years each

Source Luce and Raiffa (1957)

The prisoner's dilemma is presented as the following situation (Luce & Raiffa, 1957). Suspects 1 and 2 are arrested under suspicion of being accomplices of a crime, but they are not indicted yet because of the lack of conclusive evidence. If both of them do not confess the crime, they will be prosecuted for incidental minor crimes, so each may be sentenced to a year in prison. Given that at least one confession is necessary to prove the guilt of the two people, the prosecutor interrogates them separately and informs them that if only one person confesses, the sentence of the confessed suspect is reduced to three months. Moreover, the person who does not confess would be sentenced to ten years. If both of them confess, they will be sentenced to eight years each. Thus, each suspect has two options: denial and confession. To make this situation easier to understand, we present Table 7.1.

Let us consider this situation from the theory system called "game theory." This theory was originally proposed by von Neumann and Morgenstern (1947). Based on the theories that have evolved since then, it has been applied extensively in the fields of psychology, sociology, biology, and economics, among others. Basically, in this theory system, each individual in society estimates the other's intention and makes strategic and rational decisions.

In the situation of the prisoner's dilemma, for suspects 1 and 2, given that confessing is advantageous, regardless of whichever option the other person chooses, confessing is superior to defecting. In game theory, we call this a dominant strategy. From the viewpoint of individual rationality, it is desirable for both of them to choose the confessing option as the dominant strategy. In addition, when both persons confess by taking the dominant strategy, a kind of equilibrium state is brought about. This is because, in that situation, even if one person has changed the choice, this will not be advantageous for the decision maker. The state of such a confession and confession combination is said to be in the Nash equilibrium (Nash, 1951). The Nash equilibrium is a condition wherein, for all the members, while the option for other members is the same as the given one, even if another option has been chosen alone, no improvement of gains is observed for each member. In the prisoner's dilemma, the state of the combination of a confession and confession is the only Nash equilibrium.

However, the state of the Nash equilibrium in which both persons are betrayed and confessed is not necessarily the most desirable for suspects (while it is the most desirable for the prosecutor). Given that the state where both of them are cooperating for denial is more desirable for each other than the state where they are in the Nash equilibrium. In prisoner's dilemma, the state where both persons are defecting the allegation is said to be Pareto optimum. Pareto optimum is a state in which no one can increase his/her gain without reducing the utility of a person. In the prisoners' dilemma, Pareto optimum exists even in two states (confession, denial), (denial, confession), but the Pareto optimum state does not match the Nash equilibrium that arises based on the dominant strategies of both. Furthermore, if both persons choose denial from the Nash equilibrium, which is the consequence of individual rationality, the state of both can be improved (Pareto improvement), and this is the reason why this game situation is called a "dilemma." In this way, even if we take actions from the viewpoint of individual rationality, there may be situations that are considered irrational from a collective viewpoint.

When a prisoner's dilemma game is executed repeatedly, a cooperative relationship that makes the Pareto optimum is likely to hold true. In theory, we know that when the game is repeated infinitely, the cooperative behavior of silence can be realized as a single Nash equilibrium. Furthermore, it is theoretically derived by Kreps, Milgrom, Roberts, and Wilson (1982) that even if the game is repeated only a limited number of times, cooperative behavior can be realized if certain conditions are satisfied. In addition, Dawes and Thaler demonstrated that when players communicated regarding the situation they faced before playing the game, the feasibility of cooperation would be greatly enhanced; they concluded that the communication enhanced the sense of ethics among players (Dawes & Thaler, 1988).

Specific examples of prisoners' dilemmas appear in various problems, such as the labor relations of companies, military competition among countries, competition among companies, consumption of public goods, and declaration of taxes (Dixit & Nalebuff, 1991; Poundstone, 1993; Schelling, 1960). Under these circumstances, mutual communication enhances confidence in the other person, enhances mutual commitment, and promotes cooperative behavior.

Furthermore, in the prisoners' dilemma, there are many cases wherein the problems of the two persons are considered, but the prisoners' dilemma in a society with more participants is sometimes called a "social dilemma" (Fujii, 2003; Yamagishi, 1989). The social dilemma is a more general concept than the prisoner's dilemma. According to Dawes (1980), this is a social state under the following conditions:

1. Each individual can choose either cooperation or noncooperation.
2. Personally, choosing noncooperation over cooperation is more advantageous.
3. If everyone chooses noncooperation, the result is worse than the result generated when everyone chooses cooperation.

The social dilemma, for example, is recognized in various situations, such as the separate collection of garbage that is socially desirable but is troublesome personally, and everyone driving to the office, which causes traffic congestion of the city but is comfortable personally. Further, considering each country as an entity, the social dilemma can be recognized in the situations of the international society, where despite the fact that it is desirable for each country to respond to CO_2 reduction to overcome the global environmental problems, it remains advantageous not to reduce it from the perspective of the country's economic interests.

You cannot solve such social dilemmas by only considering personal rationality. The fact that even if what is true from the micro-viewpoint of each individual, it is not true from the macro-viewpoint of a group or society and you mistakenly apply a micro-viewpoint to a macro-event maybe called the "fallacy of composition" or "group fallacy." We have to avoid such fallacies when interpreting social phenomena. In addition, Kameda and Murata (2000) indicated that clarifying the micro–macro-relations, which are the relationships between macro- and micro-phenomena in a society, is crucial for the prediction of social behavior and the solution of various social dilemmas.

Even in such situations of social dilemmas, mutual communication between individuals contributes to the achievement of cooperation (Balliet, 2010). Communicating with each other is effective even if the result is just a conversation. In this sense, we can say that humans are drawing out mutual cooperative behavior through communication.

7.4 Theory of Economic Behavior and Interaction

7.4.1 Definition of Interaction

Through word of mouth (WoM) and other forms of negotiations, consumers are influenced by others as they make decisions under various social circumstances. The result of the decision making of individual's social actions, as described in Chap. 2, is determined by the function (mapping) from the option and state of the adopted social actions to the result, i.e., $f : A \times \Theta \rightarrow X$. When it is expressed by sets, it can be stated as $A \times \Theta = \{(a_i, \theta_k) | a_i \in A, \theta_k \in \Theta\}$. This means that the set having a pair set (a_i, θ_k) is defined as $A \times \Theta$. This set considers the order of any elements a_i of the set A of social actions and any element θ_k of the set Θ of situations as its elements. The simplest preference relation for the result is the binomial relation, such as which one of the two options is preferred. Considering the preference relation of the decision maker on the elements of the set X of results and considering the empirically observed relationships, such as which one is preferred, we can express the relationship with the symbol of $\succcurlyeq$. The preference structure can be defined by considering the set of all the ordered pairs (x_i, x_j) that make $x_i \succcurlyeq x_j$ (Ichikawa, 1983; Takemura, 2014).

From these facts, the set of sets combining five sets of the set A of social action options, set Θ of states, set X of results, mapping $f : A \times \Theta \rightarrow X$, the preference structure (X, R) can be thought of as expressing the problem of decision making, which depends on various conditions in a very simplified form. As described in Chap. 6, in the decision theory, this set $D = (A, \Theta, X, f, (X, R))$ can be expressed as an individual "decision making problem," and as an extension example of this decision making problem, we can deal with interaction among individuals.

Here, we consider the interactive behavior by further extending the individual decision making problem defined above. In the case of interaction situations and social decision making that considers multiple decision makers, let I be the set of decision makers. According to $i \in I$, the option A^i and the preference relation R^i are different, and then the decision making problem can be expressed as $\left(A^1, A^2, \ldots, A^n, \Theta, X, f, \left(X, R^1\right), \left(X, R^2\right), \ldots, (X, R^n)\right)$ and the like [see, e.g., Ichikawa (1983)]. For example, if you consider that the number of elements of the set of decision makers is two, this makes an ordinary two-person game, such as a prisoner's dilemma game, or an interaction between two persons. The interaction relationship of n people treated in social psychology and similarly the game situation of n people can be expressed simply by the above framework. Let us consider this structure as a decision making problem, taking as an example the two prisoners' dilemma (Luce & Raiffa, 1957) , which is well known in social psychology. Suspects 1 and 2 are arrested under suspicion of being accomplices of a crime, but they are not yet indicted because there is no conclusive evidence. If both of them do not confess the crime, they will be prosecuted for incidental minor crimes, so each will be sentenced to a year in prison. Given that at least one confession is necessary to prove the guilt of the two people, the prosecutor interrogates the two persons separately and informs them that if only one person confesses, the sentence of the confessed suspect is reduced to three months. In this case, the person who does not confess would be sentenced to ten years. If both of them confess, they will be sentenced to eight years each. Thus, each suspect has two options: denial and confession.

In the situation of this prisoner's dilemma, the behaviors of the other person (either silence or confession) can be considered elements of the set Θ of states, and the option of each person is also an element of option A^i. Furthermore, as the function of the set of this option and state, the result X is determined, and the so-called gain table in game theory is formed. In addition, you can think of a selection model, such as a utility theory, as a real-valued function that describes preference relations on this result X.

In this example, there are only two social actors, but considering the n-person games, and considering that there are ten actors, we know that the set of states can be the direct product of the set of ten-person options, which means that there are 1024 patterns. Furthermore, it exceeds 1 million with 20 actors, and patterns will exceed 1000 trillion with 100 actors. This is the case where there are only two options, but in the case where the number of options is four, we know that the patterns can exceed one million with only ten actors. In actual social actions, occurrence patterns may become less due to the occurrence of conformity and harmonization, but there are also enormous possibilities in combination, and statistically speaking, the probability of the occurrence of many patterns is not zero. Such enormous combinations have also occurred in an extremely simple social behavior model in which the other person's social action is the element of the set of states. Furthermore, if situational factors and knowledge are included, a more complicated world is constructed. Thus, in a realistic situation, there is a more complicated structure than that. In this abstraction, such a social situation, including a social situation and options of other social actors, among others, is treated for each social actor as shared knowledge. However, in actual social situations, the premise of such shared knowledge does not necessarily hold and there is a possibility that more complications would emerge.

7.4.2 Game Theory and Interaction

Previously, we have demonstrated that decision making in a situation without interaction can be treated by utility theory. Applying this theory to multiple decision making entities, we can consider what description is possible. Here, the utility function of any decision making entity must assume that there exists a certain function expressing its own preference relation, although it does not necessarily assume that the utility function with other persons is identical. This may be understandable. Indeed, it seems that one's own preferences and those of others are different and there is not much evidence that the utility of options is the same. That the utility function differs for each individual implies that the interpersonal comparison of utilities has fundamentally difficult problems, and this is what makes interactive decision making difficult.

Based on the idea of this utility theory, there exists a theoretical system called game theory to explain the confrontations and conflicts of multiple humans. In game theory, the development of the theory is remarkable, and recently, variations such as metagames and hyper-games have emerged (Takigawa, 1989). Research based on game theory is not limited to just theoretical research; verification by experiments is performed frequently, and these studies provide very effective ideas by which we can understand the decision making processes in light of various conflicts. Next, on the basis of the game theory, let us consider decision making in the situation where the interests are completely conflicting.

Human economic behavior is formed under the interdependence relationships among various decision making entities. Social exchange theory sees the interdependence relationship from the viewpoint of resource exchange. This theory has been adopted by some researchers in social psychology and sociology. The social psychology of Thibaut and Kelly (1959) also posits this relationship. Game theory is a mathematical theory having an affinity with such a social exchange theory.

Game theory describes the economic behavior proposed by mathematician Von Neumann and economist Morgenstern. It is also adopted in modern theoretical economics, sociology, psychology, and engineering. As this theory can describe interdependence relationships mathematically, it can express various social phenomena simply, and as a result, the predictability of theories increases (von Neumann & Morgenstern, 1944, 1947).

Here, we describe the definition of the standard-type game in game theory. A game defined by the relationship between options (strategies) and utilities (gains) of individual members (called players) involved in decision making is called a game in strategic form or a game in normal game (Okada, 2011). Given that an interdependence relationship is composed of two or more persons, when it is abstracted as an *n* persons game, such a game is defined by the following set of elements (Okada, 2011):

$$G = (N, \{S_i\}; i \in N, \{f_i\}; i \in N).$$

Here, (i) $N = \{1, 2, \ldots, n\}$ is the set of players (individual decision makers), (ii) S_i is the set of the player i's selectable behaviors (behavior options) or strategies, and (iii) f_i is a real-valued function on the direct product $S = S_1 \times S_2 \times \cdots \times S_n$ and expresses the player i's utility function (gain function). In addition, for all strategies $s = (s_1, s_2, \ldots, s_n)$, when

$$\sum_{i=1}^{n} f_i(s_1, \ldots, s_n) = 0$$

holds, it is called a zero-sum game. When a constant K exists and

$$\sum_{i=1}^{n} f_i(s_1, \ldots, s_n) = K$$

holds true, this is called a constant-sum game. In the von Neumann–Morgenstern's linear expected utility theory system, this becomes a constant-sum game. However, as the utility function has uniqueness with respect to a positive linear transformation, it results in a zero-sum game. A game that is not a zero-sum game is generally called a nonzero-sum game.

7.4.3 *Basic Concepts of Game Theory*

First, we explain the fundamental concepts of game theory, such as the best response, the Nash equilibrium, and Pareto optimum.

7.4.3.1 Best Response

First, let us define "best response." You can say that the strategy $s_i \in S_i$ of a player i is the best response to the set $s_{-i} = \left(s_{1,} s_2, \ldots, s_{i-1}, s_i, s_{i+1}, \ldots, s_n\right)$ of the strategies of other $n-1$ players when

$$f_i(s_i, s_{-i}) = \max_{t_i \in S_i} f_i(t_i, s_{-i})$$

holds true. Here, $f_i(s_i, s_{-i})$ is the utility when you adopt a strategy (option) S_i under the given condition of a strategy of others. The strategy of maximizing your utility is considered to be the best response.

7.4.3.2 Nash Equilibrium

The Nash equilibrium is the state of the combination when all the players have the best responses to their strategies. In an n persons game $G = (N, \{S_i\}; i \in N, \{f_i\}; i \in N)$, when the strategy s_i^* is the best response to the set s_i^{**} of strategies of other players for all the players $i (= 1, \ldots, n)$, the set $s^* = \left(s_1^*, \ldots, s_i^*, \ldots, s_n^*\right)$ of the player's strategies is at a Nash equilibrium point. In other words, the set s^* of strategies, for which the following relationship holds for all players $i (= 1, \ldots, n)$, constitutes the Nash equilibrium point (also referred to simply as the equilibrium point) expressed as

$$f_i\left(s^*\right) \geq f_i\left(s_i, s_i^{**}\right), \quad \forall s_i \in S_i.$$

At the Nash equilibrium point, all players have no incentive to change their strategies from that state.

7.4.3.3 Pareto Optimum

Regarding Pareto optimum, that the set $s = (s_1, s_2, \ldots, s_n)$ of the player's strategies is the Pareto optimum in a n Parsons game $G = (N, \{S_i\};\ i \in N, \{f_i\}; i \in N)$ means that for all players $i (= 1, \ldots, n)$.

There is no set $(t_1, \ldots, t_n) \in S$ of strategies that $f_i(t_1, \ldots, t_n) > f_i(s_1, \ldots, s_n)$ holds. In other words, whatever strategy is changed, if the utility of everyone does not increase, it will be the Pareto optimum. Conversely, if the change of a player's strategy does not lower everyone's utility but increases someone's utility, that state is not the Pareto optimum.

7.4.4 People's Decision Making in Prisoner's Dilemma Games

The abstract definition of a prisoner's dilemma game is as follows. The pay-off matrix of the prisoner's dilemma described above is also given below. Here, each gain can be interpreted as the utility when adopting each strategy (option). In a prisoner's dilemma game, the Nash equilibrium point (a net strategy equilibrium point with only the strategy of D or C, where strategies are not combined with probability) is the only combination of D and D. The only other combination is the Pareto optimum point (Table 7.2).

As for the prisoner's dilemma game, as we have previously explained, the qualitative property is when the relationship of $T > R > P > S, 2R > (S + T)$ holds. In the prisoner's dilemma games, the problem has to do with why the cooperative responses can be obtained. From the viewpoint of game theory, the set of the nonco-

Table 7.2 Pay-off (utility) matrix of the prisoner's dilemma games

	C	D
C	*R*, *R*	*S*, *T*
D	*T*, *S*	*P*, *P*

operative responses of confessions and denials constitutes a Nash equilibrium point, but not the Pareto optimum. All combinations, except the Nash equilibrium point, constitute the Pareto optimum state. In this prisoner's dilemma game, when considering only options C and D, the Nash equilibrium point and Pareto optimum point are completely unmatched.

In the situation of prisoners' dilemma games, what kind of decisions do people tend to make? Scodel, Minas, Ratoosh, and Liptz (1959) promised to give 20 sets of experiment participants monetary compensation (1 cent per point) for the gains in the game and had the experiment participants playing several prisoners' dilemma games from 30 to 50 trials without interactive communication among them. The researchers found that the responses of the experiment participants were likely to fall into the Nash equilibrium of the co-poverty and the tendency became stronger as the trial proceeded to the latter half. Many past studies have shown that in a situation without communication, the results of prisoners' dilemma games tend to fall into the Nash equilibrium (Colman, 1982, 1995; Davis & Holt, 1993). From these, researchers of decision making behavior in prisoners' dilemma games have examined how individuals can have the Pareto optimum state of a co-prosperity state without falling into the Nash equilibrium of a co-poverty state and found the following strategies: using cooperative motivations by teaching and interactive communication (Deutch, 1973); drawing the cooperation of the other person by tit for tat strategies in which you cooperate basically but retaliate against noncooperation (Axelrod, 1984); and changing the cognitive framework of decision making problems (Shafir & Tversky, 1992). In the social dilemma literature that extends the prisoners' dilemma to a larger number of individuals, there have been many investigations on the factors preventing the co-poverty conditions (Yamagishi, 1989). However, for the measures to solve those dilemmas, not many definitive ones have been proposed, and it has been pointed out that it is quite difficult to solve those problems by manipulating psychological factors, especially in the case of social dilemmas (Yamagishi, 1989).

In order to study decision making in conflicts, many experiments on prisoners' dilemma games have been conducted in the field of social psychology. Many studies have examined games from the viewpoint of "cooperation or competition" and investigated what kind of experimental teaching might lead to cooperation or collaboration (e.g., Deutch, 1973). However, in the first place, we must focus on the fact that in prisoners' dilemma games, the state of the zero-sum-type "competition" is not included at all. Interestingly, however, in prisoner's dilemma games, despite the fact that the competition state of having to win against the other person has not been expressed at all, people might conceive the game situation as though it is a competitive situation like a zero-sum game and then make decisions (Minas, Scodel, Marlowe, & Rawson, 1960; Saijo & Nakamura, 1995).

The situation of prisoners' dilemma games is frequently observed in social situations related to economic conflicts. For example, we can point out the substantial collapse of the Organization of Petroleum Exporting Countries' (OPEC) price agreement (Dixit & Nalebuff, 1991). In the 1970s, OPEC signed a price agreement and raised crude oil price from the level of less than $3 per barrel in 1973 to more than $30 per barrel in 1980. Some energy experts cautioned that if this situation did not change, crude oil price would rise sharply, but after that, crude oil price suddenly dropped. At the beginning of 1986, it was $10, and it seemed that the price agreement collapsed. The substantial collapse of this price agreement occurred because each OPEC member gave priority to its own interests, greatly increased the production of crude oil to lower the price, and sold large amounts to make money. In other words, although the state of the price agreement was close to the Pareto optimum, each country sought profit and eventually fell into the Nash equilibrium of co-poverty. Hence, the price agreement is desirable from the standpoint of consumers, but it is obviously undesirable for individual member countries of OPEC.

Other examples of prisoners' dilemma games appear in various problems, such as the labor relations of companies, military competition among countries, competition among companies, the consumption of public goods, and the declaration of taxes (Davis, 1970; Dixit & Nalebuff, 1991; Poundstone, 1993; Schelling, 1960). In particular, the prisoner's dilemma in a society, where the participants are not two persons but many people, is called a social dilemma. According to Dawes (1980) , the social dilemma is the situation that satisfies the following conditions:

1. Each individual can choose either cooperation or noncooperation in the situation.
2. For each individual, it is more advantageous to choose noncooperative behavior than cooperative behavior.
3. However, the result of all members choosing noncooperative behavior that is advantageous for themselves is worse than the result of all members choosing cooperative behavior.

The situation, called a "tragedy of the commons" (Hardin, 1968), is an example of a social dilemma. This refers to the situation, wherein pasture is lost and all members fall into the co-poverty relationship if too much cattle are raised in common areas. Social dilemmas, such as garbage dumping and garbage problems, sightseeing by private cars and environment pollution, commuting by car and traffic congestion, the problem of air conditioner use and electricity consumption, gold mining, and environmental pollution, are examples of situations wherein personal interests conflict with the interests of the entire group consisting of three or more people. The characteristic of the social structure in this social dilemma is that, in many cases, it is not necessarily the restricted exchange wherein the person receiving the resources matches with the person providing such resources, but the general exchange without such matching. Under such situations, considering how to solve social dilemmas is very difficult.

7.4.5 *Analysis of the Repeated Prisoner's Dilemma Games*

In the prisoner's dilemma game, in the case of a one-time game, the Nash equilibrium point constitutes only a combination of betrayal (D) and betrayal (D), and cooperation (C) cannot be derived from the viewpoint of individual rationality. What if this prisoner's dilemma game is repeated endlessly? Before thinking about this, let us consider a simpler trigger strategy.

In the trigger strategy, we first take C, after that we take C as long as the other person takes C, and if the other person takes D, then we take D. In the trigger strategy, the following theorem exists (Okada, 2011).

7.4.5.1 Theorem on Trigger Strategy

In an infinite repeating prisoners' dilemma, if a discount factor $\delta\,(0 < \delta < 1)$ for the player's future gains is $\delta \geq (T - R)/(T - P)$, the set of trigger strategies constitutes the Nash equilibrium point of the repeating game.

Proof If two players use the trigger strategy, as the discount is gained (utility), the sum $\mathrm{U_T}$ of the players is given as

$$U_T = \sum_{i=0}^{\infty} R\delta^i = R/(1 - \delta)$$

Here, the reason we considered the discount gain sum is that we assumed the discounts for future gains (utility). Consider that the discount factor δ takes a value from 0 to 1. If player 1 changes the behavior to D in the t th time ($t = 1, 2, \ldots$) of the game, the other person keeps taking D from the $t + 1$-th time. Therefore, the discount gain sum U_{D} of the player 1 from the t th time is given as

$$U_{\mathrm{D}} = \mathrm{T} + \sum_{i=1}^{\infty} P\delta^i = T + \delta P/(1 - \delta).$$

The condition of δ for U_T to be greater than or equal to the value of the U_{D}'s formula is expressed as

$$\delta \geq (T - R)/(T - P).$$

This also holds for player 2, so the theorem is proven (Q.E.D.).

7.4.5.2 Theorem on the Tit for Tat Strategy

Next, we consider the tit for tat strategy. In this strategy, you take C first, after that you take C as long as the other person takes C, and if the other person takes D, you take D; however, if the other person starts taking C, you take C. Regarding the tit for tat strategy, the following theorem holds (Okada, 2011).

In an infinite repeating prisoner's dilemma game, if the discount factor δ for the future gains is $\delta \geq \max(((T-R)/(T-P)), ((T-R)/(R-S)))$, the set of the tit for tat strategies constitutes the Nash equilibrium point of the repeating game.

Proof If the two players take a tit for tat strategy together, as the players' discount gain (utility) sum,

$$U_{\text{TFT}} = \sum_{i=0}^{\infty} R\delta^i = R/(1-\delta)$$

is given (this is the same as in the trigger strategy). If player 1 changes the behavior to D in t th time ($t = 1, 2, \ldots$), player 2 chooses D in $t+1$-th time and player 1 can choose C or D. In order for the discount gain sum to not become large regardless of how player 1 leaves from the tit for tat strategy, the following three conditions must be satisfied.

(Condition 1) Taking the behavior sequence (t and $t+1$) of length 2, the discount gain sum of player 1 when taking C in two times is equal to or greater than that when taking D only in the t-th time. This means that

$$R + \delta R \geq T + \delta S,$$

which is equivalent to $\delta \geq (T-R)/(R-S)$.

(Condition 2) When taking an infinite sequence (after the t-th time), the discount gain sum of player 1 when the players always take C is equal to or more than that when only player 1 takes the D in the t-th time and both continue taking D after the $t+1$-th time. This means that

$$R/(1-\delta) \geq T + \delta P(1-\delta),$$

which is equivalent to $\delta \geq (T-R)/(T-P)$.

(Condition 3) The discount gain sum of player 1 when player 1 always keeps taking C is equal to or more than that when player 1 only takes D in the t th time. After the $t+1$-th time, both keep taking D $m(\geq 1)$ times, and then (after $t+1+m$-th time), only player 1 takes C.

For any natural number $n \geq 2$, this is equivalent to

$$R + \delta R + \cdots + \delta^{n-1} R + \delta^n R \geq T + \delta P + \cdots + \delta^{n-1} P + \delta^n S$$

(however, $n = m + 1$).

Assuming that the n-order polynomial of δ is

$$f_n(\delta) = (R - S)\delta^n + (R - S)\delta^{n-1} + \cdots + (R - S)\delta + (R - T),$$

this proves that the above formula is nonnegative for any natural number $n \geq 2$, which is equal to proving that $R + \delta R + \cdots + \delta^{n-1} R + \delta^n R \geq T + \delta P + \cdots + \delta^{n-1} P + \delta^n S$ holds. That is, $R + \delta R + \cdots + \delta^{n-1} R + \delta^n R \geq T + \delta P + \cdots + \delta^{n-1} P + \delta^n S \Leftrightarrow f_n(\delta) \geq 0$ is given.

Initially, assuming $n = 2$, we prove whether formula (5) holds. Note that $f_n(\delta)$ is a monotonically increasing function of δ, when $\delta \geq \max(((T - R)/(T - P)), ((T - R)/(R - S)))$ is assumed.

(A) In the case of $T + S \geq P + R$, $\delta \geq (T - S)/(R - S)$ is given, so $f_2(\delta) \geq 0$.
(B) In the case of $P + R \geq T + S$, $\delta \geq (T - R)/(R - S)$ is given; therefore, $f_2(\delta) \geq 0$ holds. Next, for any $n \geq 2$ $f_{n+1}(\delta) - f_n(\delta) = (R - S)\delta^{n+1} - (P - S)\delta^n$ holds.
(C) In the case of $\delta \geq (P - S)/(R - S)$, $f_n(\delta)$ is an increasing function of n.
(D) In the case of $\delta \leq (P - S)/(R - S)$, $f_n(\delta)$ is a decreasing function of n.

In the case of (C), from $\delta \geq \max(((T - R)/(T - P)), ((T - R)/(R - S)))$, $f_n(\delta) \geq f_2(\delta) \geq 0$.

In the case of (D), for any $n \geq 2$,

$$f_n(\delta) \geq \lim_{n \to \infty} f_n(\delta) = \delta(R - P)/(1 - \delta) + R - T$$

is given. Therefore, if $\delta \geq \max(((T - R)/(T - P)), ((T - R)/(R - S)))$ holds, $f_n(\delta) \geq 0$ is given. Therefore, the theorem is proven (Q.E.D.).

In this way, in an infinite repeating prisoner's dilemma game, the cooperation-based tit for tat strategy also becomes the Nash equilibrium. However, in a finite repeating game, the opposite is true and the betrayals constitute the Nash equilibrium.

7.4.6 Computer Simulation of Repeating Prisoner's Dilemmas and Reciprocity

Numerous experimental studies have been conducted on prisoners' dilemma experiments as reported earlier (Pruitt & Kimmel, 1977), and their results indicate that the responses of all experiment participants do not fall into the Nash equilibrium point.

In a finite repeating prisoner's dilemma game, the combination of the tit for tat strategies does not constitute the Nash equilibrium point, but what is the extent of its adaptability? Axelrod (1984) conducted a computer tournament and examined which program is the strongest in a finite repeating prisoner's dilemma game. The results indicate that the tit for tat strategy gained the highest score and can widely create the state of interactive cooperation. The results of this tournament suggest that considering reciprocity is important in prisoner's dilemma games.

The behavioral principle that reciprocity shows is observed in cooperative behaviors among individuals who are not connected by blood relationships in living organisms. Gouldner (1960) cited the norm of reciprocity as the most important and fundamental social norm that makes social exchange possible. Meanwhile, evolutionary psychologist Cosmides (1989) asserted that humans are sensitive to actions that violate the norm of reciprocity and inherently have cognitive mechanisms to detect the existence of such violations. For example, in the inference tasks called "four-card problems" that examine logical relationships, the reason why a person is easily able to solve the problem with the type of normative violation even if it has the same theoretical structure is that humans possess inherent cognitive mechanisms that can detect when others violate the norm of reciprocity. According to a certain technique, when one anticipates a huge request being refused initially, then he/she makes a small request instead. It is a request technique that expects reciprocity for the requester's concession and which, for example, is sometimes used for requesting donations.

7.4.7 Framing of Decision Making Problems in Game Behavior

Thus far, the experimental results of the prisoner's dilemma game, so far, show that people's choices are likely to fall into a co-poverty state, which can easily lead to the Nash equilibrium. Does this mean that the Nash equilibrium has been realized because the participants of the game are selfish and rational like homo economicus assumed in economics? In the field of economics, predicting interactive behavior from the viewpoints of the Nash equilibrium and dominant strategy in game theory has been established (e.g., Davis & Holt, 1993; Kandori, 1993), but experimental facts that are contrary to such an assumption have been presented.

Minas et al. (1960) showed the experiment participants a gain table, shown in Table 7.3, promised to give them monetary compensation (1 cent per point) based on the results of their choices, and conducted 30 trial experiments. As shown in Table 7.3, if the two persons choose dominant strategies A1 and B1, their gains will be 4 cents each. This choice pattern constitutes the Nash equilibrium and is a Pareto

Table 7.3 Pay-off matrix of the nonzero-sum game with Pareto optimum and Nash equilibrium by the dominant strategies of two persons

		Choice of player 2	
		B1	B2
Choice of player 1	A1	4 cents each	1 cent for 1 and 3 cents for 2
	A2	3 cents for 1 and 1 cent for 2	0 cent each

Source Minas et al. (1960)

optimum. Therefore, such a choice pattern is pretty obvious from the perspective of an egoist who hopes to maximize his own gains. Surprisingly, the choice patterns of A2 and B2 of the worst cases that are neither the Nash equilibrium nor Pareto optimum occupied 47% of the total trials. Furthermore, in 15 trials of the latter half, the choice patterns of A2 and B2 exceed half, suggesting that people are not mere egoists but simply hope to win even if they suffer a loss by comparing themselves with others.

Saijo and Nakamura (1995) named the phenomenon a "spite dilemma," which occurs when the decision makers do not adopt the dominant strategy by behaving with emphasis on the ranking among decision makers and not for their own gains. In their study, they had seven experiment participants playing games and gave them monetary compensation according to the results. They found that this phenomenon of the spite dilemma appeared frequently. In this experiment, many of the participants in the experiment reported that "they wanted to get money, but did not want to lose to other participants."

From the viewpoint of game theory, these results are very strange phenomena. Assuming that individual utility functions are different, and even assuming an ordinal utility function that allows some monotonic increasing transformation for their own gains, we can neither predict this experimental result from the viewpoints of the Nash equilibrium and dominant strategy nor predict it from the Pareto optimum criteria. If some kind of gain transformation has been done, this transformation should be a function that covers both the gains of others and their own gains, such as the "ranking" that participants answered in the experiment by Saijo and Nakamura (1995). Now, let us examine the problem of the psychological transformation of such gains.

The framing effect, also described in Chap. 5, is a phenomenon that deviates from the fundamental assumption of this game theory (Tversky & Kahneman, 1981). The framing effect deviates from the descriptive invariance, such that if the subjects described by a theory are identical, then their theoretical descriptions and prediction results are identical as well.

In decision making, the psychological reconstruction of decision making problems is called framing and the so-called framing effect refers to the phenomenon wherein the results of the decision making depend on the manner of framing (Tversky & Kahneman, 1981, 1986). For example, in the case where a patient is considering whether or not to undergo surgery, the patient's decision when the doctor tells him that the probability of success is 80% differs from the decision when he is told that the probability of failure is 20%. The phenomena of such framing effect can be observed frequently in personal decision making (Takemura, 1994; Tversky & Kahneman, 1981, 1986). We can assume that such framing is one of the causes of the spite dilemma.

According to interpersonal interdependence theory, first proposed by Kelly and Thibaut (1978), given a pay-off matrix in the experiment game, people transform this matrix psychologically to make it an effective one and base their decision on such a matrix. In interpersonal relationships, there are various types of transformations from the given matrix to the effective matrix. Moreover, they assert that the problem of the transformation from a given matrix to an effective one can be considered as a framing problem in interpersonal decision making. The existence of the spite dilemma shows that there is a kind of bias in the manner of framing.

Decision makers' way of framing has a major impact on decision making in both zero-sum and nonzero-sum games (Colman, 1995). For example, Eiser and Bhavnani (1974) examined how decisions are made by changing the teaching of the situation content of prisoner's dilemma games with the same structure. They taught experiment participants of each condition the situation settings as a problem of economic negotiation, negotiation between countries, and human relations, respectively, and in experiments the other persons of the game (cooperators) dealt with them using a tit for tat strategy. Their results indicate that the rate of cooperation choices realizing the Pareto optimum is higher for the cases where they are taught as a problem of the negotiation between countries or human relations than the cases where they are taught as a problem of economic negotiation. In the case of this experiment, the utility function is highly likely to be different among the situations due to the differences in units of a gain between situations. However, even if it is true, in the case of a prisoner's dilemma game, it cannot be predicted from game theory that different results occur from different situations. Therefore, the results of this experiment indicate the framing effect.

The problem of framing is extremely important when thinking about decision making in conflicts. This is well understood when considering the disarmament issues between the countries (Tversky, 1994) . Let us assume that two countries are negotiating the reduction of the number of missiles. Reducing the number of missiles in one country is understood as losses from the present situation, and the reduction of the number of missiles in the other country is understood as gains. According to Tversky (1994) , the impact of perceived losses is about twice as much as the impact of perceived gains, so one country's reduction of two missiles is balanced with reduction of one missile of the other country. As countries think in such a way, reaching an agreement becomes very difficult. Although one country gaining or losing from the current point may be generally acceptable, how to grasp the current point depends largely on the framing of the problem situation.

This framing is amplified by emotion or may affect emotions. Many researchers have pointed out that the feelings of anger have an important influence on decision making in conflict situations (Frank, 1988; Heitler, 1990; Ohbuchi, 1993). From a common sense perspective, it is difficult to think of a conflict situation without emotions, such as anger. In the process of conflict accompanied by emotions like anger, one can assume a chain reaction that the emotion of anger is influenced by the framing of the situation and the emotion of anger affects the way of framing. For example, we daily observe that when we perceive the other person's selection behavior as a betrayal or attack, anger is intensified; from such anger, the interpretation of the situation is distorted in a specific direction and the situation gets bogged down further.

However, framing is an important factor in decision making in economic behavior and this poses a serious problem to us. This is because, the fact that the interpretation of the decision problem, even if it is the same decision problem, differs between individuals and even within the individual, when we negotiate or make social decisions, it is difficult for us to think rationally about the methods of mediation and agreement. Several methods have been proposed for mediation in situations where interests are conflicting (Saeki, 1980). For example, regarding negotiation issues, Nash (1950) proves that there is a rational mediation scheme if you accept a group of axioms, even though individuals' utility functions are not the same. As a result, it is a rational negotiation solution to choose the point that maximizes the product of the utility of both sides from the negotiation origin. However, the negotiation solution of the negotiation experiment deviates from a rational solution, such as the Nash solution (Miyajima, 1970). Thus, it is highly likely that the framing effect is involved in this experimental result. If the framing effect always occurs, it will be almost impossible for the negotiation origin to remain unchanged.

Therefore, in order to think about satisfactory mediation and agreement, it will be necessary to undertake some sharing of cognition in consideration of the framing problem. The problem of the sharing of cognition, which is the basis of comparison with others, is considered important not only for negotiation problems, but also for thinking about the problem of social decision making widely (Sen, 1982; Takemura, 1993). Many people may think that discussions may be the most appropriate method for the sharing of cognition.

Then, can we really share cognition by discussion? The situation is not so easy. Kameda (1994) reviewed the research of the simulated trial experiment of the jury he and others had conducted and concluded that group discussion does not actually facilitate the sharing of cognition. Kameda's research does not deal with decision making in the situation of conflicts of interest but decision making in which the legitimacy of a trial matters. However, past studies have pointed out that the sharing of cognition is difficult even in the situation of conflicts of interest. Takemura and Eguchi (1995) examined what kind of information each individual considered during the discussion in joint decision making and found that as the discussion progresses, the ratio of the information considered and shared between the two sides tends to increase slightly, but its value is quite low. These findings suggest the difficulty of sharing cognition by discussion. Then, can we resolve the conflict rationally? This problem is our future task. Regarding the problem of the sharing of cognition and agreement, it seems that there is a clue in elucidating sympathy phenomena and its mechanism as considered by modern thinkers, such as Hume and Smith (Takemura, 1993). In order to elucidate the problem, we will have to wait for further developments in this area.

Box 1: Kenneth Joseph Arrow

was born in 1921 and deceased in 2017. He received Ph.D. at Columbia University in 1951. He was awarded the Nobel Prize in economics in 1972 with John Hicks. He contributed to social choice theory, especially the so-called Arrow's impossibility theorem, and general equilibrium theory of economics.

Photograph: Wikipedia

7.5 Word of Mouth and Diffusion Processes

As described earlier, decision making behaviors change through communication in social processes. Especially in consumer behavior research, the term "word of mouth" (WoM) is used as an important concept of consumer interaction. WoM information is transmitted not only in interpersonal communication but also on the Internet. Recently, some companies have provided services through which WoM information is systematically posted on the Internet. With the emergence of such services, it is expected that consumers' WoM information will have a greater influence on consumer behavior and carry a significant weight on corporate management as well.

Rogers (2003) published a theory on the diffusion process and stated that ordinary advertisements affect the publicity of products, but that WoM has an influence on the selection stage of products. In 1999, the Japan Advertising Agencies Association "Advertising Function and Role Subcommittee" conducted in-depth interviews with twelve men and women in their 20s–50s who live in the Tokyo metropolitan area. They found that the WoM information, such as advice from friends, acquaintances, and salespersons, influences one's final judgment (Japan Advertising Agencies Association "Advertising Function and Role Subcommittee," 2000). According to this survey, new products are recognized mainly through articles and advertisements posted in newspapers and television, traffic advertisements, mail advertisements, and detailed information on products of interest is usually provided in catalogs, specialized magazines, and Internet Web sites. The survey findings indicate that the final judgment is influenced by WoM.

Consumers also conduct some kind of evaluation after decision making, and cognitive dissonance may be involved in this process. In addition, consumers who conducted evaluation will influence other consumers through WoM, among others. In the research of Sundaram and Webster (1999), who examined how consumer behavior is influenced by the WoM based on the hypothesis that the effect of WoM changes depending on familiarity with the brand, they examined which is more effective on the purchase intention involving WoM and attitude to the brand: the case where the degree of familiarity is low or the case where it is high. Their experiment plan was a 3×2 design with two factors: three levels for the factor of WoM and two levels for the factor of the brand familiarity. The factor of WoM has three levels, namely positive, negative, and no WoM. In addition, the factor of the brand familiarity has two levels, namely brands with high familiarity and brands with low familiarity. For this factor of familiarity, in the preliminary research by 34 experiment participants, rating tests for brand familiarity was conducted, and two brands were selected: Whirlpool as a brand with high familiarity and Electrohome as a brand with low familiarity. The experiment participants were 120 college students studying business relations in the Middle West (however, 72. 4% of them had jobs in some form, 40% were men, and the average age was 26.4 years old). Each experiment participant was randomly assigned to one of the six conditions, and each was given a scenario of a fictitious situation from the experimenter. In that scenario, the experiment participant was in a situation wherein he/she was collecting information on various brands at retail stores, etc., to purchase an air conditioner. As the manipulation of the brand, the experiment participant was shown an advertisement with only its brand name changed (stuff actually used in the market), and then the WoM information was given.

The positive WoM is as follows: "You gathered various pieces of information and told a friend that you were going to buy a new air conditioner and mentioned the brand you found (specify the actual name) to that friend. The friend said that the brand is quite reliable, easy to install, durable, and its maintenance and energy costs are low."

The negative WoM is as follows: "You gathered various pieces of information and told a friend that you were going to buy a new air conditioner and mentioned the brand you found (specify the actual name) to that friend. The friend said that the brand's reliability is low, it easily breaks down, is difficult to install, is not durable, and its maintenance and energy costs are high."

Finally, on the condition without WoM, they were given only the name of the brand and not presented any kind of WoM information.

After these manipulations, the experiment participants carried out brand evaluation over 11 items with the nine-stage scale method. For example, they evaluated items, such as the possibility of buying the brand, how much the brand is liked, and the quality of the brand. In order to check the manipulation, the experiment participants subsequently answered the items, such as the brand's familiarity, evaluation of the WoM (evaluation on whether it was a positive or negative WoM), and to what extent this fictitious scenario was realistic. Checking answers to such items is to check whether the experimental factors targeted by the experimenter are really manipulated. Results of the manipulation check indicate that the manipulation is effective and that the reality is relatively high (for the reality, the average value is 6. 28 out of 9 points).

As a result of analyzing 11 evaluation items by the multivariate statistical analysis of the principal component analysis, 83.5% of the data dispersion was explained by two factors. When the axis transformation called the varimax rotation was performed, the first factor was interpreted as purchase intention, and the second factor was interpreted as attitudes to brands. Therefore, we calculated the average value of items showing a high load on each of these factors and plotted the average value in each condition about purchase intention and attitudes to brands, as shown in Figs. 7.3 and 7.4.

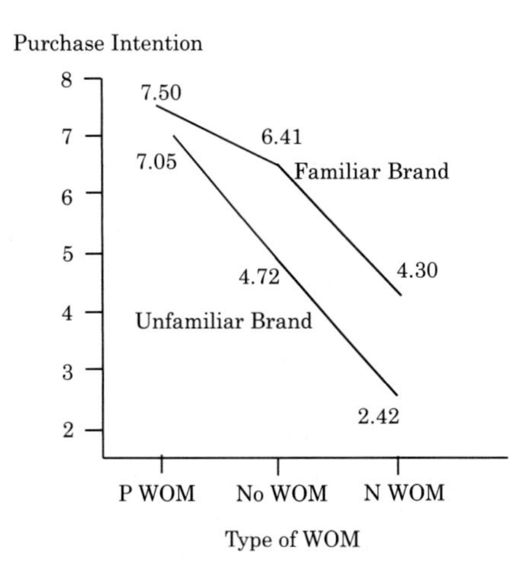

Fig. 7.3 Interaction effect of brand familiarity and WoM on purchase intention. *Source* Sundaram & Webster (1999)

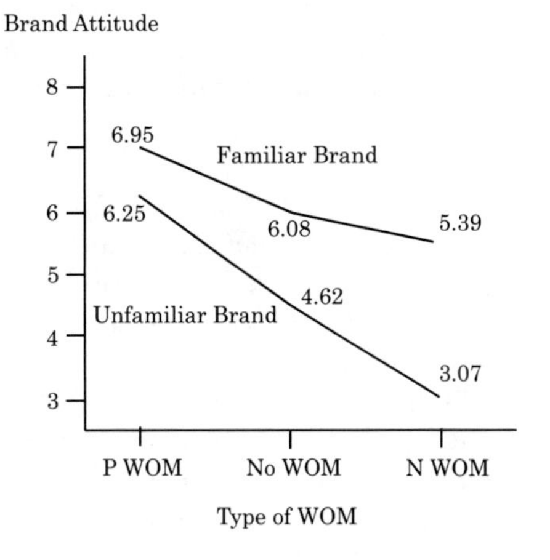

Fig. 7.4 Interaction effect of brand familiarity and WoM. *Source* Sundaram & Webster (1999)

From these figures, we can see the impact of WoM. The positive WoM leads to the highest purchase intention of brands, attitudes to brands are also favorable, and the negative WoM performs the worst in purchase intention of both brands and attitudes. As for familiarity, the brands with high familiarity are linked to a higher purchase intention and are more favorable in attitudes to brands than those with low familiarity. Notably, the results differ significantly depending on the combination of the type of WoM and familiarity. That is, as for a positive WoM, regardless of whether the brand‘s familiarity is high or low, it does not have a huge impact on purchase intention or attitudes. For a negative WoM, the purchase intention of the brand and attitude does not decrease much when the brand‘s familiarity is high, but if its familiarity is low, they tend to decline considerably. Hence, if a brand‘s familiarity is high, regardless of a positive or negative WoM, this can still contribute to sales promotion. In particular, even if a negative WoM occurs, the degree suppressing sales is lower than the case where its familiarity is low. Conversely, it will be fatal for a brand with low familiarity if a negative WoM occurs.

In experiments and surveys on the effect of the WoM, Herr et al. (1991) conducted psychological experiments on the product evaluation of personal computers and found that the effect on product evaluation is greater when people are told experience stories by the WoM than when they are told with printed media. This indicates that a bad WoM is more effective than a good WoM.

So, is there a person who is likely to be an information sender or the receiver of WoM? In their two-step theory of communication, Katz and Lazarsfeld (1955) stated that there are two kinds of participants: those who are likely to be opinion leaders with huge influence on others and those who are likely to be followers. According to this theory, in public opinion formation on political content, ordinary people (followers) are not influenced directly by media-derived information, but they are likely to be influenced through the mediation of opinion leaders.

Rogers describes the theory based on a more detailed category than the two-step theory of Katz and Lazarsfeld; the former describes how innovation, such as new concepts and ideas of technology, can spread and diffuse along with products. In the survey, he divided the adopters in the process of diffusion and spreading of innovation into standard five categories and found that when plotting the number of these adopters over the time axis, a bell-shaped distribution appeared (see Fig. 7.5), and the curve of cumulative frequency distribution became an S curve. Each category is labeled as follows: "innovators," "early adapters," "early majority," "late majority," and "laggards," in order of adoption. Innovators are a group of people who adopt new ideas and technologies first, have a high ability to deal with risks and uncertainty, are tolerant of new ideas, and interact with other innovators. They are not respected much compared with the members of other groups, but they play an important role in the diffusion process. Early adapters are equivalent to opinion leaders, and they have the highest degree of influence on the surroundings compared with the members of other categories; they are respected, conduct choice and adoption in socially more adaptive manner than innovators do, and maintain their position as opinion leaders. Meanwhile, the members who are early adapters are cautious about innovation, only adopt ideas after a certain period of time, and interact frequently with their friends.

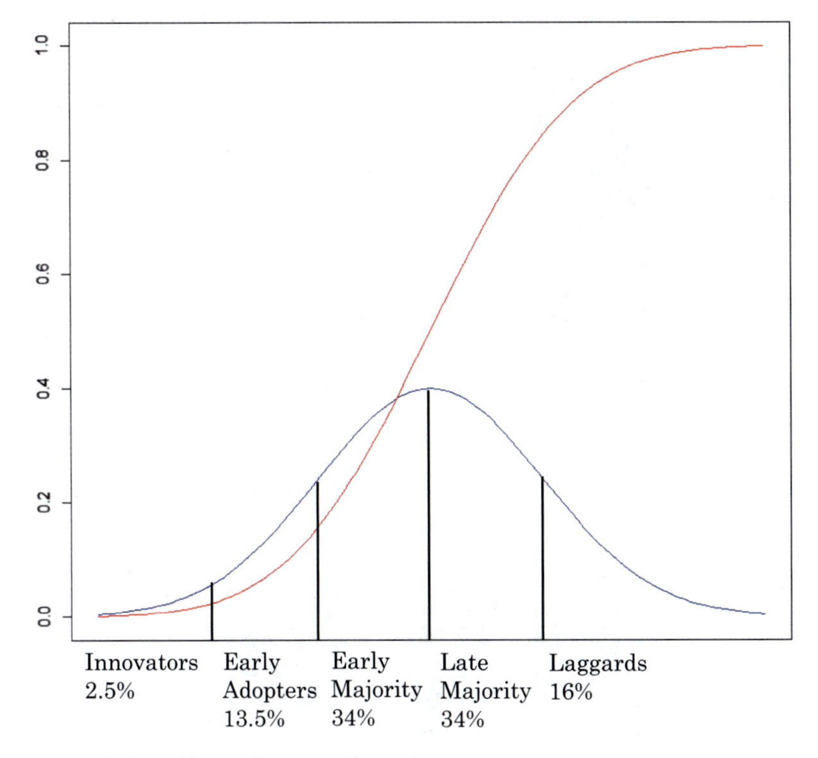

Fig. 7.5 Diffusion of innovations according to Rogers

Members in the category of late majority, who adopt ideas after the adoption by average people, are wary of innovations even though innovation is already spreading midway. Compared with other categories, laggards' social influence is extremely small, and they dislike change, prefer tradition, and adopt innovations only at the end.

7.6 Decision Making Process in Diffusion

Rogers (2003) proposed a conceptual model in which the diffusion of innovation is based on the five decision stages of individuals (Fig. 7.6). According to this model, the first stage is the knowledge stage, where individuals become aware of the existence of innovation and understand how it works. The second is the attitude or persuasion stage, where individuals form favorable or nonfavorable attitudes to innovation. The third is the decision stage, where individuals decide whether they should use innovation or not. The fourth is the implementation stage, where individuals use innovation. The final fifth is the assurance or confirmation stage, where individuals re-evaluate and confirm the decision on innovation that has already been made.

In addition to the above conceptual classification, let us consider these stages of decision making in more detail. People's decision making processes represented

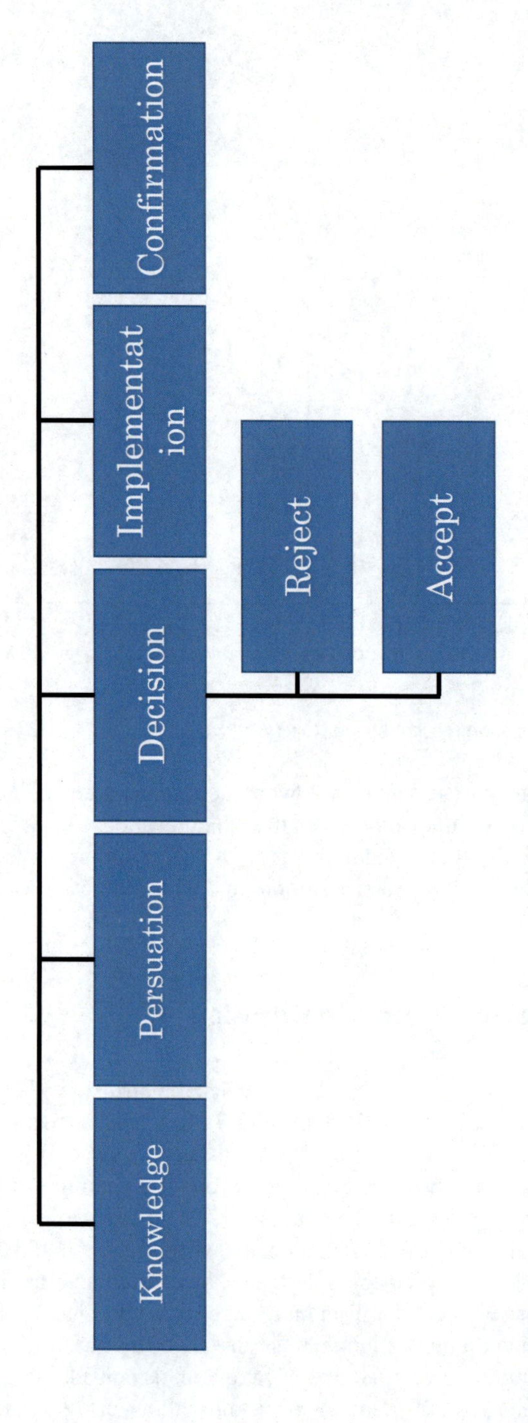

Fig. 7.6 Five stages in the decision innovation process according to Rogers

by economic behavior, among others, can be understood as a kind of problem-solving process, and they search for related information to achieve the target state in the problem solving (Takemura, 2009). Therefore, information search is performed before Rogers' knowledge stage and attitude stage. In general, the information search starts from an internal information search, during which the relevant information is retrieved in one's memory, and if there is not enough information in the memory, an external information search is made from external information sources.

The external information search may be performed not only for information search for decision making, such as purchase behavior of a specific product, but also for the purpose of satisfying knowledge desire or satisfying curiosity. In the information search of the WoM, there are cases where a consumer who is thinking about buying a personal computer asks friends and clerks about the reputation of manufacturers or brands, or asks what kind of points to consider, and in daily conversation, there are cases where he hears of the reputation of manufacturers and brands and rumors. The effect of WoM is expected to become particularly noticeable with respect to a bad reputation. For example, news reports on agricultural crop damage and poisoning caused by environmental pollution and defects of automobiles, etc., greatly affect the relevant purchase psychology and purchase behavior of consumers.

According to Sugimoto (1997), the occurrence of WoM is based on the situation where the information sender is strongly involved in a certain product, has a special interest in the product, and enjoys talking about it to others, and on his desire to show his knowledge or to help others. On the one hand, that person may have a desire to gain psychological stability by telling others about the defects of products or the poor quality of services. For the senders of such WoM information, there is no information on their purchase or objective gains, but it seems that there is a function to satisfy the psychological desire as shown above. On the other hand, for the receiver of the WoM information, the WoM has the function of supporting purchase decision making in the sense that it provides knowledge and information about the product. Sugimoto (1997) points out that when the receiver of the WoM information does not have enough knowledge of the product or when the judgment of product selection is not easy or not convincing even though he has some information, purchase decisions become easier by using such WoM information.

What kind of influence does interpersonal interactions, such as WoM, have on decision making? What kind of decision making is influenced by it? Takemura and Eguchi (1995) conducted experiments on the cooperative decision making related to package tours using information monitoring methods that record the process of information acquisition in decision making. In the experiments, the following three conditions were set for 106 male and female university students: a condition of making decisions in face-to-face situations, a condition of making decisions while communicating with one another using video phones while in rooms away from others, and a condition of making decisions while communicating using a telephone. The basic experiment procedure was that experiment participants of each condition freely searched for information on package tours, using a personal computer, and made decisions while talking with each other. After making decisions, the experiment participants were tested how much they remembered the options, and they also answered questionnaires on the psychological state during the decision making process. Experimental results indicated no noticeable differences between the conditions regarding the information search behavior, but the participants revealed that they felt the least risk of decision in the face-to-face situation. Meanwhile, the subjects with the telephone condition felt the highest risk of decision. This suggests that face-to-face, nonverbal communication affects the perception of risks. In addition, subjects under the face-to-face condition had a tendency to feel that they had made less concession on their own opinions in cooperative decision making and that they had asserted their opinion, compared with subjects with other conditions. The fact that facing the other person has some effect indicates that, apart from the related information of purchase decision making, nonverbal information can also influence interaction.

7.7 Economic Behavior and Social Welfare

7.7.1 *Does Money Make People Happy?*

What is it about life that makes you happy? Many people are interested in this problem. "Living" is a series of decisions, so this problem leads us to the question of what kind of decision making makes us happy. Thus, what kind of decision making can make us happy? In this chapter, we consider this problem based on the results of recent decision making research.

In traditional economics, the problem of happiness is not usually handled directly, but there seems to be an implicit premise that an increase in wealth leads to happiness, but is this implicit premise really right?

Certainly, according to survey studies so far, the super-rich, whose annual income exceeds ten million dollars, shows somewhat stronger feelings of happiness than the ordinary people, and extremely few of them report negative emotions (Diener, Horowitz, & Emmons, 1985). However, according to Diener et al. (1999), the correlation between income and happiness was extremely weak with the result from 0.10 to 0.15. Diener and Oishi (2000) analyzed the results of GNP and life satisfaction survey in 15 countries from 1965 to 1990 and found that GNP and life satisfaction had few relationships and Brazil and Argentina with GNPs that were less than half of Japan had higher satisfaction than Japan. Considering such things, we can say that an increase in income and wealth does not necessarily lead to happiness. Indeed, people may be unhappy if they have no money, but money does not guarantee happiness.

7.7.2 *Can Rational People Be Happy?*

What kind of decision making in daily life can lead to happiness? We can find many articles on what kind of decisions we should make. The normative decision theory explaining what kind of decision making is desirable recommends that when making a decision, we should look for decision making options as much as possible, evaluate them one by one, and adopt an option with the best evaluation. Utility theory, which is a representative theory of economics and decision theory, also suggests that such decision making is rational. On the contrary, selecting an option that is not given the highest evaluation among all the options is "irrational." We may say that rational people look for an option given the highest evaluation and select it.

In addition, people who make rational decisions make consistent decisions. This is because if one chooses from noticeable options, he/she is making inconsistent haphazard decisions and it cannot be said that such a person is doing rational decision making.

According to previous studies, the decision making processes of many people rarely go through rational procedures in the sense described above (Abelson & Levi, 1985; Gigerrenzer & Selten, 2001; Simon,1957). Especially, Simon (1957) pointed out that humans do not make decisions based on the principle of maximization or optimization, which helps humans pick out the best choice from as many options available as possible. Rather, because of the limitations in their information processing capabilities, humans make decisions based on the principle of satisfaction wherein they seek options that can satisfy them at a certain level. He points out that humans only have "bounded rationality." Making decisions based on the principle of satisfaction only with this limited rationality means that if all the necessary conditions of several conditions are satisfied, humans are likely to adopt that option on the spot, thus making an inconsistent decision.

In this way, many people do not find the best option. Rather, they tend to make inconsistent and irrational decisions. However, even if that is true, are people who make rational decisions happier than those who do not?

Schwartz et al. (2002) have sociopsychologically expanded Simon's theory and clarified the relationship between happiness and clinical adaptation through surveys. They created a psychological measure using the questionnaire of "Regret and Maximization Scale" and named those who use the decision method based on Simon's (1957) satisfaction standard in decision making "satisficer" and those who collect information on options as much as possible and try to adopt the best option among them "maximizer." They then investigated the psychological tendencies of both. As a result of this survey, they found that those with maximizer tendencies had positive correlations with depression, perfectionism, and the degree of regret and had negative correlations with happiness, optimism, satisfaction with life, and self-respect. Schwartz (2004) reported that maximizers regretted options they had not adopted, so they were less satisfied with their decisions than the satisficers. Such a trend is also revealed in the research of Hisatomi et al. (2005), who surveyed Japanese participants from Hokkaido to Okinawa.

The research done by Schwartz and others suggests that a rational "homo economicus," who makes optimal decisions by considering as many options as possible, has an unhappy sensation in the modern society or may become clinically maladaptive. Such findings are very interesting. That the process of searching for options that might bring the greatest evaluation and making consistent rational decisions does not necessarily lead to happiness may also cast doubt on the concept of "freedom of choice," a fundamental value in modern capitalism. In a capitalist society, there are as many options (goods, etc.) as possible and one can freely choose from those many options, but it can also be said that the "freedom of choice" can also make humans unhappy. In addition, the sense of values of "choosing the best" may also make people unhappy. For example, choosing the best option for going to a university, finding employment, or getting married is a natural value in modern society. However, in thinking about "happiness," it may be necessary to re-examine these values.

7.7.3 Does Strategically Having Relationships with People Lead to Happiness?

In modern society, it seems that not only the value of seeking rationality of individual decision making is admired, but also strategic negotiation and interaction with others are recommended. For example, we know some enlightening books that encourage strategic interpersonal interaction based on game theory, and some scholars and business persons have argued based on game theory that "it is important for Japanese people in the future to live strategically without getting caught in conventions or fixed human relationships." Will this lead to happiness than doing strategic interactions following game theory?

In considering this, we first discuss the prisoners' dilemma commonly used in game theory. According to the experimental results of the game expressing the prisoner's dilemma, from actual human decision making results, one can hardly fall into the Nash equilibrium predicted by game theory (Sanfey, 2007). From game theory, confession is recommended and keeping silent may be foolish, but does following the normative instruction of game theory lead to the happiness of both persons? Recent neuroscientific research on decision making suggests that taking cooperative behavior like silence leads to people's happiness in prisoners' dilemma games. For example, Ringling, Sanfey, Aronson, Nystrom, and Cohen (2004) used the prisoner's dilemma game and examined the brain activity when the game result was presented using a brain image device of fMRI. Their findings suggested that positive emotion derived from cooperative behavior led to the activity of a site that was deeply involved in the reward of the striatum of the brain in the same manner as actual remuneration. Furthermore, according to the recent review of the brain scientific research of game behavior, cooperation and acting fairly without being moved by gains activate the brain's reward system (Sanfey, 2007; Tabibnia, & Liberman, 2007; Takemura, Ideno, Okubo & Matsui, 2008). In this way, it seems that people feel joy and happiness more while cooperating and behaving fairly with others rather than when they are trying to maximize their gains by behaving strategically.

Furthermore, among the empirical studies of the basic assumption of game theory by economists, it is the final proposal negotiation game (ultimatum bargaining game) that has recently attracted the most attention and has been debated on vigorously. In general, this game is also applicable to games wherein negotiation is conducted in finite times alternately. However, for the sake of simplicity, we explain it below with a game to negotiate only once.

First, there is a certain amount of C yen (e.g., 1000 yen) that can be divided by two persons. Two players are assigned either of two roles: the side (player A) who presents the distribution amount or the side (player B) who decides whether or not to accept the presentation. Player A presents k yen to B as a distribution amount. For player A, C yen–K yen is the amount to be distributed to him. After this presentation, player B decides whether or not to accept A's proposal. After accepting A's proposal, A can obtain C yen–K yen and B can obtain K yen. However, if B refuses A's proposal, neither of them will receive anything.

In noncooperative games theory by using the backward induction to reach a subgame perfect equilibrium, it can be easily derived that as the optimal response player A presents the lowest amount (1 yen), close to 0 yen, to player B, and player B accepts it. It is clearly financially better for B to accept the presentation rather than refuse A's presentation with the acquisition amount becoming 0 yen. As a result, A can acquire almost 100%. However, the results of the experimental studies by Güth, Schmittberger and Schwarz (1982) showed that the average of the proposed amount of player A was about 35% of the total amount, which was considerably different from almost 0% of the prediction of game theory. Moreover, when player A proposed an extremely small amount of money, player B refused the proposal (about 20% of the total). Hence, the behavior of players in this game was far from a rational human image seeking opportunistic self-interests in game theory. Güth et al. interpreted this result as the indication of the subjects emphasizing fairness in distribution rather than economic benefits and thought that the explanation based on theory of distribution fairness was more appropriate than that based on game theory. After their research, numerous experimental studies were conducted, and controversies between the group of researchers defending the basic assumptions of game theory and the group of researchers defending distribution fairness arose (Roth, 1995). After many such studies have been conducted, it is now accepted that at least in the final proposal negotiation game, fairness and other social psychological factors are important determinants in addition to economic interests (Heap & Varoufakis, 1995; Fukuno and Obuchi, 1997). However, even if people emphasize factors other than the size of money, the basic assumption of game theory holds.

7.7.4 Happiness and Peak–End Rule

The normative decision theory to guide correct rational decision making explains that we should grasp the decision making problem, evaluate what kind of final result arises if choosing a certain option, and choose the option with the highest evaluation. However, according to Kahneman (1999), humans are unable to predict the evaluation concerning the results of choosing an option beforehand. Aside from the low human ability of predicting results, this means that it is very difficult to predict subjective conditions of one's own. According to Kahaneman, the evaluation of the experienced result is determined by the difference between the two points of the average of the evaluation of "the highest or the lowest moment (peak)" to the evaluation at "the end." This is called the peak–end rule.

The authors conducted very interesting experiments, the results of which suggest the peak–end rule (Redelmeier, Katz, & Kahneman, 2003). They asked 682 patients who underwent colonoscopy treatment to report the momentary feeling during the examination and the feeling after the examination. Colonoscopy is an examination conducted by inserting a tube with a small camera at the tip into the rectum from the anus. They carried out the usual examination for half of the patients; for the other half, they left the instrument in the body for about one minute in addition to the usual examination. Both groups had the same discomfort at the peak time, but the former normal examination group had a lower degree of discomfort at the end than the latter group and the discomfort time was shorter. Nevertheless, the former examination group, who showed a large difference between the peak and the end, showed more discomfort as a whole in the retrospective evaluation. In addition, the latter group had a higher proportion of agreeing to the notification of endoscopic re-examination after 5 years than the normal examination group. In general, one may think that the normal examination is more desirable than the examination leaving the instrument in the body for an additional one minute in the preliminary evaluation. However, in this way, the happiness or unhappiness that people felt after the experience varied from what was actually experienced at that time and what was evaluated beforehand.

7.7.5 Decision Making and Multi-attribute Value

How can people live happily? Certainly, it seems that some financial resources are necessary to achieve a happy life. However, no matter how much money we earn, it is not necessarily true that we can be happy because of it. The relationship between money and happiness is not so strong.

In order to be happy, as it is assumed in traditional economics, should we instead find an option that gives the greatest value and make consistent and rational decisions? Such actions of seeking rationality seem to lead one to unhappiness rather than happiness. It also does not seem true that strategically having relationships with others as assumed in game theory leads you to happiness. Rather, people seem to search for a moderate option and adopt it if they can be satisfied to a certain extent and feel happiness by behaving cooperatively and fairly with the other people. Other research results reveal that happy individuals become happier by being kind to others (Otake, Shimai, Tanaka-Matsumi, Otsui, & Fredrickson, 2006).

In the field of psychology, critical thinking had been temporarily admired, and critical consideration had been emphasized. Critical consideration of things may be necessary to some extent to prevent us from being deceived or outdone by other people. However, thinking too deeply or critically can have us doubt people or, indeed, make our lives unpleasant. If we make decisions without thinking sufficiently, we may have to regret being deceived or making too irrational decisions. Having a regretful experience could make us unhappy. However, even if such an unhappy event has been experienced, if the difference between the peak and the end does not change drastically, the happiness eventually remembered is not so low.

Computer simulation results indicate that even if we make decisions contrary to rational decision making as suggested here, the final result is not so different from the result of rational decision making (Gigerenzer, Todd, & The ABC Research Group, 1999; Payne, Bettman, & Johnson, 1993; Takemura et al., 2015). Although it may seem like irrational decision making, I do not think that it is a bias or an error that should be corrected immediately. We may find happiness in life by associating ourselves with others based not on strategies but on personality, and with a moderate desire and experiences of various failures, reflect on them as we try to live a positive and joyful life.

Box 2: Gerd Gigerenzer

was born in 1947. He received Ph.D. from the University of Munich in 1977. He studies human judgment and decision making in uncertainty. He proposes some heuristics that make us smart and rational in an uncertain world using the terms of the adaptive toolbox. He is Director Emeritus of the Center for Adaptive Behavior and Cognition (ABC) at the Max Planck Institute for Human Development.

Phograph: Franz Johann Morgenbesser

https://www.flickr.com/photos/vipevents/14010756670/in/photolist-nm5MQb-nCyXP1-nEmzht

Summary

- People are communicating in an economic society. Communication between two people consists of a two-way interaction, but as the number of members increases from tripartite to quadripartite, the combination of interactions increases dramatically and the way of communication becomes dramatically complicated as well.
- The interaction in economic behavior is complicated, and the theory dealing with this interaction is game theory.
- In game theory, concepts like the Nash equilibrium point and the Pareto optimum are important.
- Communication can be done strategically in circumstances where the interests of each person are conflicting and they are assumed to behave strategically as in game theory. Nonetheless, mutual communication increases trust, brings about cooperation, and suppresses the emergence of social dilemmas to a certain extent.
- In communication in society, there exist not only the so-called WoM between people but also large-scale information transmission by newspapers, magazines, television programs, and so on (also called mass communication) and information transmission by other forms of media, such as the Internet. Regarding the influence of mass communication, Katz and Lazarsfeld proposed a two-step theory, which states that communication passes through two steps from opinion leaders to followers.
- Regarding the diffusion process of innovations, such as ideas and technologies, Rogers (2003) proposes that there are decision making processes, such as knowledge, attitudes, decision, execution, and beliefs, and there are five groups of people involved: innovators (those who adopt new innovations early), the early adapters, the early majority, the late majority, and the laggards, who finally introduce innovations and popular things.

Recommended Books and Reading Guides for More Advanced Learning

- Luce, R. D., & Raiffa, H. (1957). *Games and decisions: Introduction and critical survey*. New York, NY: Wiley (Dover Publications; Reprint version).

This book is a classical text of game theory dealing with interaction in economic behavior carefully explained from an introductory point of view.

- Rogers, E. M. (2003). *Diffusion of innovation* (5th ed). New York, NY: Simon and Schuster.

This book explains how innovations, such as new technology, new products, and new lifestyles, spread and become popular. According to the author, innovation increases uncertainty by increasing the people's decision making options.

- Camerer, C. F. (2003). *Behavioral game theory: Experiments in strategic interaction.* In Roundtable Series in Behavioral Economics. Princeton, NJ: Princeton University Press.

This book provides the first substantial and authoritative effort to integrate game theory and psychological experimental study. The authors use psychological principles and hundreds of experiments to develop mathematical theories of interactions, and clarifying relationships with game theory and traditional economics, and also explain the findings of many behavioral game experiments.

- Ogaki, M., & Tanaka, S. (2017). *Behavioral economics: Toward a new economics by integration with traditional economics*. Tokyo: Springer.

This book explains the basic concepts of behavioral economics in an easy-to-understand manner, clarifying relationships with traditional economics, and also explains the findings of recent neuroeconomics.

References

Abelson, R. P., & Levi, A. (1985). Decision making and decision theory. In G. Lindzey, & E. Aronson (Eds.), *The handbook of social psychology* (3rd ed., Vol. 1, pp. 231–309). New York, NY: Random House.

Axelrod, R. (1984). *The evolution of cooperation*. New York, NY: Basic Books.

Balliet, D. (2010). Communication and cooperation in social dilemmas: A meta-analytic review. *Journal of Conflict Resolution, 54*(1), 39–57.

Colman, A. M. (1982). *Game theory and experimental games* (1st ed.). Oxford, UK: Pergamon Press.

Colman, A. M. (1995). *Game theory and its applications in the social and biological sciences* (2nd ed.). Oxford, UK: Butterworth-Heinemann.

Cosmides, L. (1989). The logic of social exchange: Has natural selection shaped how humans reason? *Studies with the Wason selection task, Cognition, 31,* 187–276.

Davis, M. D. (1970). *Game theory: A nonthechnical introduction*. New York, NY: Basic Books.

Davis, D. D., & Holt, C. A. (1993). *Experimental economics*. Princeton, NJ: Princeton University Press.

Dawes, R. M. (1980). Social dilemmas. *Annual Review of Psychology, 31,* 169–193.

Dawes, R. M., & Thaler, R. H. (1988). Anomalies: Cooperation. *Journal of Economic Perspectives, 2*(3), 187–197.

Deutch, M. (1973). *The resolution of conflict: Constructive and destructive processes*. New Haven, CT: Yale University Press.

Diener, E., Horowitz, J., & Emmons, R. A. (1985). Happiness of the very wealthy. *Social Indicators Research, 16,* 263–274.

Diener, E., & Oishi, S. (2000). Money and happiness: Income and subjective well-being across nations. In E. Diener & E. M. Suh (Eds.), *Culture and subjective well-being* (pp. 185–218). Cambridge, MA: MIT Press.

Diener, E., Suh, E. M., Lucas, R. E., & Smith, H. E. (1999). Subjective well-being: Three decades of progress. *Psychological Bulletin, 125,* 276–302.

Dixit, A. K., & Nalebuff, B. J. (1991). *Thinking strategically: The competitive edge in business, politics and everyday life*. New York, NY: W. W. Norton & Company.

Dunbar, R. I. (1998). The social brain hypothesis. *Evolutionary Anthropology, 6,* 178–190.

Eiser, J. R., & Bhavnani, K. K. (1974). The effect of situational meaning on the behavior of subjects in the prisoner's dilemma game. *European Journal of Social Psychology, 4,* 93–97.

Frank, R. H. (1988). *Passion within reason: The strategic role of the emotions*. New York, NY: W. W. Norton & Company.

Fujii (2003). *Shakaiteki jirenma no shohōsen: Toshi, Kōtsū, kankyō mondai no shinrigaku [Prescription for social dilemma: Psychology of urban, traffic, and environmental problems]*. Tokyo, JP: Nakanishiya Shuppan.

Fukuno, M., & Ohbuchi, K. (1997). Kōshōji no ninchiteki baiasu: Kotei shigen chikaku to kōsei baiasu no kiteiin no kentō [Cognitive biases in negotiation: The determinants of fixed-pie assumption and fairness bias]. *Shakai shinrigaku kenkyu [Japanese Journal of Social Psychology], 13*(1), 43–52.

Gigerenzer, G., Todd, P. M., & The ABC Research Group (Eds.). (1999). *Simple heuristics that make us smart*. New York, NY: Oxford University Press.

Gigerenzer, G., & Selten, R. (Eds.). (2001). *Bounded rationality: The adaptive toolbox*. Cambridge, MA: MIT Press.

Gouldner, A. W. (1960). The norm of reciprocity: A preliminary statement. *American Sociological Review, 25,* 161–178.

Güth, W., Schmittberger, R., & Schwarz, B. (1982). An experimental analysis of ultimatum bargaining. *Journal of Economic Behavior & Organization, 3,* 367–388.

Hardin, G. (1968). The tragedy of the commons. *Science, 162,* 1243–1248.

Heap, S. P. H., & Varoufakis, Y. (1995). *Game theory: A critical introduction*. London, UK: Routledge.

Heitler, S. (1990). *From conflict to resolution: Skills and strategies for individual, couple, and family therapy*. New York, NY: W. W. Norton & Company.

Herr, P. M., Kardes, F., & Kim, J. (1991). Effects of word-mouth and product attribute information on persuation: An accessibility-diagnocity perspective. *Journal of Consumer Research, 17,* 454–462.

Hisatomi, T., Isobe, A., Oniwa, T., Matsui, Y., Ui, M., Takahashi, N., & Takemura, K. (2005). Anshin to fuan no shakai shinri (IV): Ishikettei sutairu to shinrai tono kanrensei [Social psychology of security and insecurity (IV): Relevance between decision-making style and trust]. *Paper presented at the 46th Conference of the Japanese Society of Social Psychology* (pp. 230–231).

Ichikawa, A. (1983). *Ishiketteiron: Enjiniaringu saiensu kōza 33 [Decision-making theory: Engineering science course 33]*. Tokyo, JP: Kyoritsu Publishing.

Kahneman, D. (1999). Objective happiness. In D. Kahneman, E. Diener, & N. Schwarz (Eds.), *Well-being: The foundations of hedonic psychology* (pp. 3–25). New York, NY: Russel Sage.

Kameda, T. (1994). Shūdan ishikettei to shakaiteki kyōyūsei [Group decision making and socially shared cognition]. *Shinrigaku hyoron [Japanese Psychological Review], 37,* 367–385.

Kameda, T., & Murata, K. (2000). *Fukuzatusa ni idomu shakai shinrigaku: Tekiō ējento to shiteno ningen kaiteiban [Social psychology from a complex-system perspective: Humans as adaptive agents* (2nd ed.), Tokyo, JP: Yuhikaku Publishing.

Kandori, M. (1993). Gēmu riron ni yoru keizaigaku no shizukana kakumei [Quiet revoluciton by game therory]. In K. Iwai, & M. Ito. (Eds.), *Gendai no keizai riron [Modern Economic Theory]* (pp. 15–56). Tokyo, JP: University of Tokyo Press.

Katz, E., & Lazarsfeld, P. F. (1955). *Personal influence: The part played by people in the flow of mass communications*. New York, NY: Free Press.

Kelly, H. H., & Thibaut, J. W. (1978). *Interpersonal relations: A theory of interdependence*. New York, NY: Wiley.

Kreps, D. M., Milgrom, P., Roberts, J., & Wilson, R. (1982). Rational cooperation in the finitely repeated prisoners' dilemma. *Journal of Economic Theory, 27*(2), 245–252.

Luce, R. D., & Raiffa, H. (1957). *Games and decisions: Introduction and critical survey*. New York, NY: Wiley.

Minas, J. S., Scodel, A., Marlowe, D., & Rawson, H. (1960). Some descriptive aspects of two-person non-zero-sum games, II. *Journal of Conflict Resolution, 4*(2), 193–197.

Miyajima, M. (1970). Kōshō gēmu [Bargaining game]. In M. Suzuki (Ed.), *Kyōsō shakai no gēmu riron [Competitive society and game theory]* (pp. 169–225). Tokyo, JP: Keiso Shobo.

Nash, J. F. (1950). The bargaining problem. *Econometrica, 18,* 155–162.

Nash, J. F. (1951). Non-cooperative games. *The Annals of Mathematics, 54*(2), 286–295.

Ohbuchi, K. (1993). *Hito wo kizutukeru kokoro: Kōgekisei no shakai shinrigaku [Mind of hurting people: Social psychology of aggressiveness]*. Tokyo, JP: Saiensu Sha Co., Ltd.

Okada, A. (2011). *Gēmu riron shinban [Game theory (New edition)]*. Tokyo, JP: Yuhikaku Publishing.

Otake, K., Shimai, S., Tanaka-Matsumi, J., Otsui, K., & Fredrickson, B. (2006). Happy people become happier through kindness: A counting kindness intervention. *Journal of Happiness Studies, 7,* 361–375.

Payne, J. W., Bettman, J. R., & Johnson, E. J. (1993). *The adaptive decision maker*. New York, NY: Cambridge University Press.

Poundstone, W. (1993). *Prisoner's dilemma: John von Neumann, game theory, and the puzzle of the bomb*. New York, NY: Anchor Books.

Pruitt, D. G., & Kimmel, M. J. (1977). Twenty years of experimental gaming: Critique, synthesis, and suggestions for the future. *Annual Review of Psychology, 28,* 363–392.

Redelmeier, D. A., Katz, J., & Kahneman, D. (2003). Memories of colonoscopy: A randomized trial. *Pain, 104,* 187–194.

Ringling, J. K., Sanfey, A. G., Aronson, J. A., Nystrom, L. E., & Cohen, J. D. (2004). Opposing BOLD responses to reciprocated and unreciprocated altruism in putative reward pathways. *NeuroReport, 15,* 2539–2543.

Rogers, E. M. (2003). *Diffusion of Innovation* (5th ed.). New York, NY: Simon and Schuster.

Roth, A. E. (1995). Bargaining experiments. In J. H. Kagel & A. E. Roth (Eds.), *The handbook of experimental economics* (pp. 253–348). Princeton, NJ: Princeton University Press.

Saeki, Y. (1980). *"Kimekata" no ronri [Logic of decision]*. Tokyo, JP: University of Tokyo Press.

Saijo, T., & Nakamura, H. (1995). The "spite" dilemma in voluntary contribution mechanism experiments. *Journal of Conflict Resolution, 39,* 535–560.

Sanfey, A. G. (2007). Social decision making: Insights from game theory and neuroscience. *Science, 318,* 598–602.

Schelling, T. C. (1960). *The strategy of conflict*. Cambridge, MA: Harvard University Press.

Schwartz, B. (2004). *The paradox of choice: Why more is less*. New York, NY: Harper Collins.

Schwartz, B., Ward, A., Monterosso, J., Lyubomirsky, S., White, K., & Lehman, D. R. (2002). Maximizing versus satisficing: Happiness is a matter of choice. *Journal of Personality and Social Psychology, 83,* 1178–1197.

Scodel, A., Minas, J. S., Ratoosh, P., & Liptz, M. (1959). Some descriptive aspects of two-person non-zero-sum games, I. *Journal of Conflict Resolution, 3,* 114–119.

Sen, A. (1982). *Choice, welfare and measurement*. New York, NY: Basil Blackwell.

Shafir, E., & Tversky, A. (1992). Thinking through uncertainty: Nonconsequential reasoning and choice. *Cognitive Psychology, 24,* 449–474.

Simon, H. A. (1957). *Administrative behavior: A study of decision making process in administrative organization*. New York, NY: McMillan.

Sugimoto, T. (1997). *Shōhisha rikai no tame no shinrigaku [Psychology for understanding consumer]*. Tokyo, JP: Fukumura Shuppan.

Sundaram, D. S., & Webster, C. (1999). The role of brand familiarity on the impact of word of mouth communication on brand evaluation. *Advances in Consumer Research, 26,* 664–670.

Tabibnia, G., & Liberman, M. D. (2007). Fairness and cooperation are rewarding: Evidence from social cognitive neuroscience. *Annals of the New York Academy of Sciences, 1118,* 90–101.

Takemura, K. (1993). The effect of decision frame and decision justification on risky choice. *Japanese Psychological Research, 35,* 36–40.

Takemura, K. (1994). Furemingu koka no rironteki setsumei: Risukuka deno ishikettei no jokyo izonteki shoten moderu [Theoretical explanation of the framing effect: Contingent focus model for decision-making under risk]. *Japanese Psychological Review, 37,* 270–293.

Takemura, K. (2009). *Kōdō ishiketteiron: Keizai kōdō no shinrigaku [Behavioral decision theory: Psychology of economic behavior]*. Tokyo, JP: Nippon hyoron sha Co. Ltd.

Takemura, K. (2014). *Behavioral decision theory: Psychological and mathematical representations of human choice behavior*. Tokyo, JP: Springer.

Takemura, K., & Eguchi, A. (1995). Kotonaru tsūshin media de no sōdan ni yoru kyōdō ishikettei no katei: Terebi denwa, denwa, taimen jokyo no hikaku [Dyadic decision making process using different communication medias: Video phone, telephone, and face to face communication]. In *The Proceedings of the 36th Annual meetings of the Japanese Society of Social Psychology* (pp. 280–283).

Takemura, K., Haraguchi, R., & Tamari, Y. (2015). Tazokusei ishikettei katei ni okeru kettei houryaku no ninchiteki doryoku to seikakukusa: Keisanki simyurēshon ni yoru kōdō ishiketteiron teki kentō (Effort and accuracy in multi-attribute decision making process: A behavioral decision theoretic approach using computer simulation technique). *Ninchi kagaku [Cognitive Studies], 22,* 368–388.

Takemura, K., Ideno, T., Okubo, S., & Matsui, H. (2008). Shinkei keizaigaku to zentōyō [Neuroeconomics and frontal lobe]. *Bunshi seisin igaku [Japanese journal of molecular psychiatry], 8*(2), 35–40.

Takigawa, T. (1989). Meta gēmu, sūpā gēmu, haipā gēmu [Meta game, super game, and hyper game]. *Shinrigaku hyōron [Japanese Psychological Review], 32*(3), 333–350.

Thibaut, J. W., & Kelly, H. H. (1959). *The social psychology of groups*. New York, NY: Wiley.

Tversky, A. (1994). Contingent preferences: Loss aversion and tradeoff contrast in decision making. *Japanese Psychological Research, 36,* 3–9.

Tversky, A., & Kahneman, D. (1981). The framing of decisions and the psychology of choice. *Science, 211,* 453–458.

Tversky, A., & Kahneman, D. (1986). Rational choice and the framing of decisions. *Journal of Business, 59,* 251–278.

von Neumann, J., & Morgenstern, O. (1944/1947). *Theory and games and economic behavior*. Princeton, NJ: Princeton University Press.

Yamagishi, T. (1989). Shakaiteki jirenma no shuyō na rironteki apurōchi [Major theoretical approaches in social dilemmas research]. *Shinrigaku hyoron [Japanese Psychological Review], 32,* 262–294.

Chapter 8
Consumers' Preference Construction, Affects, and Neuroscientific Research: Research on Consumer's Preference and Neuromarketing

Keywords Preference construction · Mere exposure · Gaze cascade · Decision making process · Affect · Mood · Emotion · Prospect theory · Neuromarketing · Neuroeconomics

"Why did you choose that option?" When asked to explain their choice, consumers are likely to offer the following explanation: "I made the choice because I prefer it more than the other options." This basic explanation of the decision making process involved in making a choice or preference is widely accepted in real-life situations. Moreover, it is also commonly accepted in social science research on decision making. In this chapter, we provide an in-depth discussion of the preference construction. In this chapter, I explain the consumer's preference construction, the effects of feelings such as moods and emotions on the decision making process from the viewpoint of prospect theory, which has been studied vigorously, especially in the fields of behavioral economics and neuroeconomics. I explain the relation of prospect theory with the problems of affect, along with the related neuroscientific findings. Furthermore, I briefly describe the definition of affect and the development of affect research and explain the methods of affect manipulation and affect measurement. Next, I explain the influence of affects on the decision making process related to economic behaviors based on prospect theory and present the experiment examples related to it. Then, I theoretically examine the influence of affect on decision making based on the findings of neuroeconomics related to prospect theory and affect. Finally, I discuss the implications of these findings on consumer behavior research and marketing.

K. Takemura, *Foundations of Economic Psychology*,
https://doi.org/10.1007/978-981-13-9049-4_8

8.1 Preference Construction and Economic Psychology

Traditional economicsposits that economic behavior is predicted by the utility function that expresses preference relations (Ideno & Takemura, 2018). Hence, gaining a better understanding of the preference relations inferred from the choice relations can help determine the utility function. However, we have yet to fully examine the problems related to preference construction, namely, how preferences are formed and how such preferences are related to options. In fact, even though Nishibe (1975) already emphasized in the 1970s the importance of the problem of preference construction in his work entitled "Socio-Economics", this problem has only gained attention in the field of behavioral economics and behavioral decision theory in recent years.

According to psychological studies, preference construction depends on multiple factors, including an inherent fondness for something, past experiences as shown in operant conditioning, and the current situation. Thus far, the psychological research on preference construction has centered on the idea that one's current situation influences preference and preference relations (Lichtenstein & Slovic, 2006). According to Fujii and Gärling (2003), preference is made up of two components: (1) a part (contingent preference) formed depending on the situation, and (2) a stable part (core preference) formed by experiences until the time of decision making. The current research is biased toward the contingent preference.

Although our decision making has an aspect of continuity as it is repeated in our daily lives, the notion that preference is formed by a past choice has yet to be fully examined in the literature. Therefore, we aim to re-evaluate the meaning of choice in this chapter by focusing on the idea that preference is developed through repeated daily decision making processes. The aim of this chapter is to experimentally evaluate the influence on subsequent preference relations of the choosing actions that are not linked with preference.

8.2 Situation Dependence in the Preference Construction Processes

Situation dependence has been emphasized in psychological research on preference construction processes. The "preference reversal effect" (Lichtenstein & Slovic, 1971) and the "framing effect" (Tversky & Kahneman, 1981) are often cited as main examples of decision making that demonstrate situation dependence. In the context of behavioral decision theory and behavioral economics, these phenomena are not necessarily treated as instances of preference construction; nevertheless, both have important implications concerning preference construction.

According to the preference reversal effect, the preference relations vary according to the statement procedures of such relations. As shown in Table 8.1, the typical procedure of the preference reversal effect is a set featuring matching and preference tasks (Tversky, Sattath, & Slovic, 1988). In the matching task, the participants are

Table 8.1 Task examples used in the preference reversal effect

Choice tasks			Matching tasks		
	Traffic accident deaths	Cost		Traffic accident deaths	Cost
Measure *X*	500	$550,000	Measure *X*	500	?
Measure *Y*	570	$120,000	Measure *Y*	570	$120,000

Source Takemura (2014) and Tversky et al. (1988)

required to estimate the amount of measure *X*, which is equivalent to measure *Y* (570 people, $12 M). Assuming that the participant estimated $200,000, based on the results of the matching task, we can say that the preferences for measure *X* (500 people, $200,000) and measure *Y* (570 people, $120,000) are equivalent, so the choice of measure *Y* is predicted in the choice task. However, in that study, many participants actually chose measure *x*. Hence, the preference reversal effect is considered as a deviation of procedural universality, in which preference is reversed by the procedure of preference statement (Lichtenstein & Slovic, 1971).

Meanwhile, the framing effect is a phenomenon, wherein the decision making results vary depending on how differently the decision making problems are presented even for the same problem. Such an effect is considered a phenomenon that deviates from the descriptive universality, which indicates that the same decision making problems have the same result. For example, in the disease problem in Asia, which is a representative task, we have reported that choice results change depending on the inclusion of expressions, such as "saved" (or "died"), in the description of the presented options. Regarding these various phenomena on the situation dependence of decision making, theoretical developments include prospect theory (Tversky & Kahneman, 1992) and the contingent focus model (Fujii & Takemura, 2001; Takemura, 1994; Takemura & Fujii, 2015). By contrast, in decision research literature, not many studies have recently investigated the construction process by which stable preference such as liking is formed.

Although the preference reversal effect and framing effect, along with various theories presented, help explain the process of preference construction, they do not deal directly with such a problem. Therefore, instead of investigating studies on the problem of the situation dependence of decision making, we examine those studies dealing with the following problems: preference change after making a choice, the relationship between physical contact and the psychological gaze (viewing), and the notion that preference is formed by choice. These problems conflict with the commonly accepted premise in decision making research.

8.3 Exposure and Preference Construction

In making a decision, the basic premise is that options are perceived and actions are chosen. Even when choices are made in some vague way, or even when the attributes of different options are carefully examined, exposure to options is a prerequisite. In the following, we list the phenomena exemplifying the idea that being exposed to options, such as viewing and touching, facilitate preference construction.

8.3.1 Mere Exposure Effect

The "mere exposure effect" states that experiencing an object (or being exposed to it) helps a person form a preference and influences that person's subsequent choice of action (Zajonc, 1968). This effect describes a phenomenon, wherein the preference for an object increases by repeatedly experiencing that specific object. Using meaningless words and kanji for stimuli, among others, Zajonc (1968) demonstrated that the evaluation value of favorableness for objects increased after the repeated presentation of approximately five times. In addition, even in the two options in the forced choice task, which used a stimulus presented for a short time and another stimulus seen for the first time, the choice rate for the former stimulus shown in advance is significantly higher than the choice rate for the latter (Kunst-Wilson & Zajonc, 1980). One hypothesis employed to theoretically explain the mere exposure effect is perceptual fluency, which posits that processing becomes more "fluent" when objects are perceived through repeated exposure, resulting in one's increased affinity for that object. In addition, Zajonc (2001) attempted to explain the mere exposure effect in relation to conditioning, stating that a correspondence relationship is developed through the repeated experience of a certain stimulus.

The mere exposure effect attracted increased attention when many studies reported the preference construction under the condition of subconscious exposure to a stimulus (Bornstein, 1989; Bornstein, Leone, & Galley, 1987; Monahan, Murphy, & Zajonc, 2000; Murphy & Zajonc, 1993; Zajonc, 2001). Bornstein et al. (1987), for example, reported that preference is formed even with the stimulus presentation of about 4 ms. Given that the mere exposure effect is also observed even with a subliminal stimulus, he proposed affect–cognition theory (Zajonc, 1980), which states that the occurrence of affect is a prerequisite before cognition is established. However, the important point in relation to the current chapter is that preference is formed even with passive exposure to a stimulus.

8.3.2 *Gaze Cascade Effect*

Meanwhile, other studies examined the "gaze cascade effect," which hypothesizes the relationship between active gaze and preference construction based on the measurement of eye movements in two options under forced choice situations (Shimojo, Simion, Shimojo, & Scheier, 2003). The gaze cascade effect is substantially different from the mere exposure effect in that the former argues that preference is formed by an active gaze. As hypothesized in the preferential gaze method, gazes reflect two factors: seeing a more preferable object and liking it by seeing it (Fantz, 1963). Shimojo et al. (2003) used preference judgment tasks in examining the interrelation between looking at an object of interest and liking it by seeing it. Figure 8.1 (left) shows the bias of gazing when conducting the choice task of two face photographs. Figure 8.1 (right) presents the task of choosing a face photo with a rounder outline for the two face photographs. In their experimental results shown in Fig. 8.1 (left), when performing the preference judgment task, the likelihood of the line-of-sight pointing toward the chosen object increases just before a choice to draw a cascaded curve. Such findings lend support to the gaze cascade effect hypothesis. Under their experimental conditions, they reported that such an effect only occurs when a preference judgment is made as a result of setting conditions for the object (e.g., the roundness of facial contours in this case).

In their second experiment, Shimojo et al. (2003) examined the influence of active gazes on choice by manipulating the presentation time and position of the given stimulus. Specifically, using the presentation time and position of the face photographs as variables, the authors manipulated the participants' active gazes in the task of choosing which of the two face photographs was more preferable over the other. In terms of position, three conditions were set; Condition 1: the left and right, Condition 2: top and bottom, and Condition 3: center on the screen. In terms of presentation time, each face photograph was presented in two levels: 300 and 900 ms. In Condition 1, the face photographs were alternately presented one by one at the left and right positions of the screen. In Condition 2, the images were alternately presented one by one at the top and bottom positions. In Condition 3, the photographs were alternately

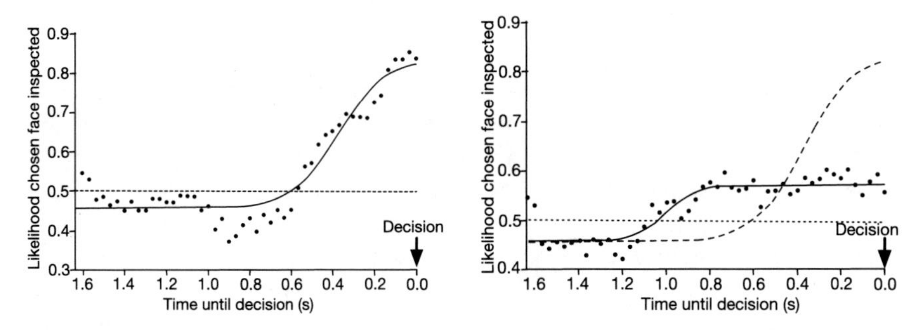

Fig. 8.1 Gaze ratio until choice. *Source* Shimojo et al. (2003)

presented one by one at the center of the screen. To manipulate the active gazes, the photographs were alternately presented at the positions of the left and right or top and bottom.

For the manipulation of gaze time, they used procedures in which one face photograph was presented for a short time (300 ms) and the other face photograph was presented for a long time (900 ms). They set three levels for the number of repetitions of the photograph presentation (2, 6, and 12 times, respectively). A sample experimental procedure is given as follows: In the case of the left and right presentation condition, the experiment participants gazed at the face photographs one by one, as presented on the left and right of the computer screen; on the final choice screen, on which two face photographs are presented, they expressed their preference by pushing either the left or right key. Their results indicated that the choice rate (approximately 60%) of the face photograph presented for a longer time was higher when the photographs were presented one by one under Conditions 1 and 2. When the photographs were presented under Condition 3, because the choice rate of the face presented for a longer time did not increase, they concluded that the active gazes were determinants of preference construction (Shimojo et al., 2003).

8.4 Effect on Preference of Tactile Exposure and Making Things Yourself

Exposure, apart from seeing (viewing) an object, involves touching an object with one's bare hand. Several studies have shown that touching or holding an object is a main contributor to preference construction. For instance, using mugs, Peck and Shu (2009) reported that the ownership and preference for an object increased after actually touching the object, compared with the situation, wherein the object was not touched. In their research background, Peck and Shu (2009) cited the research on the endowment effect (Kahneman, Knetsch, & Thaler, 1990). According to the "endowment effect," one's assessment of the value of what he/she owns has a close relationship with whether he/she has touched or held that object (Chark & Muthukrishnan, 2013; Wolf, Arkes, & Muhanna, 2008), such that both ownership and preference increase upon exposure to an object.

Another concept that suggests the influence of exposure on preference is the so-called IKEA effect reported by Norton, Mochon, and Ariely (2011). They reported that one's preference for a product increases when he/she has made that product him/herself, as in the case of a self-assembled IKEA furniture, origami, DIY creations, and so on. Norton and colleagues called this the "IKEA effect." Through tasks such as assembling IKEA's products and creating origami themselves, the authors reported that the participants' preference for products they made themselves actually increased. Based on such findings, they proposed the hypothesis that labor forms preference. In that work, Norton et al. (2011) also discussed the relationship with sunk cost.

On the basis of such studies, we can say that various phenomena, such as "seeing (viewing) an object," "touching an object," and "making an object" all contribute to preference construction. In the subsections below, we outline the previous studies dealing with preference construction by choice experiences.

8.5 Cognition of Choice and Preference Construction

A person makes numerous decisions throughout the day, and his/her previous choice experiences are likely to determine the choices he/she subsequently makes. In this subsection, we explain previous research, which reported the idea that preference is formed after experiencing a choice.

8.5.1 Cognitive Dissonance Theory and Preference Construction by Choice

The choices we make do not necessarily allow us to end up with a high-preference option (Ideno & Takemura, 2018). For example, in the "sour grapes" story found in Aesop's Fables, if one cannot choose an option with a high preference, after making a certain choice, he/she can then evaluate that option as a high-preference option and the option not chosen as a low-preference option. The attitudinal changes that occur when choice and preference do not match are explained in the literature using cognitive dissonance theory.

In a state of cognitive dissonance, discomfort and tension increase when an inconsistency (dissonance) among cognitive factors occurs and a person is motivated to reduce such dissonance; hence, in this theoretical framework, a change in attitude is predicted by cognition and behavior (Festinger, 1957, 1964). For example, if the option for which one has shown the same degree of preference is chosen once, the preference for the subsequently chosen option increases and the preference for the other option not chosen decreases. In other words, cognitive dissonance theory posits that "preference is formed by choice."

Recently, some studies have re-investigated the phenomenon of preference change induced by one's choice (Chen & Risen, 2010; Izuma et al.,2010; Lieberman, Ochsner, Gilbert, & Schacter, 2001; Simon, Krawczyk, & Holyoak, 2004). This is an experimental approach initiated by Brehm (1956), in which he used an experimental paradigm called the "free choice paradigm." Figure 8.2 presents the flow of the experiment. In this paradigm, the preferences are measured before and after the measurement of preference relationships (Stage 2 choice tasks). The procedures that are used to measure the preferences in Stages 1 and 3 are favorableness evaluation (or ranking evaluation). The evaluation objects used to include products and painting prints. Here, evaluation is required for numerous products, including products used

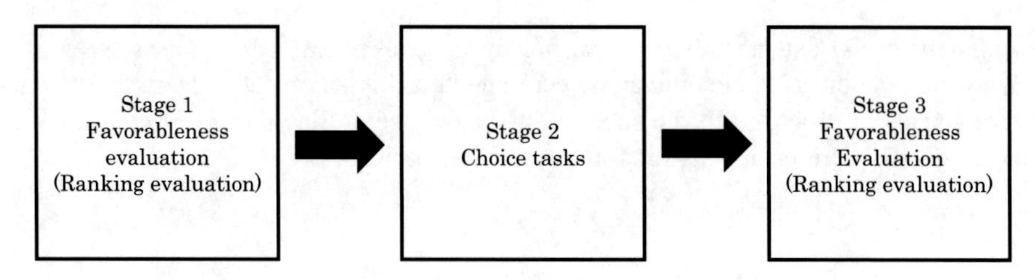

Fig. 8.2 Experimental flow of the free-choice paradigm

as options in Stage 2. In order to estimate the effect of choice on preference, the amount of change is calculated based on the evaluation values obtained in Stages 1 and 2 for the objects used as options in Stage 2, which are then used as indexes. Previous studies have shown that the evaluation values for objects adopted at Stage 2 further increased at Stage 3 than at Stage 1, and that the evaluation values for options not adopted at Stage 2 further declined at Stage 3 than at Stage 1 (Brehm, 1956; Festinger, 1964; Gerard & White, 1983; Lieberman et al., 2001).

Regarding the free-choice paradigm, some authors have argued for the opposite notion: preference is not formed by choice (Chen & Risen, 2010). Many options used for choice tasks at Stage 2 are objects that do not show significant difference in the evaluation results at Stage 1. In recent years, researchers have examined the possibility that a difference can be seen at Stage 3 because the difference between the objects has been recognized based on the resulting choice at Stage 2.

Izuma et al. (2010) used an image of the brain function during the tasks in Stages 1–3 and analyzed the participants' decisions based on the free-choice paradigm. In this case, they used striatal activity as an index of preference. Many studies have shown that the corpus striatum is activated when the prediction of favorable results and rewards is presented. The experimental results of Izuma et al. (2010) indicated that, in the comparison between Stages 1 and 3, the evaluation value of the object not chosen at Stage 2 changed, and the activation of the corpus striatum was reduced. They concluded that preference is formed by choice not only at the behavior level, such as favorableness evaluation, but also at the activity level of the brain (Izuma et al., 2010).

Thus far, we have dealt with various phenomena, such as increased and decreased preference after the measurement of preference relations. The above-mentioned studies are considered as research on changes in preferences for objects by stating preference relations and their influence on subsequent choices. In addition to this research, we also examine the commitment effect (i.e., the evaluation of held objects increases during the decision making process) as a phenomenon related to the preference construction after making a choice (Brockner, 1992; Kiesler, 1971; Klinger, 1975).

8.6 Experiments Dealing with the Preference Construction Process

The various phenomena we have discussed above exemplify the favorableness evaluation after stating the preference relationship (Ideno, & Takemura, 2018). So far, we have yet to discuss the preference construction for the object itself. Thus, we examined the possibility of preference construction by choice behavior while controlling the exposure experience of options (Ideno et al., 2011a, 2011b; Takemura et al., 2012). The choice behavior used in the following experiments is designed to control, as much as possible, the influence of the statement of preference relations in order to examine the influence on preference construction of choice behavior itself. Specifically, we investigate whether choice behavior based on visual judgment, such as size judgment, has an influence on subsequent preference relations. Each experiment consists of three phases: (1) the perceptual judgment task, (2) the preference judgment task (two options choice), and (3) the favorableness evaluation. The perceptual judgment task performed was a choice task unrelated to a preference statement. This task was created for the purpose of manipulating the number of choices of a specific stimulus. In Phase 2, the preference relations were measured according to the choice task of the two options. In Phase 3, the favorableness evaluation of objects was used for choice.

8.6.1 Experiment Using Meaningless Figures

In the perceptual judgment task in Phase 1, two meaningless figures were presented to the left and right; by seeking the judgment related to size, the number of choices for the stimulus was manipulated (Ideno & Takemura, 2018). In this research framework, the preference judgment in Phase 2 and the influence of figures in Phase 3 on the favorableness evaluation were examined based on the number of choices of meaningless figures targeted in Phase 1. The experimental hypothesis states that the preference for the stimulus chosen many times in the perceptual judgment task increases. In this choice task, the multi-choice stimulus was chosen.

We conducted a preliminary survey to sample random figures with similar preferences (Fig. 8.3). They used the random figures presented by Attneave and Arnoult (1956) and Vanderplas and Garvin (1959). We sampled 10 octagonal meaningless figures in the preliminary experiments. These figures have low association values with concrete things and equivalent favorableness values. In the choice task, we used two stimuli as the target stimuli, and for the perceptual judgment tasks, they used four stimuli as fillers.

Twenty-two university students were asked to respond accurately and as fast as possible. Specifically, they were asked to choose which of the stimulus displayed on the left and right parts of the screen was larger (or smaller). The characters ("large" or "small") displayed at the center of the screen were task assignments, and

Fig. 8.3 Two stimuli used as targets

the participants were asked to respond with the left and right keys (e.g., "a," "l"). In addition, we manipulated the number of choices by preparing stimuli that were larger (or smaller) than the targets; this was done by expanding or contracting the four filler stimuli by 10% in advance. A total of 96 trials were conducted. We adjusted the number of presentations of the stimuli so that all the stimuli would be equalized; in addition, we counterbalanced the stimuli presentation positions and the positions of the response keys. We designed the task so that participants would do the following: choose one of two targets 32 times (hereinafter called "multi-choice target"), choose the remaining targets 0 times (hereinafter called "non-choice target"), and choose each filler 16 times. We did not conduct trials using two targets. This was because we wanted to prevent the choice of both targets from affecting the subsequent preference judgment.

8.6.1.1 Preference Judgment Tasks

In this part of the experiment, we used two targets and four fillers different from the perceptual judgment task. We also changed the positions of the left and right for all combinations and asked the participants to choose their preferred stimulus. We conducted 30 trials, 2 of which used the two stimuli of the targets. We randomly assigned the presentation order.

A total of 19 participants were analyzed after excluding three participants with a mis-response rate exceeding 10% in the perceptual judgment task. Table 8.2 presents the preference judgment task in Phase 2 for the two targets and the favorableness evaluation using seven points scale results of Phase 3.

Regarding the preference task, the results indicate that the number of choices of the multi-choice target of the perceptual judgment task was significantly larger than that of the non-choice target (65.8 vs. 34.2%), as indicated by the direct probability

Table 8.2 Result of preference judgment tasks and evaluation task

	Multi-choice target	Non-choice target
Number of choice (%)	25 (65.8%)	13 (34.2%)
Average evaluation value	3.11	3.26

Note The stimuli used in the experiment were evaluated using the seven-stage scale method (from 1: "like" to 7: "dislike")

test ($p = 0.037$). This finding suggests that preference is formed by a choice that is not related to preference.

8.6.1.2 Evaluation Task

Once the above two tasks were completed, the preferences of 30 stimuli, including the stimuli used in the experiment, were evaluated using the seven-stage scale method (From 1: "like" to 7: "dislike"). We randomly assigned the presentation order. Next, we performed the t-test based on each participant's given evaluation value. We observed no significant mean difference [3.11 for high-frequency object vs. 3.26 for low-frequency object; $t(18) = 0.33$, n.s.].

In this experiment procedure, we measured the favorableness evaluation after conducting the preference tasks. Compared with the free-choice task paradigm, the Stage 1 favorableness evaluation was implemented, assuming that it corresponded with the choice tasks and the favorableness evaluation in Stage 2 and Stage 3, respectively. Therefore, based on the average evaluation value in this experiment, we predicted a difference between the non-choice and multi-choice targets; however, the results indicated no such difference.

Using the target option in the choice task, in this experiment, two trials were conducted for each choice. The choice results can be divided into two groups: (1) choosing the same option twice, and (2) choosing different options. Next, we checked the difference between the average evaluation value (3.0, SD = 1.7) of the meaningless figures chosen in Phase 2 by the experiment participants (14 people) who chose the same option twice and the evaluation value (3.5 SD = 1.2) of the meaningless figures that were not chosen. This was done in order to examine the favorableness evaluation after making a choice. Based on the results, no significant difference can be found [$t(13) = 0.82$, n.s.].

8.6.1.3 Interpretation of the Experimental Result

Using meaningless figures in this experimental condition, results indicate that preference can be formed by a choice that is unrelated to preference. However, the problem with this experiment was that we did not control the gazes while the participants were performing the perceptual judgment task. Thus, when performing the perceptual judgment task, each participant's preference might have increased by seeing more stimuli. Shimojo et al. (2003) reported a choice rate by the gaze cascade effect of approximately 60%, which is just about near the choice rate (65.8%) shown in our result. This finding suggests that preference could be formed by choice.

Furthermore, in this experiment, we adopted a procedure of conducting favorableness evaluation after making a choice. This procedure corresponds to those followed in Stage 2 and Stage 3, which have been used in previous similar studies on preference change by choice. Past studies on preference change by choice have shown that chosen options are evaluated more favorably after a choice is made, and that products that have not been chosen are evaluated as having lower preference value. However, in the current research, no difference exists in the favorableness evaluation between the target adopted in the choice task of Phase 2 and the target that has not been adopted. This result is different from the prediction made by studies on preference change by choice. Regarding this experiment, we may be able to say that the influence of using meaningless figures to control the degree of preference for stimuli.

8.6.2 Experiment Using Mineral Water Bottles

In this experiment, we also used the experimental paradigms of measuring preference relations and evaluating favorableness after conducting perceptual judgment tasks (Ideno & Takemura, 2018). In this task, a discriminative stimulus was presented in the form of an image of a mineral water bottle, after which the participants would press a key corresponding to the bottle presented with a specific discriminative stimulus.

We conducted a preliminary survey so that we can identify the mineral water brands, with which the experiment participants had no previous choice experience. In the preliminary survey, we asked 13 university students three questions while showing images of 20 kinds of mineral water bottles one by one. All 20 kinds of bottles were made in foreign countries, and the experiment group members gathered products they had never seen before. The evaluation average of two items ("favorableness" and "happiness") was calculated for each product. Among them, we selected a pair of products with a small difference in the evaluation average for any item as the experimental stimulus. The final products obtained from this preliminary survey included the following: the Italian-made BrioBlu (hereafter called "Product B") and American-made Penta (hereafter called "Product P").

This experiment involved 133 undergraduate and graduate students. For the presentation of the stimuli, we used personal computers and monitors. In addition, we used the EyeLink CL Illuminator TT-890 (SR Research Ltd.) for measuring each participant's eye movements. The experiments were conducted individually.

8.6.2.1 Perceptual Judgment Task

The flow of one trial is shown in Fig. 8.4. After a gaze point was presented, a discriminative stimulus (i.e., a triangle and an inverse triangle) was given, and the images of Product P and Product B were simultaneously presented on the left and right parts of the screen, respectively. The presentation times of the screen on which both bottles were presented were set to four time periods, namely, 400, 800, 1200, and 1600 ms, respectively. Four kinds of presentation times were provided in order to prevent the prediction of such presentation times. Next, we presented a discriminative stimulus to both images. The participants were asked to press the button on the presented side of the same triangle mark as the discriminative stimulus, when a triangle mark was presented on a bottle. The key corresponding to the left bottle was labeled "A," and the key corresponding to the left bottle was labeled "L." After choosing a product, the chosen product image disappeared from the screen, and the image of the unchosen product was presented at the same duration (800 ms) and position. In order to control the gaze time for each product, we had to present the unchosen product.

As the presentation conditions of the stimulus, equal-frequency conditions were provided, wherein Product B and Product P were chosen with an equal frequency, along with a high-frequency condition, where each of Product B and Product P was chosen more by the participants. For example, in the high-frequency condition of Product B, a discriminative stimulus was presented for Product B at a frequency of 80 trials out of a total of 88 trials. After completing the practice trial of a perceptual

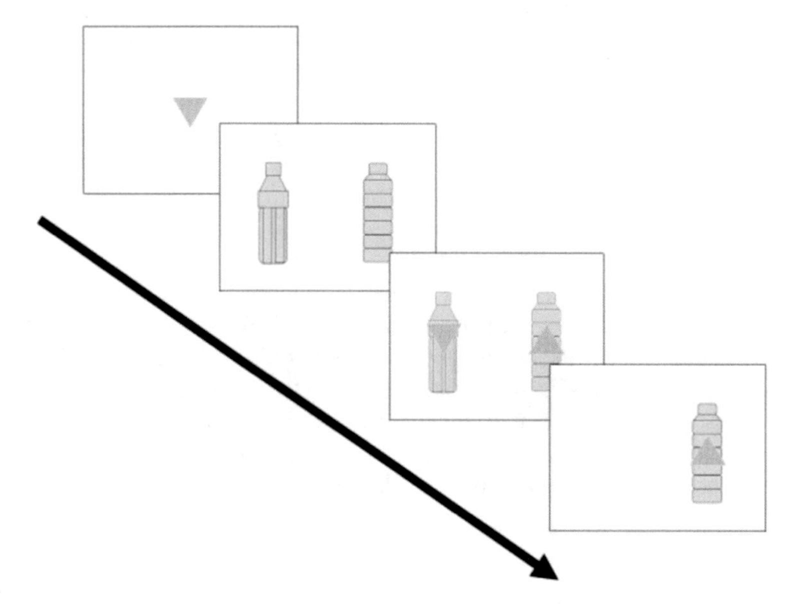

Fig. 8.4 Trial image of the perceptual judgment tasks for experiment 2

judgment task, we set the eye camera and proceeded to conduct the trial of the perceptual judgment task. Next, the choice task and favorableness evaluation were carried out. Upon completion, we gave the participants the products they chose in the choice task as their reward for participating in the experiment.

The experiments were conducted individually. After completing the practice trial of a perceptual judgment task, we set the eye camera and proceeded to conduct the trial of the perceptual judgment task. Next, the choice task and favorableness evaluation were carried out. Upon completion, we gave the participants the products they chose in the choice task as their reward for participating in the experiment.

8.6.2.2 Choice Task

For this task, we presented two products on the left and right sides of the monitor and requested the participants to choose which product they would bring home. We counterbalanced the presentation positions of the two products for each experiment participant. The number of choice for product B was 80 for high-frequency condition, 44 for equal-frequency condition, and 8 for low-frequency condition, respectively. The number of choice for product P was 8 for high-frequency condition, 44 for equal-frequency condition, 80 for low-frequency condition, respectively. The results of the choice task for each of the low-frequency and high-frequency choice targets of the perceptual judgment task indicated the number of choices of the preference judgment task of the low-frequency choice stimulus was 45 times (43%) and that of the high-frequency choice stimulus was 60 times (57%). The choice probability of the high-frequency choice products tended to be higher than that of the low-frequency choice products. However, this value is lower than the 65.8% rate of the multi-choice condition and the 60% rate reported by Shimojo et al. (2003). No significant difference can be found ($p = 0.17$, n.s.) after performing the exact probability test based on the number of choices of the preference judgment task.

8.6.2.3 Evaluation Task

Here, we presented each product on a monitor and asked the participants to use the nine-stage scale in evaluating the favorableness of each product. We counterbalanced the presentation order of the two products for each experiment participant. Next, we tested the difference between the average values on the basis of the evaluation value for each product. The results indicate no difference in the average evaluation value of high-frequency choice products [5.97 for high-frequency product vs. 5.74 for low-frequency product, $t(104) = 0.97$, n.s.].

8.6.2.4 Gaze Time

We also calculated the average gaze time for the high-choice and low-choice products in the perceptual judgment task of the experiment participants who had a high capture rate of eye movements (equal-frequency condition: 11 people, *B* high-frequency condition: 16 people, *P* high-frequency condition: 16 people).

In this experiment, we aimed to predict the longer gaze time for products with many choices in the perceptual judgment task. Thus, we presented non-choice targets for 800 ms after the choice. Our aim was to increase the gaze time for products with a few choices. The average gaze time of each product until the choice for each trial indicates a large amount of gazing at multiple options. The gaze time is longer for products with a few choices than for products with many choices. After testing the difference between the average gaze time of products with many choices and that of products with a few choices, we found a significant difference [$t(31) = 3.05$, <0.01]. Furthermore, the products with a few choices had a higher number of gazes than the products with more choices.

8.6.2.5 Interpretation of the Experimental Result

In this experiment, we examined the influence of choice behavior on preference construction by controlling the products during the perceptual judgment tasks. From the examination of the gaze time, we found that the gaze time for products with a few choices was significantly longer than that for products with many choices. We controlled the gaze time for products with many choices in the perceptual judgment task. Furthermore, in this experiment, the products were used for stimuli, and the preferences were measured with the choice task of giving away products. The results of the preference judgment task revealed that the choice rate of products with many choices is 57.1%. This value is lower than that recorded in the former experiment and the value reported by Shimojo et al. (2003). The analysis of products with many choices in the perceptual judgment task depending on the type showed a tendency to be different depending on the kind of product presented. This finding suggests that the effect of a preceding choice task is not necessarily large in the choice task of products that are actually given away.

8.7 Consumer's Preference Construction and Future Perspective

In this section, after reviewing the psychological findings on preference construction processes, we presented our experimental study focusing on the causal relation, in which choice leads to preference. Contrary to the almost implicit common idea of choice by preference, we focused on the aspect of preference by choice by performing related laboratory experiments and then analyzing the results. In Experiment 1, the stimuli used in the experiment were meaningless figures, and the choice objects were objects hardly encountered in daily life. The results indicate that the choice itself could contribute to the formation of preference. By contrast, we also examined the relationship between choice and preference using familiar products (i.e., mineral water bottles). The effect of choice in the mineral water experiment is not recognized as much as in the meaningless figure experiment, suggesting that the degree of the influence of choice actions varies depending on the choice objects. Given that the preference judgment of mineral water bottles is a kind of decision making that is repeatedly carried out on a daily basis, we can point out the possibility that preference criteria can be formed in advance. In addition, we examined the favorableness evaluation of chosen objects, no difference was found between the favorableness evaluations of the chosen and unchosen options.

Indeed, decision making is a continuous process, and even a simple situation shows that a certain choice influences a person's subsequent choice. When we look at the decision making process in detail, we need to perceive a relevant option before we choose it. In the process of perceiving options, we must first be exposed to an object by seeing and touching it. The relationship between sensory information at the time of exposure to an object and the judgment and decision making that follows is a theme that has attracted much attention in the area of sensitivity marketing or embodied cognition, especially in recent years. Hence, the relationship between preference construction and sensory information, which is the participant of this research, should be examined further in the future.

Moreover, in the process after choice, we may touch an object and gaze on it. Then, after the choice, the next decision making stage follows. In psychological research, so far, regarding the process after choice, we have hardly seen any development other than the research on preference change by choice. If the preference for the object continues to rise after making a choice, the reason why our mind chooses a different option from the one we have chosen previously cannot be easily explained. Hence, we consider the process that happens after making a choice and subsequent attitude changes as the remaining issues that must be studied.

Finally, research on the relationship between preference and choice, as explained in this section, involves a methodological issue as well: how to measure preference, "favorableness evaluation," or "choice." This corresponds to the fact that the problem of whether either method is justified in measuring preference in the matching tasks or choice tasks has yet to be sufficiently elucidated in preference reversal research. The research discussed in this chapter operationally defines preference and choice, but this essential problem has yet to be fully explained. Thus, this is another problem to be resolved by future studies.

8.8 Consumers' Affects and Economic Behaviors

Many studies have shown that affect affects decision making related to economic behaviors (Chaudri, 2006; Ishibuchi, 2003; Takemura, 1996, 1997; Turley & Miliman, 2000). For example, Donovan and Rossiter (1982) conducted an interview survey to assess the affective value of stores. They found that when consumers evaluated the environment in the store as comfortable, their purchasing intention for products was stronger. Moreover, they revealed that when consumers evaluated the environment in the store as comfortable, purchasing intention increased with the arousal level. However, when consumers thought that the store environment was uncomfortable or had a neutral opinion (i.e., the store was neither uncomfortable nor comfortable), the purchasing intention was not related to the arousal level.

Consumers' affects influence not only their purchase behavior in stores but also the acceptance effect of advertisements. Batra and Stayman (1990) experimentally examined the influence of positive affects on the psychological effect of advertisements. They had participants evoke positive affects by letting them read stories, after which they examined the effect on brand evaluation using advertisements of printed media. The results indicated that participants who evoked positive affects evaluated brands more favorably than those in the control group. In addition, they asked the participants to freely describe what they were thinking about while reading advertisements to examine the acceptance process of advertisements. They analyzed the data and found that positive affects suppressed deep thinking (refinement) about advertising messages, suggesting that positive affects caused favorable brand evaluation. Batra and Stayman (1990) also found that the suppressive effect on refinement by positive affects became noticeable when the participants' cognitive desire was weak or when advertisements comprised weak messages.

Many other findings about such an affective effect have been published in the fields of marketing and consumer behavior research over the past 30 years. Recently, studies on psychological mechanisms and neuroscientific foundations have also become more popular. This chapter explains the influence of affect on the decision making of economic behavior from the viewpoint of prospect theory (Kahneman & Tversky, 1979; Tversky & Kahneman, 1992), which has been vigorously studied in the fields of behavioral economics and neuroeconomics. Furthermore, I explain the relation of prospect theory and the problems of affect, along with the relevant neuroscientific findings.

This chapter first describes the definition of affect and the development of affect research. It explains the methods of affect manipulation and affect measurement. Next, I outline prospect theory in explaining the decision making phenomena, discuss the influence of affect on the decision making related to economic behavior based on this theory, and further examine the influence of affect on decision making based on the findings of neuromarketing using neuroscientific methods of decision making. Finally, I consider the implications of these findings on consumer behavior research and marketing.

8.9 Definition of Affect and the Development of Affective Effect Research

8.9.1 Definition of Affect and the Trend of Affect Research

Although affects comprise a central research participant in psychology, researchers have yet to agree on a standard definition (Endo, 2015; Ohira, 2010). The relatively broad concept of affect, as explained by Ohira (2010), is defined by Ortony, Clore, and Collins (1988) as "evaluation responses to agents, things, events and circumstances, which are information processing that one conducts in the mental process."

On the one hand, relatively strong feelings accompanied by physiological arousal, such as anger, fear, sadness, surprise, and joy, are called emotions. On the other hand, the feelings that are not strong but lasts for a relatively long period of time are called temper or mood. Some of these affects are grasped in terms of pleasure and discomfort, arousal, and so on, while some are related to the fundamental form of affects such as joy and sadness. Basically, affects cannot be clearly defined like cognition, but they are considered as mental activities that differ from ordinary cognitive activities.

Affects have always been considered in the history of psychology, and according to Endo (2015), two major trends have emerged from the ancient history of thought to modern times. One is the way of thinking that regards affects as factors that disturb reason. This thought was proposed by philosophers such as Plato, Descartes, and Kant. The other trend is the notion that affects are factors leading reason conversely, as proposed by philosophers Aristotle, Smith, and Hume. These two major trends exist in psychology in which a school of thought argues that cognitive evaluation is an antecedent factor of affects (proposed by Lazarus), whereas another argues that affects are antecedent factors of cognition (proposed by Zajonc and Damasio). In the era when cognitivism was the mainstream, the standpoint of the latter was not the mainstream. Nevertheless, the latter notion has attracted increased attention with the recent recognition of the neuroscience of decision making.

8.9.2 Methods of Affect Manipulation and Measurement of Affects

When studying the influence of affect on social behavior, causal relationships from affect to economic behavior cannot be clarified by simply conducting correlation analysis research. In general, in outdoor and laboratory experiments, researchers conduct experimental manipulation for one group of participants (experimental group) to evoke some affects and not for the other group (control group) to examine the influence of affect on economic behavior by comparing the occurrence rates in both groups.

Various methods have been adopted in the literature to manipulate affects. Among these, the simplest method is to give some gifts. For example, researchers may choose to give free gifts, such as notebooks, clips, or cookies, or deliberately place a coin in the change slot of a public phone. Previous studies have shown that such slight stimulation evokes participants' affects and affects their social behaviors. Other methods include the hypnotic method, which manipulates affects by hypnotism; the task performance feedback method, which makes false feedback, such as success or failure after some tasks; the Velten method, which requires participants to evoke affects by asking them to read a series of cards; and the recall interview method, which evokes affects by interviews and manipulates affects with music, movies, or a certain kind of smell.

Do such various methods that manipulate affects evoke the same affects? Is it possible for a participant who had a positive affect by being told the success with the task performance feedback method and another participant who had a positive affect by listening to music to have the same affect? When using the former method, it is highly likely that a sense of pride has also increased at the same time, whereas in the latter case, a psychological process completely different from the former, such as evaluation of music, might have occurred at the same time. Owing to those possibilities, maintaining the reliability of the result would not be possible by merely using one type of affect manipulation method; hence, in many cases, various affect manipulation methods are used for experiments.

Using many methods to manipulate one affect is also effective in checking convergent validity. Convergent validity ensures that if the measure is appropriate, then the results of the two methods are quite similar even if a characteristic or concept is measured with different methods. Previous studies have shown that similar behavioral results can be obtained using various different affective manipulation methods, which guarantee convergent validity (Isen, Means, Patrick, & Nowicki, 1982).

Even if different types of affective manipulation cause the same behavioral results like these, is there evidence that the participants have actually experienced a particular affect? Without the evidence, we cannot simply assume that the manipulation of affect was successful. In relation to these issues, measurement methods based on physiological indicators concerning affect have been examined in previous researches (Davidson, 1984; Murakami, 2010; Tucker, 1981; Tucker, Vannatta, & Rothlind, 1990). For the traditional decision making related to affective economic behavior, the most commonly used method is the measurement method, such as the questionnaire method. In this method, affective conditions are measured by having participants reflect on their affective conditions, evaluate the affective conditions with adjective pairs, and evaluate ambiguous neutral stimuli. Most studies, so far, have shown that such measurement by self-evaluation has been effective for the manipulation of affect. However, such an evaluation method based on self-report may also have certain problems. For example, participants may falsely report their affects or are oblivious to their affects. To avoid these self-evaluation problems, a method of analyzing the facial expression and voice of the participant by making video and audio recording is also adopted. Masters, Barden, and Ford (1979) confirmed that the method of evaluating facial expressions was effective; similarly, Bugental and Moore (1979) confirmed that evaluating the quality of the participant's voice was effective.

8.9.3 *Affective Decision Making Research by Neuroscientific Methods*

Along with the development of the fields incorporating the neuroscientific methods (e.g., neuroeconomics and neuromarketing) in recent years, physiological methods and neuroscientific methods on affect and decision making are also being increasingly used for measuring affects. Several reasons are cited as to why researchers use these methods. First, the noninvasive brain activity measurement methods, such as the functional magnetic resonance imaging (fMRI) and the positron emission tomography apparatus (PET), have been developed. A system has been established in which psychologists and economists can examine the findings previously handled only with behavior experiments in cooperation with neuroscientists. In addition to fMRI and PET, which measure blood flow in the brain, other methods measure brain waves and skin electrical activities, eye movement, and so on, in neuromarketing, and the measurement devices for these have already been developed since a few decades before. However, in the measurement research of affect, these physiological indicators are also being used widely together with neuroscientific methods.

Second, many economists and psychologists doubt the human model of "rational homo economics" that has been assumed in economics so far, leading to the development of behavioral decision theory and behavioral economics, which describe actual human affects and decision making (Ogaki & Tanaka, 2014; Takemura, 2009a, 2009b; Takemura, 2014). These movements have already appeared in the research by Simon, who won the 1978 Nobel Prize in Economics, as well as in the research of Kahneman, who won the same award in 2002. In addition, the research groups of Farrell and Camerer and others, who have been actively conducting research in neuroeconomics on affects and decision making, seem to have a large influence.

Third, in marketing research, practitioners and researchers are becoming aware of the problem that they cannot sufficiently predict behaviors or gain objective data from the consumer behavior research by merely depending on the questionnaire method, web surveys, interview methods, and behavior observation methods. Accordingly, new ways of measuring affects and decision making by neuroscientific methods have also been used. Meanwhile, the traditional affect-evoking method and the subjective measurement method shown earlier have also been used together.

Examples of such neuroeconomic research are shown below. Takemura, Ideno, Okubo, Kodaka, and Takahashi (2009c) examined the effect of the background information of products on the product evaluation by selecting products heavily consumed in the summer and in the winter. Using fMRI and the questionnaire method together, they examined how consumers evaluated the products under the conditions that summer and winter background images were consistent with the products and that the background and products were not consistent with the measurement of brain function images. This experiment comprised 48 trials, with each trial having two phases: the priming phase and the selection phase. The experiment was carried out, as shown in Fig. 8.5. At the questionnaire level, more positive affects and evaluations were observed when the background stimuli were consistent than when they were inconsistent.

In their group analysis, no significant difference was observed in brain function images among the background consistency, inconsistency, and control groups. However, a participant showed differences in brain function images between the conditions below. Figure 8.6 shows the brain regions with high activation under the consistent background condition compared with the controlled condition. With respect to the right caudate nucleus, a reward-related site, and the inside part of the orbitofrontal cortex (OFC), significant activation was seen in the consistent background condition compared with the controlled condition. Moreover, significant activation was seen at the dorsolateral prefrontal cortex (DLPFC) and supplementary motor area, among others. In this way, based on the difference from the controlled condition, the consistent background effect was analyzed by the individual or by the group.

Studies on such affective decision making initially gained much interest among neuroscience researchers when the research group of Montague, a neuroscientist from the Baylor College of Medicine, reported their experimental results on the preference of Coca-Cola and Pepsi-Cola in Neuron, a neuroscience magazine, in 2004. For participants who liked Coca-Cola and Pepsi-Cola, they used fMRI to measure the blood flow of the brain while drinking under two conditions: The brand names were hidden and not hidden (McClure et al., 2004). Under the condition that

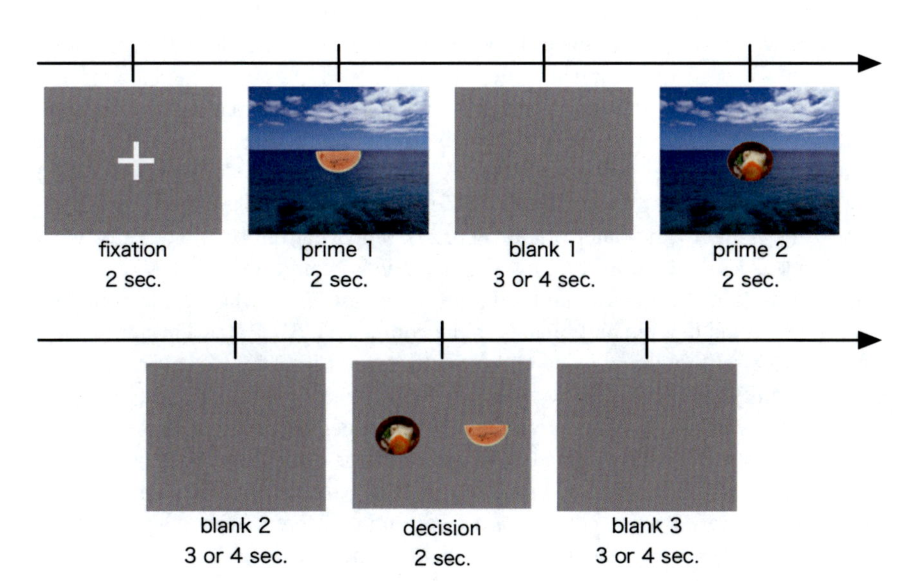

Fig. 8.5 Flow of one trial of product choice task in the fMRI experiment. *Source* Takemura et al. (2009c)

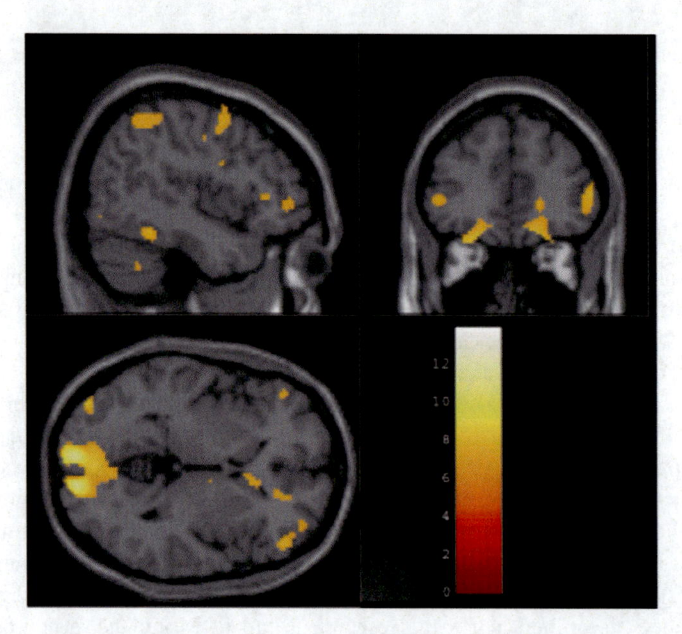

Fig. 8.6 Brain images of a participant who showed differences in brain function images between the conditions. *Source* Takemura et al. (2009c)

the brand names were hidden, the ratios of choosing Coca-Cola and Pepsi-Cola were almost the same. The results of the fMRI experiments indicated that when the brand names were hidden, each of the times of choosing Coca-Cola and Pepsi-Cola was significantly correlated with the brain activity of the ventromedial prefrontal cortex of the frontal lobe. This finding indicates that activities of the ventromedial prefrontal cortex are genuinely expressing choices by individual taste for cola when the brand names are hidden. Furthermore, in the condition wherein one brand label was shown in one cup and the other cup was not labeled (containing either Coca-Cola or Pepsi), they found that the cups with the Coca-Cola label were chosen more than those bearing the name of the other brand (Pepsi). Comparing the brain activities when they drank Coca-Cola after looking at Coca-Cola's picture to the ones when they drank Coca-Cola after a stimulation that they did not know what would come, the researchers found that the participants' hippocampus and DLPFC were significantly more active after looking at Coca-Cola's picture (see Fig. 8.7). However, there was no significantly intensive activity after looking at Pepsi's picture. This suggests that there were at least two systems explaining the desire occurrence of consumers: a system for evaluating a taste itself and a system for image evaluation based on recalling brands. Moreover, depending on brand information based on communication strategies, such as advertisements, there was a preferential difference from the taste based on the natural physiology reaction, suggesting that the consumers' desire system was strongly influenced by advertisements and the manipulation of brand images.

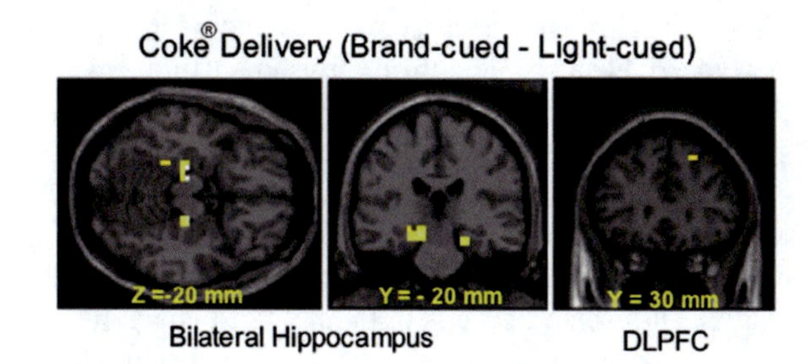

Fig. 8.7 Brain images of participants who underwent the preference experiments of Coca-Cola and Pepsi-Cola. *Source* McClure et al. (2004)

On the basis of that research, other marketing researchers have directed their attention to neuromarketing as an objective method to measure the communication effect of marketing. At present, neuroscientists recognize that the neuroscientific method can be used to solve the problems of real-world marketing.

8.9.4 Methodology of Neuroscientific Research and Significance as Consumers' Decision Making Research

Neuromarketing has been increasingly regarded in recent years as a subdomain of neuroeconomics, which clarifies the neuroscientific basis of human decision making behavior (Fugate, 2007; Hubert & Kenningy, 2008; Lee, Broderick, & Chamberlain, 2007). Currently, neuroeconomics is considered to be a neuroscientific study on human economic decision making; hence, neuromarketing can be positioned in the field of neuroscientific research of economic decision making. Under such circumstances, neuromarketing is being studied by a wide range of researchers and marketing practitioners, including business scholars, economists, psychologists, and neuroscientists.

According to Sanfey and Stallen (2015), the methods of neuromarketing include imaging studies such as fMRI, near-infrared spectroscopy, PET magnetoencephalography, and electroencephalogram. Meanwhile, transcranial magnetic stimulation (TMS) has also been used in recent years. In this interventional method, which affects the brain activities of participants, the researchers can observe the influence on cognitive judgment tasks and decision making tasks. Regarding TMS, several approaches depend on the stimulus method. Single-pulse TMS and paired-pulse TMS depolarize the neuronal cell population in the cerebral neocortex by the pulse stimulus to cause action potential. Repetitive TMS (rTMS) increases or decreases the excitability of the corticospinal tract or transcortical pathway by the intensity of the stimulus, the direction of the coil and the frequency of the stimulus, among others. Every method has advantages and disadvantages, and studies combining various methods are under way. For the features of these methods, refer to Sanfey and Stallen (2015).

As a reason why neuroeconomics and neuromarketing (as its subdomain) have developed, I can consider the establishment of the research system from the development of neuroscience-related methodologies. Moreover, in consumers' decision making research and marketing research, practitioners and researchers are becoming increasingly aware of the problem that they cannot sufficiently predict behaviors or gain objective data by merely depending on the questionnaire method, web surveys, interview methods, and behavior observation method. In addition, the Swiss and American research groups in the field of behavioral economics have been actively conducting research in neuroeconomics, and their findings seem to have a large influence. In addition, in the field of neuroscience, researchers have been better able to

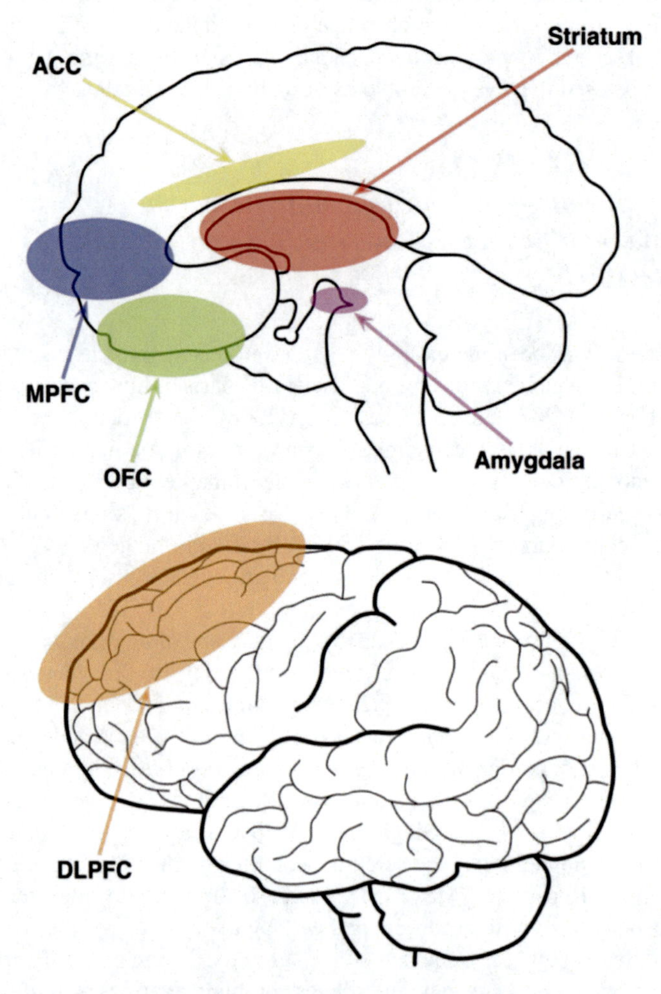

Fig. 8.8 Schematic illustrations of the brain regions considered relevant to decision making. *Source* Takemura, Ideno, Okubo, and Matsui (2008)

predict and explain people's decision making processes compared with the questionnaire method by language reports.

The neuroscientists consider the following areas of the brain as important parts: the medial prefrontal cortex (MPFC), located behind the forehead (Fig. 8.8), and the dorsolateral prefrontal cortex (DLPFC) , an area set back in the prefrontal cortex (Fig. 8.8). Knutson et al. (2007) performed an experiment in which an actual product was presented, and subsequently, the price was presented to the test subjects using fMRI, who were, then, to decide whether they would purchase the product. The result was that the more attractive the product, the more active the nucleus accumbens (NAcc) in the ventral striatum. When high prices were shown, the insula was active and the activity of the medial prefrontal cortex (MPFC) decreased (Fig. 8.8). This result suggests that the brain activities in the areas of loss and gain differ and can be interpreted as being consistent with prospect theory, which uses different value functions in the areas of loss and gain. Hsu et al. (2005) compared decision making under ambiguity and decision making under risk and observed activation of the orbitofrontal cortex (OFC) (presumably related to the integration of affects and cognitive input: Fig. 8.8), the amygdale (which presumably reacts to affective information), and the dorsomedial prefrontal cortex (DMPFC) (which presumably adjusts the activity of the amygdale) in decision making under ambiguity and activation of the caudate nucleus in decision making under risk. Hsu et al. (2005) also demonstrated that the more active the OFC, the stronger the tendency of ambiguity aversion.

Box 1: Colin Farrell Camerer

Born in 1959. He received Ph.D. from the University of Chicago in 1981. He investigates cognitive, neural, and behavioral aspects of human decision making and economic behavior. He is considered to be an important contributor to neuroeconomics with Ernst Fehr. He is Professor of Behavioral Finance and Economics at the California Institute of Technology.

Photograph by Universitat Pompeu Fabra:

https://www.flickr.com/photos/universitatpompeufabra/44649077635/in/photolist-ePtPbf-pPQpYd-2cqgoiH-NDYwWe-22D6Gdw-29DE4z9-NDYCrH-29DDRaE-QhpHZo-2ckPj8A-2b2uf9R-QhpjoL-QhpkWq-2b2ucen-2ckP257-2ckNUs9-2b2udGc-2ckPij1

8.10 Affective Effect and Prospect Theory

8.10.1 Affective Effect and Framing

As shown in Chaps. 4 and 5, prospect theory advocated by Kahneman and Tversky (1979) is a comprehensive theory based on the previous findings of behavior decision theory and nonlinear utility theory (or generalized expected utility theory). Prospect theory was initially proposed as a descriptive theory dealing with decision making under risk (Kahneman & Tversky, 1979), but this later developed into a theory that can explain decision making under uncertainty (Tversky & Kahneman, 1992).

In the editing phase, where framing is conducted and the problem of decision making is psychologically constructed, each prospect is reconstructed and the prospect with the highest evaluation value is selected in the evaluation phase. In this phase, evaluation is made with the value function (i.e., utility function) and with the weighting function to the probability. In the editing phase, the reference point, which is the origin of the value function, is determined. The method of evaluation in this phase is basically the same as the rank-dependent type utility theory in the nonlinear utility theory.

When analyzing decision making, the method of tracing the verbal protocol and the information search process in the decision process is often used. I can also examine how the decision makers have structured the decision problem based on the decision result. Tversky and Kahneman (1981) considered how I can understand the decision problem by decision makers, that is, the phenomenon of decision framing. Using the method used by Tversky and Kahneman (1981), Takemura (1988a) examined how affects affect decision framing. In this experiment, affective manipulation was performed by giving false feedback on the results of mental examinations (Uchida/Kraepelin test). The factors of decision framing comprised two conditions, namely, the ticket loss condition and the cash loss condition, which appear in the ticket problem used by Tversky and Kahneman (1981) . The task on the ticket loss condition is as follows: "Imagine the scene below. You decide to go see a movie, and after purchasing a ticket for 1500 yen, go to a movie theater. When you enter the movie theater, you have noticed that you have lost the ticket. Will you buy the ticket again?" The task on the cash loss condition is as follows: "Imagine the scene below. You decide to go see a movie and go to a movie theater. The price for the ticket is 1500 yen. When you enter the movie theater, you have noticed that you have lost 1500 yen in cash. Will you buy the ticket?"

According to Tversky and Kahneman (1981), these two problems are structurally identical, but because the way of framing is different, the ticket becomes more difficult to buy under the former ticket loss condition than under the latter cash loss condition. In the research by Takemura (1988a), participants under controlled conditions tended to show the effect of framing in the direction predicted by the prospect theory, whereas participants with positive affective conditions and those with negative affects did not show such a tendency. In addition, the purchase intention of the ticket tended to be higher in participants with positive affective conditions than in those with negative affective conditions. Whether it is a positive affect or a negative affect, can it suppress the cognitive bias such as the framing effect? Stroessner, Hamilton, and Mackie (1992) reported that positive and negative affects caused cognitive bias, such as illusory correlations to disappear, corresponding with Takemura's research results. Thus, positive affects promote the use of simple but efficient decision strategies in decision making, reduce framing effects, and enhance purchase intention.

People's affects are known to affect behavior and judgment in uncertain and risky situations. If the likelihood of suffering loss is low or the extent of the loss is small, participants with positive affects expressed by comfort tend to adopt high-risk options; on the contrary, if there is large potential loss or loss actually exists, participants with positive affects tend not to adopt high-risk options compared with participants under controlled conditions (Arkes, Herren, & Isen, 1987; Isen & Geva, 1987; Isen, & Patrick, 1983). For example, Isen and Patrick (1983) devised a bet-type gamble in which participants can obtain high credit if they win that gamble and examined whether participants would bet on that gamble. The results indicated that the participants who received the coupon ticket of hamburgers (positive affect group) showed lower betting rate on a gamble with a high risk (winning percentage of 17%) than those who did not receive the coupon ticket (control group).

Preference under risk is explained by two aspects, a belief in a future event occurrence (usually expressed by subjective probability) and a value (utility) brought about by the occurrence of that event. Is the effect on risk preference of a positive affect an influence of the subjective probability or of influenced utility?

First, positive affects affect perceived utility. Isen et al. (1988) measured the utility with the probability kept constant and found that the participants who received free gifts (positive affect group) gave a high negative utility value to the potential loss compared with the participants who did not receive the free gifts (control group). Therefore, positive affects make the subjective utility of loss more negative.

8.10.2 *Influence of Affects on Subjective Probability*

Next, positive affects also affect subjective probability. Johnson and Tversky (1983) found that participants with positive affects gave a low estimation of the subjective probability of the occurrence of negative incidents. The researchers manipulated the affects of participants by letting them read the story and asking them to identify the frequency of various incidents. As a result, the group who read the story causing positive affects perceived a lower risk compared with the control group. In contrast, the group who read the story causing negative affects perceived a higher risk. Thus, the positive affect lowers the subjective probability that negative incidents may occur, and the negative affect seems to increase the subjective probability that negative incidents may occur. These findings suggest that affect affects the probability weighting function in prospect theory.

In the research on the cumulative prospect theory, various metric models have been proposed in the evaluation formula of the probability weighting function, and their comparisons have been made (e.g., Gonzalez & Wu, 1999; Murakami et al., 2014; Takemura, 2014; Takemura, Murakami, Tamari, & Ideno, 2014; Wu & Gonzalez, 1996).

In the original prospect theory, the probability weighting function showed non-additivity, the overestimation of low probability events, nonproportionality, and discontinuity near the end point. The cumulative prospect theory, as shown above, is formulated as

$$W(p) = \frac{p^{\gamma}}{(p^{\gamma} + (1-p)^{\gamma})^{\frac{1}{\gamma}}},$$

where p expresses a probability, $W(p)$ is a subjective weight to the probability p, and γ is a parameter that takes values from 0 to 1. Here, only the probability weighting function in the gain region is shown. The shape for each parameter of this model is shown in Fig. 8.9.

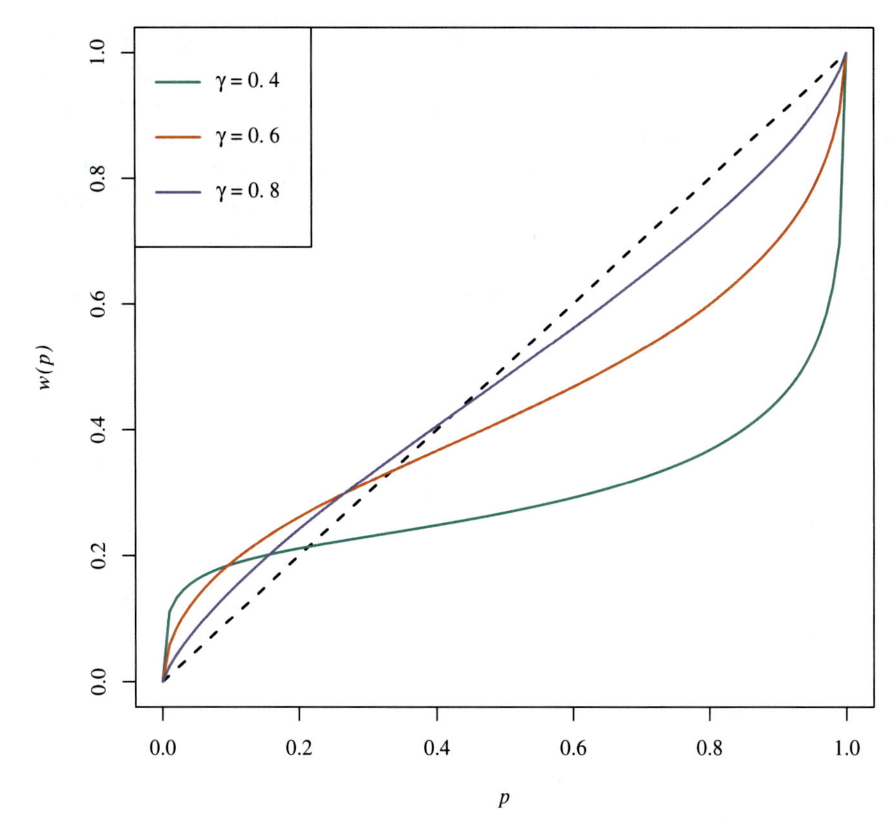

Fig. 8.9 Probability weighting function in cumulative prospect theory (constructed based on the Tversky and Kahneman (1992) model)

Rottenstreich and Hsee (2001) showed that the overestimation and underestimation of objective probability became more noticeable by increasing the distortion of the probability weighting function from evoking pleasant and unpleasant affects. This is indicated by the fact that the value of γ in the metric model of the probability weighting function above becomes small based on affect.

Suter, Pachur, and Hertwig (2015) also examined the applicability of prospect theory by conducting experiments of affective manipulation. Indeed, when affect was evoked, the parameter γ of the metric model of the probability weighting function decreased, suggesting that affect increased the distortion of the probability weighting function. However, when affect was evoked, it was often decided by the method that ignored the probability information and decided only by the result rather than by prospect theory, which integrated the probability and result information. Moreover, according to another study, when affect is evoked, the probability information is ignored and a decision is made only by the result (Pachur, Hurtwig, & Wolkewitz, 2014).

In the neuroscientific research on the probability weighting function in prospect theory, Takahashi et al. measured the dopamine D1 and D2 receptors in the brain using PET. They examined the relation between the probability weighting function and the D1 and D2 receptors of the corpus striatum in the brain (Takahashi et al., 2010: see Fig. 8.11). They made decision making tasks under risk to estimate the degree of the nonlinear weighting of probability. In this research, they used Prelec's (1998) formula in which axiomatic foundations are being considered as a metric model. The simple formula of this model is often used in the fields of behavioral economics, behavior decision making research, and neuroscience. This simple formula is given by

$$W(p) = \exp\{-(-\ln p)^{\alpha}\},$$

where p is a probability, $W(p)$ is a subjective weight to p, and α is a free parameter that takes the interval (0, 1). In addition, as shown in Fig. 8.10, the fixed point is 1/e or about 0.36 regardless of the value of the parameter α.

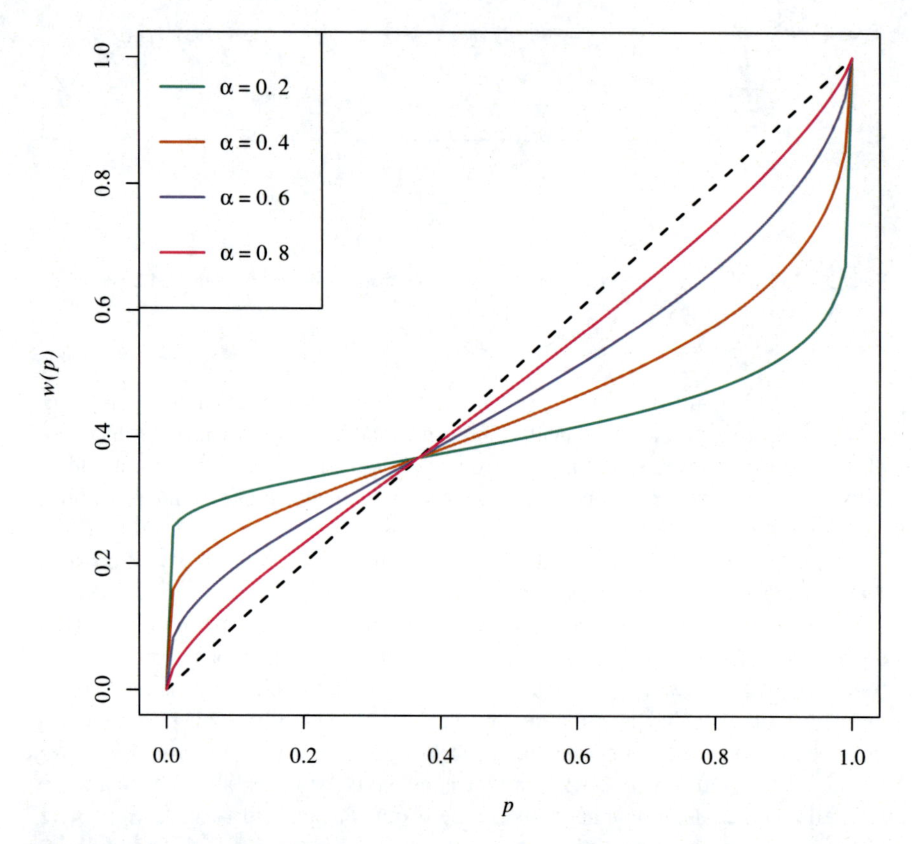

Fig. 8.10 Probability weighting function by Prelec (1998) (constructed based on the formula)

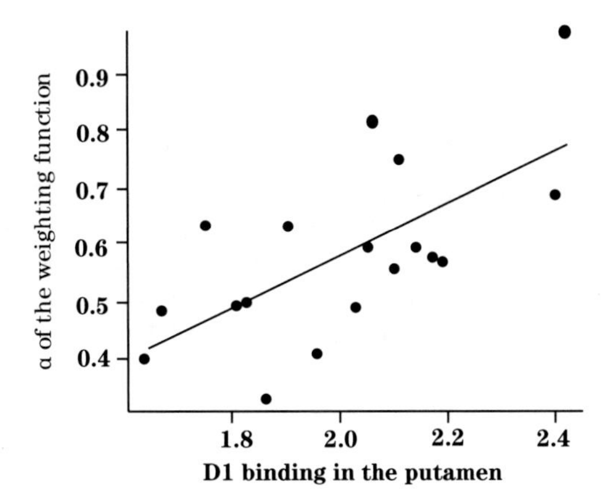

Fig. 8.11 Relations between the variables expressing the degrees of estimating low probability to be higher (and high probability to be lower) and the dopamine receptor of the corpus striatum. *Source* Takahashi (2013). *Note* The density of dopamine receptor is shown in *x* axis. The smaller the variable value of *y* axis, the tendencies estimating low probability to be higher (and high probability to be lower) become stronger

Takahashi et al. (2010, 2013) also estimated the probability weighting function based on the simple formula of Prelec (1998) and sought α in defining the probability weighting function. They obtained an average of 0.5–0.6, which is in agreement with past reports. However, at the same time, individual differences were recognized as well. In addition, when they investigated the D1 and D2 receptor binding abilities of the corpus striatum measured by PET, they found positive correlations between the D1 receptor binding ability of the corpus striatum and α, which defines the probability weighting function (see Fig. 8.11). In other words, people with lower D1 receptor density of the corpus striatum have a higher nonlinearity of the probability weighting function, and they are more likely to estimate low probability as high and high probability as low. From this result, if we infer relations with affect, the D1 receptor-binding ability of the corpus striatum may be related to human affects.

Finally, Takahashi et al. (2013) examined the neural basis corresponding to the value function in prospect theory. They examined the relation between the degree of loss aversion, which the participants judge by placing emphasis on loss, and the density of the thalamus noradrenaline transporter. They found that participants with low density of the thalamus noradrenaline transporter have a higher degree of loss aversion in decision making with risk. They found a tendency that those who had a low density of the thalamus noradrenaline transporter did not choose an option with risk unless the amount of gain was expected to be much higher than the amount of loss. Takahashi (2013) inferred that people with a low density of the thalamus noradrenaline transporter tend to give higher affectal attention to loss. Past studies also reported that participants with positive affects tend not to choose risky options (Arkes et al., 1987; Isen & Geva, 1987; Isen & Patrick, 1983). Overall, positive affects and the efficiency of noradrenaline uptake may be related. However, as for noradrenaline uptake and the D1 receptor binding ability of the corpus striatum, it remains unclear what kind of causal relationship they have with affect. Hence, there is room for further research in this area in the future.

Meanwhile, Suter et al. (2015) examined the influence of affect on the probability weighting function and value function using fMRI experiments. In the case of the decision making of results with affect, they found that the amygdala, posterior cingulate cortex, and the thalamus became more activated. In contrast, in the case of decision making involving less affect, they found that the supramarginal gyrus and occipital lobe became more activated. The results of these analyses are also based on an estimation combining mathematical models and metric models. Future studies may need to examine the stability of these results.

8.11 Implications for Consumer Behavior Research and Marketing Practices

Previous studies have reported that affect distorts the probability weighting function and ignores the numerical information of the probability; then, decisions are made from the viewpoint of the result (Suter et al., 2015). Furthermore, such effects of affect affect the value function and the probability weighting function in prospect theory. This also corresponds to the experimental result of the influence of affect carried out in multi-attribute decision making tasks.

Isen and Means (1983) examined the effect of positive affects on the decision making process. They showed that when participants received a fake feedback that they succeeded in perceptual movement tasks, compared with participants who did not receive a feedback, their time for choosing a fictitious car was short and they did not explore the information much when making a decision. They revealed that in the analysis of the protocol using the verbal protocol method in which participants utter what they are thinking of in the decision making process, compared with participants in the control group, those with positive affects tended to make decisions by the decision strategies based on attributes, such as the elimination by aspects (EBA)-type decision strategies of eliminating options sequentially according to the attributes of interest. The EBA-type decision strategy does not necessarily lead to the optimal decision but is a strategy with a low cognitive load, so positive affects are thought to promote the use of strategies of this property.

In addition, As shown in Chap. 6, Takemura (1988b, 1988c) showed that participants who were given false feedback (top 4% with a deviation value of 67.5) that they obtained high grades in mental examinations (actually, Uchida and Kraepelin tests) took a shorter time in decision making than those in the control group. Moreover, the numbers of their searched information and behaviors for information reexamination were fewer than those of the control group in actual radio-cassette recorder choice (high-involvement condition) and in virtual radio-cassette recorder choice (low-involvement condition). These findings suggest that positive affects induce participants to adopt attribute-based decision strategies with low information processing loads during the decision making process. Forgas (1991), who conducted a similar study on choice tasks between opposite sexes, also found that positive affects induce participants to shorten the time for decision making and adopt a simple information search strategy while making a decision. Even in consumer behavior research, examining the psychological mechanism of the effect of such affects not only facilitates a better understanding of the decision making process of consumers but also produces practical benefits for them.

In addition, as shown above, the affectal effect on decision making is considered to influence the decision strategy—both in decision making under risk and in multiattribute modeldecision making—and we can understand that this influence of affect can also be predicted with the model equation of prospect theory. From this, it is also possible to simulate the influence of affect on purchase decision making from a practical point of view. Further, the measurement and prediction of affectal effect may also be practically possible in the future based on neuroscientific methods.

As shown in Chapter 6, according to the computer simulation of the decision strategy of Takemura et al. (2015), the EBA-type and lexicographic-type decision strategies, which are likely to appear due to the influence of affect, have relatively high decision accuracy in spite of the low cognitive efforts required. From this finding, consumers' decision making may be simplified and somewhat impulsive if the influence of affect merely influences decision strategies, but the matter of which option to choose will not change much. In comparison, affects not only influence decision strategies but may also change the weights of attributes; in that case, significant differences will result in terms of the chosen brands and products. Such an investigation in basic research will be useful for practical applications.

Chaudri (2006) considered the methods of new marketing strategies for consumers while studying affects of consumers by organizing affect research. The examination of the affectal effect on decision making may be useful not only in consumer behavior research but also in practical marketing. As an application of consumer affect research, Gardner (1985) reviewed the influence of affect on consumer behavior 30 years ago and showed the influence process of affect in consumer behavior. Gardner assumed that the manner of communication, such as response in the service category, storefront stimulation, and advertisements, affected the affects of consumers, which in turn, influenced their memories, evaluations, and behaviors. According to this framework, it is practically effective to evoke positive affects of consumers in each marketing category (Gardner, 1985). Regarding the affects leading to the purchase behavior of consumers, I can think of various affects, including positive ones. Further, the idea that affects are preparatory behaviors for behavior exists in traditional psychology. An important contribution is possible from the practical point of view by studying the influence of various affects on decision making and behavior from the basic level and examining it from the viewpoint of prospect theory and new psychological models of decision making.

Box 2: Ernst Fehr

Born in 1956. He was doctorate from 1980 to 1986 at the University of Vienna. He investigates the human altruism and the interplay between social preferences, social norms, and strategic interactions. He also studies on the role of bounded rationality in strategic interactions and on the neuroscientific foundations of social and economic behavior. He is considered to be an important contributor to neuroeconomics. He is Professor of Microeconomics and Experimental Economics at the University of Zürich.

Photograph by Universitat Pompeu Fabra:

https://www.flickr.com/photos/universitatpompeufabra/22113262766/in/photolist-XoGYvZ-yMmc6q-zrKHgC-zG5qxS-zHg7ks-zG5fB9-zG5hQY-zKg7Dk-zG3MC1-zrQNke-yMt81M-zrRoHT-zHfGgW-23PLyA4/

Summary

- Preference is widely accepted in real-life situations as the cause of making a choice. We provided discussions of the preference construction process illustrating some experimental findings and theoretical explanations.
- According to psychological studies, preference construction depends on multiple factors, including an inherent fondness for something, past experiences as shown in operant conditioning, and the current situation.
- We listed the phenomena exemplifying the idea that being exposed to options, such as viewing and touching, facilitate preference construction. The "mere exposure

effect" states that experiencing an object helps a person form a preference and influences that person's subsequent choice of action. The "gaze cascade effect," which hypothesizes the relationship between active gaze and preference construction based on the measurement of eye movements in two options under forced choice situations. The gaze cascade effect is substantially different from the mere exposure effect in that the former argues that preference is formed by an active gaze. We also proposed the preference construction process by choice behavior while controlling the exposure experience of options.
- Many studies have shown that affect influences decision making related to economic behaviors. Consumers' decision making and economic behavior are affected by their affects, such as moods and feeling. Various methods of measuring affects have been introduced, such as the questionnaire, observation, and neuroscientific methods. Various methods for affect manipulation have been proposed, such as the method of task feedback and Velten method. Affects influence consumers' decision making strategies, probability weighting, and value functions.
- The relatively broad concept of affect is defined as "evaluation responses to agents, things, events and circumstances, which are information processing that one conducts in the mental process." Relatively strong feelings accompanied by physiological arousal, such as anger, fear, sadness, surprise, and joy, are called affects. On the other hand, the feelings that are not strong but lasts for a relatively long period of time are called temper or mood.
- In terms of neuroscientific research on consumers' decision making process, neuromarketing and neuroeconomics have been established, and studies using brain function images such as fMRI have been introduced in recent years. Positive affect influences perceived utility and subjective probability. Some fMRI studies found that the amygdala, posterior cingulate cortex, and the thalamus involved in affective process of decision making. A PET study suggested that people with a low density of the thalamus noradrenaline transporter tend to give higher affective attention to loss.

Recommended Books and Reading Guides for More Advanced Learning

- Lichtenstein, S., & Slovic, P. (Eds.) (2006) . *The construction of preference.* New York, NY: Cambridge University Press.

This book provides the view that people's preferences are often constructed in the process of elicitation. This book introduces not only the historical perspective of preference construction but also the important concepts of construction of preference in psychology, law, marketing, philosophy, environmental policy, and economics. This book is considered as a classic book of the preference construction.

- Glimcher, P. E., & Fehr, E. (Eds.) (2013). *Neuroeconomics: Decision Making and the Brain* (2nd ed.), London, UK: Academic Press.

This book is the second edition of the textbook of neuroeconomics. This book provides useful introductions to the disciplines of microeconomics, the psychology of judgment and decision, computational neuroscience, and anthropology, risk, time preferences, social preferences, emotion, pharmacology, and the foundations of neuroeconomic thought.

– Delgado, M. R., Phelps, E. A., & Robbins, T. W. (Eds.) (2011). *Decision Making, Affect, and Learning (Attention and Performance)*, Oxford, UK: Oxford University Press.

This book focused on decision making and emotional processing. This book investigates the psychological and neural systems underlying decision making, and the relationship with reward, affect, and learning, and neuroeconomics. This book would be also beneficial for graduate students, researchers, and practitioners in this field.

– Reuter, M., & Montag, C. (Eds.) (2016). N*euroeconomics (Studies in Neuroscience, Psychology and Behavioral Economics)*, Berlin, DE: Springer.

This book provides several series studies in neuroscience, psychology, and behavioral economics. This book starts with an introduction to neuroeconomics followed by an overview of frequently applied experimental frames in neuroeconomics research. Moreover, this book addresses the molecular basis of decision making, as well as developmental and clinical approaches to neuroeconomics. This book would be useful for researchers and practitioners and may also be beneficial for graduate level students in this field.

References

Arkes, H., Herren, L. T., & Isen, A. M. (1987). The role of potential loss in influence of affect on risk-taking behavior. *Organizational Behavior and Human Decision Process, 99,* 1–13.

Attneave, F., & Arnoult, M. D. (1956). The quantitative study of shape and pattern perception. *Psychological Bulletin, 53,* 452–471.

Batra, R., & Stayman, D. M. (1990). The role of mood in advertising effectiveness. *Journal of Consumer Research, 17,* 203–214.

Bornstein, R. F. (1989). Exposure and affect: Overview and meta-analysis of research. 1968–1987. *Psychological Bulletin, 106,* 265–289.

Bornstein, R. F., Leone, D. R., & Galley, D. J. (1987). The generalizability of subliminal mere exposure effects: Influence of stimuli perceived without awareness on social behavior. *Journal of Personality and Social Psychology, 53,* 1070–1079.

Brehm, J. W. (1956). Postdecision changes in the desirability of alternatives. *The Journal of Abnormal and Social Psychology, 52,* 384–389.

Brockner, J. (1992). The escalation of commitment to a failing course of action: Toward theoretical progress. *Academy of Management Review, 17,* 39–61.

Bugental, D. B., & Moore, B. S. (1979). Effect of induced moods on voice affect. *Developmental Psychology, 15,* 664–665.

Chark, R., & Muthukrishnan, A. V. (2013). The effect of physical possession on preference for product warranty. *International Journal of Research in Marketing, 30,* 424–425.

Chaudri, A. (2006). *Emotion and reason in consumer behavior*. Burlington, MA: Elsevier.

Chen, M. K., & Risen, J. L. (2010). How choice affects and reflects preferences: Revisiting the free-choice paradigm. *Journal of Personality and Social Psychology, 99,* 573–594.

Davidson, R. J. (1984). Affect, cognition, and hemispheric specialization. In C. E. Izard, J. Kagan, & R. B. Zajonc (Eds.), *Emotions, cognition, and behavior*. New York, NY: Cambridge University Press.

Donovan, R., & Rossiter, J. (1982). Store atmosphere: An environmental psychology approach. *Journal of Retailing, 58,* 34–57.

Endo, T. (2015). Moroha naru jōdō: gōrisei to higōrisei no awai ni aru mono [Multi-edged emotion: Existence between rationality and irrationality]. In M. Watanabe & S. Funahashi (Eds.), *Jodo to ishikettei [Emotion and decision making]* (pp. 93–131). Toyo, Japan: Asakura Shoten.

Fantz, R. L. (1963). Pattern vision in newborn infants. *Science, 140*(3564), 296–297.

Festinger, L. (1957). *A theory of cognitive dissonance*. Stanford, CA: Stanford University Press.

Festinger, L. (1964). *Conflict, decision, and dissonance*. Stanford, CA: Stanford University Press.

Forgas, J. P. (1991). Affective influence on partner choice: Role of mood in social decisions. *Journal of Personality and Social Psychology, 61,* 708–720.

Fugate, D. L. (2007). Neuromarketing: A layman's look at neuroscience and its potential application to marketing practice. *Journal of Consumer Marketing, 24*(7), 385–394.

Fujii, S., & Gärling, T. (2003). Application of attitude theory for improved predictive accuracy of stated preference methods in travel demand analysis. *Transport Research A: Policy & Practice, 37,* 389–402.

Fujii, S., & Takemura, K. (2001). Risuku taido to chūi: Jōkyō izonteki shōten moderu ni yoru furēmingu kōka no keiryō bunseki [Risk attitude and attention: A psychometric analysis of framing effect by contingent focus model]. *Kodo keiryogaku [The Japanese journal of behaviormetrics], 28,* 9–17.

Gardner, M. P. (1985). Mood states and consumer behavior: A critical review. *Journal of Consumer Research, 12*(3), 281–300.

Gerard, H. B., & White, G. L. (1983). Post-decisional reevaluation of choice alternatives. *Personality and Social Psychology Bulletin, 9,* 365–369.

Gonzalez, R., & Wu, G. (1999). On the shape of the probability weighting function. *Cognitive Psychology, 38*(1), 129–166.

Hsu, M., Bhatt, M., Adolphs, R., Tranel, D. & Camerer, C. F. (2005). Neural systems responding to degrees of uncertainty in human decision-making. *Science, 310*, 1680–1683.

Hubert, M., & Kenningy, P. (2008). A current overview of consumer neuroscience. *Journal of Consumer Behaviour, 7,* 272–292.

Ideno, T., Hayashi, M., Sakagami, T., Fujii, S., Okubo, S., Tamari, Y., et al. (2011a). Chikaku handan kadai wo mochīta senkō keisei katei no kentō [An experimental study of preference construction using a perceptual judgment task]. In *Proceedings of the 75th Annual Convention of the Japanese Psychological Association* (p. 88).

Ideno, T., Hayashi, M., Sakagami, T., Fujii, S., Okubo, S., Tamari, Y., et al. (2011b). Sentaku kodo ni yoru senko keisei katei no kisoteki kenkyu [A basic study of preference formation process by choice]. In *Paper presented at the 15th Conference of Experimental Social Science*.

Ideno, T., & Takemura, K. (2018). Senko no keisei katei ni kansuru jikkenteki kento [An experimental study of preference construction process]. In K. Takemura (Ed.), *Senkō keisei to ishikettei [Preference construction and decision making]* (pp. 99–122). Tokyo, Japan: Keiso Shobo.

Isen, A. M., & Geva, N. (1987). The influence of positive affect on acceptable level of risk: The person with a large canoe has a large worry. *Organizational Behavior and Human Decision Processes, 39*(2), 145–154.

Isen, A. M., & Means, B. (1983). The influence of positive affect on decision making strategy. *Social Cognition, 2,* 18–31.

Isen, A. M., Means, B., Patrick, R., & Nowicki, G. P. (1982). Some factors influencing decision-making strategy and risk taking. In M. S. Clark & S. T. Fiske (Eds.), *Affect and cognition: The 17th annual carnegie symposium on cognition* (pp. 243–261). Hillsdale, NJ: Erlbaum.

Isen, A. M., Nygren, T. E., & Ashby, F. G. (1988). The influence of positive affect on the subjective utility of gains and losses: It's not worth the risk. *Journal of Personality and Social Psychology, 55*, 710–717

Isen, A. M., & Patrick, R. (1983). The effect of positive feelings on risk taking: When the chips are down. *Organizational Behavior & Human Performance, 31*(2), 194–202.

Ishibuchi, J. (2003). Kaimono kōdō to kanjō [Buying behavior and affect]. *Maketingu janaru [Japan Marketing Journal], 22*(4), 109–116.

Izuma, K., Matsumoto, M., Murayama, K., Samejima, K., Sadato, N., & Matsumoto, K. (2010). Neural correlates of cognitive dissonance and choice-induced preference change. *Proceedings of the National Academy of Sciences, 107,* 22014–22019.

Johnson, E. J., & Tversky, A. (1983). Affect, generalization, and the perception of risk. *Journal of Personality and Social Psychology, 45*(1), 20–31.

Kahneman, D., Knetsch, J. L., & Thaler, R. H. (1990). Experimental tests of the endowment effect and the Coase theorem. *Journal of Political Economy, 98,* 1325–1348.

Kahneman, D., & Tversky, A. (1979). Prospect theory: An analysis of decision under risk. *Econometrica, 47,* 263–292.

Kiesler, C. A. (1971). *The psychology of commitment*. New York, NY: Academic Press.

Klinger, E. (1975). Consequences of commitment to and disengagement from incentives. *Psychological Review, 82,* 1–25.

Knutson, B., Rick, S., Wimmer, G. E., Prelec, D., & Loewenstein, G. (2007). Neural predictors of purchases. *Neuron, 53*(1), 147–156.

Kunst-Wilson, W. R., & Zajonc, R. B. (1980). Affective discrimination of stimuli that cannot be recognized. *Science, 207*(4430), 557–558.

Lee, N., Broderick, A. J., & Chamberlain, L. (2007). What is 'neuromarketing'? A discussion and agenda for future research. *International Journal of Psychophysiology, 63,* 199–204.

Lichtenstein, S., & Slovic, P. (1971). Reversals of preference between bids and choices in gambling decisions. *Journal of Experimental Psychology, 89,* 46–55.

Lichtenstein, S., & Slovic, P. (Eds.). (2006). *The construction of preference*. New York, NY: Cambridge University Press.

Lieberman, M. D., Ochsner, K. N., Gilbert, D. T., & Schacter, D. L. (2001). Do amnesics exhibit cognitive dissonance reduction? The role of explicit memory and attention in attitude change. *Psychological Science, 12,* 135–140.

Masters, J. C., Barden, R. C., & Ford, M. E. (1979). Affective states, expressive behavior and learning in children. *Journal of Personality and Social Psycholgy, 37,* 380–390.

McClure, S. M., Li, J., Tomlin, D., Cypert, K. S., Montague, L. M., & Montague, P. R. (2004). Neural correlates of behavioral preference for culturally familiar drinks. *Neuron, 44*(2), 379–387.

Monahan, J. L., Murphy, S. T., & Zajonc, R. B. (2000). Subliminal mere exposure: Specific, general, and diffuse effects. *Psychological Science, 11,* 462–466.

Murakami, I. (Ed.). (2010). *Irasuto rekuchā ninchi shinkei kagaku: Shinrigaku to nō kagaku ga toku kokoro no shikumi [Lecture on cognitive neuroscience using illustrations: Mechanism of mind studied by brain science and psychology]*. Tokyo, Japan: Ohmsha Ltd.

Murakami, H., Ideno, T., Tamari, Y., & Takemura, K. (2014). Kakuritsu kajū kansū ni taisuru moderu no teian to sono hikaku kentō [Proposal of probability weighting function model and psychological experiment for the comparisons]. In *The Proceedings of the 78th Annual Convention of the Japanese Psychological Association* (p. 515).

Murphy, S. T., & Zajonc, R. B. (1993). Affect, cognition, and awareness: Affective priming with optimal and suboptimal stimulus exposures. *Journal of Personality and Social Psychology, 64,* 723–739.

Nishibe, S. (1975). *Soshio economikkusu: Shūdan no keizai kōdō [Socio-economics: Economic behavior of groups]*. Tokyo, Japan: Chūōkōronsha.

Norton, M. I., Mochon, D., & Ariely, D. (2011). The 'IKEA effect': When labor leads to love. *Journal of Consumer Psychology, 22,* 453–460.
Ogaki, M., & Tanaka, S. (2014). *Kōdō keizaigaku: Dentōteki keizaigaku tono tōgō ni yoru atarashī keizaigaku wo mezashite [Behavioral Economics: Aiming for New Economics through Integration with Traditional Economics].* Tokyo, Japan: Yuhikaku Publishing.
Ohira, H. (Ed.). (2010). *Kanjō shinrigaku nyūmon [Introduction to Psychology on Emotions].* Tokyo, Japan: Yuhikaku Publishing.
Ortony, A., Clore, G., & Collins, A. (1988). The cognitive structure of emotion. *Contemporary Sociology, 18*(6), 957–958.
Pachur, T., Hertwig, R., & Wolkewitz, R. (2014). The affect gap in risky choice: Affect-rich outcomes attenuate attention to probability information. *Decision, 1*(1), 64–78.
Peck, J., & Shu, S. B. (2009). The effect of mere touch on perceived ownership. *Journal of Consumer Research, 36,* 434–447.
Prelec, D. (1998). The probability weighting function. *Econometrica, 66*(3), 497–527.
Rottenstreich, Y., & Hsee, C. K. (2001). Money, kisses, and electric shocks: On the affective psychology of risk. *Psychological Science, 12*(3), 185–190.
Sanfey, A. G., & Stallen, M. (2015). Neuroscience contribution to judgement and decision making: Opportunities and limitations. In G. Keren & G. Wu (Eds.), *The Wiley Blackwell handbook of judgement and decision making* (pp. 268–294). West Sussex, UK: Wiley.
Shimojo, S., Simion, C., Shimojo, E., & Scheier, C. (2003). Gaze bias both reflects and influences preference. *Nature Neuroscience, 6,* 1317–1322.
Simon, D., Krawczyk, D. C., & Holyoak, K. J. (2004). Construction of preferences by constraint satisfaction. *Psychological Science, 15,* 331–336.
Stroessner, S. J., Hamilton, D., & Mackie, D. (1992). Affect and stereotyping: The effect of induced mood on distinctiveness-based illusory correlations. *Journal of Personality and Social Psychology, 62*(4), 564–576.
Suter, R. S., Pachur, T., & Hertwig, R. (2015). How affect shapes risky choice: Distorted probability weighting versus probability neglect, *Journal of Behavioral Decision Making*, Published online in Wiley Online Library https://doi.org/10.1002/bdm.1888
Takahashi, H. (2013). Shakai shinkei kagaku to seishin igaku [Social Neuroscience and Psychiatry]. *Seishin shinkeigaku zasshi [Psychiatry and Clinical Neurosciences], 115*(10), 1027–1041.
Takahashi, H., Fujiie, S., Camerer, C., Arakawa, R., Takano, H., Kodaka, F., et al. (2013). Norepinephrine in the brain is associated with aversion to financial loss. *Molecular Psychiatry, 18,* 3–4.
Takahashi, H., Matsui, H., Camerer, C., Takano, H., Kodaka, F., Ideno, T., et al. (2010). Dopamine D1 receptors and nonlinear probability weighting in risky choice. *Journal of Neuroscience, 30,* 16567–16572.
Takemura, K. (1988a). Ishikettei katei no kenkyū (XI) Kanjō oyobi kettei furēmingu ga ishikettei katei ni oyobosu kōka [Study of decision-making process (XI): The influence of affect and framing on decision making process]. In *Proceedings of the 5th Annual Meeting of the Japanese Cognitive Science Society* (pp. 120–121).
Takemura, K. (1988b). Ishikettei Katei No Kenkyū (XII): Kanjō oyobi kanyo ga ishikettei katei ni okeru jōhō kensaku kōdō ni oyobosu kōka [Study of decision-making process (XII): The influence of affect and involvement on the information search behavior in decision-making process]. In *Proceedings of the 36th Annual Meeting of the Japanese Group Dynamics Association* (pp. 23–24).
Takemura, K. (1988c). Ishikettei katei no kenkyū (XIII) Kanjō oyobi kanyo ga ishikettei katei ni okeru jōhō no saikentō kōdō ni oyobosu kōka [Study of decision making process (XIII): Influence of affect and involvement on decision making process]. In *Proceedings of the 52th Annual Convention of the Japanese Psychological Association* (p. 198).
Takemura, K. (1994). Furemingu Koka no Rironteki Setsumei: Risukuka Deno Ishikettei no Jokyo Izonteki Shoten Moderu [Theoretical explanation of the framing effect: Contingent focus model for decision-making under risk]. *Japanese Psychological Review, 37,* 270–293.

Takemura, K. (1996). Ishikettei to sono shien [Decision-making and support for decision-making]. In S. Ichikawa (Ed.), *Ninchi shinrigaku 4kan Shikou [Cognitive psychology thoughts]* (Vol. 4, pp. 81–105). Tokyo, Japan: University of Tokyo Press.

Takemura, K. (1997). Shōhisha no jōhō tansaku to sentakushi hyōka [Alternative evaluation and consumer buying decision]. In T. Sugimoto (Ed.), *Shōhisha rikai no tame no shinrigaku [Psychology for understanding consumer]* (pp. 56–72). Tokyo, Japan: Fukumura Shuppan.

Takemura, K. (2009a). *Kōdō ishiketteiron: Keizai kōdō no shinrigaku [Behavioral decision theory: Psychology of economic behavior]*. Tokyo, Japan: Nippon Hyoron Sha Co. Ltd.

Takemura, K. (2009b). Ishikettei to shinkei keizaigaku [Decision making and neuroeconomics]. *Rinshō seishin igaku [Japanese journal of clinical psychiatry], 38,* 35–42.

Takemura, K. (2014). *Behavioral decision theory: Psychological and mathematical representations of human choice behavior*. Tokyo, Japan: Springer.

Takemura, K., & Fujii, S. (2015). *Ishikettei no shohō [Prescription for decision making]*. Tokyo, Japan: Asakura Shoten.

Takemura, K., Haraguchi, R., & Tamari, Y. (2015). Tazokusei ishikettei katei ni okeru ketteihōryaku no ninchiteki doryoku to seikakukusa: Keisanki simyurēhshon ni yoru kōdō ishiketteiron teki kentō (Effort and accuracy in multi-attribute decision making process: A behavioral decision theoretic approach using computer simulation technique). *Ninchi kagaku [Cognitive Studies], 22,* 368–388.

Takemura, K., Ideno, T., Hayashi, M., Sakagami, T., Fujii, S., Okubo, S., et al. (2012). *Hannō ga senkō oyobi hisenkō no keisei katei ni oyobosu kōka* [Effect of response on preference and non-preference construction process]. In *Paper presented at the Joint Conference of the Association of Bevioral Econmics and Finance (the 6th Conference), and the Experimental Social Science (the 16th Conference).*

Takemura, K., Ideno, T., Okubo, S., Kodaka, F., & Takahashi, H. (2009c). *Shōhisha no senkō ni kansuru shinkei keizaigaku teki kenkyū: Ninchi hannō to nōgazō kaiseki* [A neuroscientific study of consumer preference]. In *Paper presented at the 39th Conference of Japan Association of Consumer Studies.*

Takemura, K., Ideno, T., Okubo, S., & Matsui, H. (2008). Shinkei keizaigaku to zentōyō [Neuroeconimics and frontal lobe]. *Bunshi seishin igaku [Japanese journal of molecular psychiatry], 8*(2), 35–40.

Takemura, K., Murakami, H., Tamari, Y., & Ideno, T. (2014). Probability weighting function models and psychological experiment for the comparisons. In *Paper presented at Soft Science Workshop of Japan Society for Fuzzy Theory and Intelligent Infomatics.*

Tucker, D. M. (1981). Lateral brain function, emotion and conceptualization. *Psychological Bulletin, 89,* 19–46.

Tucker, D. M., Vannatta, K., & Rothlind, J. (1990). Arousal and activation systems and primitive adaptive controls on cognitive priming. In N. Stein, B. Leventhal, & T. Trabasso (Eds.), *Psychological and biological approaches to emotion* (pp. 145–166). Hillsdale, NJ: Lawrence Erlbaum.

Turley, L. W., & Miliman, R. E. (2000). Atmospheric effects on shopping behavior: A review of the experimental evidence. *Journal of Business Research, 49*(2), 193–211.

Tversky, A., & Kahneman, D. (1981). The framing of decisions and the psychology of choice. *Science, 211,* 453–458.

Tversky, A., & Kahneman, D. (1992). Advances in prospect theory: Cumulative representation of uncertainty. *Journal of Risk and Uncertainty, 5,* 297–323.

Tversky, A., Sattath, S., & Slovic, P. (1988). Contingent weighting in judgment and choice. *Psychological Review, 95,* 371–384.

Vanderplas, J. M., & Garvin, E. A. (1959). The association value of random shapes. *Journal of Experimental Psychology, 57,* 147–154.

Wolf, J. R., Arkes, H. R., & Muhanna, W. A. (2008). The power of touch: An examination of the effect of duration of physical contact on the valuation of objects. *Judgment and Decision Making, 3,* 476–482.

Wu, G., & Gonzalez, R. (1996). Curvature of the probability weighting function. *Management Science, 42*(12), 1676–1690.

Zajonc, R. B. (1968). Attitudinal effects of mere exposure. *Journal of Personality and Social Psychology*, *9*(2, Pt.2), 1–27.

Zajonc, R. B. (1980). Feeling and thinking: Preferences need no inferences. *American Psychologist, 35,* 151–175.

Zajonc, R. B. (2001). Mere exposure: A gateway to the subliminal. *Current Directions in Psychological Science, 10,* 224–228.

Author Index

K. Takemura, *Foundations of Economic Psychology*,
https://doi.org/10.1007/978-981-13-9049-4

G

H

I

J

K

L

M

N

O

P

Q

Index

K. Takemura, *Foundations of Economic Psychology*,
https://doi.org/10.1007/978-981-13-9049-4